VOLUME ONE HUNDRED AND THIRTY SEVEN

PROGRESS IN MOLECULAR BIOLOGY AND TRANSLATIONAL SCIENCE

The Molecular Basis of Drug Addiction

VOLUME ONE HUNDRED AND THIRTY SEVEN

PROGRESS IN MOLECULAR BIOLOGY AND TRANSLATIONAL SCIENCE

The Molecular Basis of Drug Addiction

Edited by

SHAFIQUR RAHMAN
Department of Pharmaceutical Sciences,
South Dakota State University,
Brookings, South Dakota, USA

ELSEVIER

AMSTERDAM • BOSTON • HEIDELBERG • LONDON
NEW YORK • OXFORD • PARIS • SAN DIEGO
SAN FRANCISCO • SINGAPORE • SYDNEY • TOKYO
Academic Press is an imprint of Elsevier

Academic Press is an imprint of Elsevier
50 Hampshire Street, 5th Floor, Cambridge, MA 02139, USA
525 B Street, Suite 1800, San Diego, CA 92101-4495, USA
125 London Wall, London EC2Y 5AS, UK
The Boulevard, Langford Lane, Kidlington, Oxford OX5 1GB, UK

ISBN: 978-0-12-803786-7
ISSN: 1877-1173

For information on all Academic Press publications
visit our website at http://store.elsevier.com/

CONTENTS

CONTRIBUTORS

Richard L. Bell
Department of Psychiatry, Indiana University School of Medicine, Indianapolis, Indiana, USA

Thomas P. Beresford
Department of Veterans Affairs Medical Center, Laboratory for Clinical and Translational Research in Psychiatry, Denver, Colorado, USA
Department of Psychiatry, School of Medicine, University of Colorado, Denver, Colorado, USA

Patrick Chan
Department of Pharmacy and Pharmacy Administration, Western University of Health Sciences, College of Pharmacy, Pomona, California, USA

Howard J. Edenberg
Departments of Biochemistry and Molecular Biology and Medical and Molecular Genetics, Indiana University School of Medicine, Indianapolis, Indiana, USA

Eric A. Engleman
Department of Psychiatry, Indiana University School of Medicine, Indianapolis, Indiana, USA

Sheketha R. Hauser
Department of Psychiatry, Indiana University School of Medicine, Indianapolis, Indiana, USA

Simon N. Katner
Department of Psychiatry, Indiana University School of Medicine, Indianapolis, Indiana, USA

Kabirullah Lutfy
Department of Pharmaceutical Sciences,College of Pharmacy, Western University of Health Sciences, Pomona, California, USA

William J. McBride
Department of Psychiatry, Indiana University School of Medicine, Indianapolis, Indiana, USA

Jeanette McClintick
Departments of Biochemistry and Molecular Biology and Medical and Molecular Genetics, Indiana University School of Medicine, Indianapolis, Indiana, USA

Bethany S. Neal-Beliveau
Department of Psychology, Purdue School of Science, Indiana University-Purdue University Indianapolis, Indianapolis, Indiana, USA

Pamela M. Quizon
Department of Drug Discovery and Biomedical Sciences, South Carolina College of Pharmacy, University of South Carolina, Columbia, South Carolina, USA

Shafiqur Rahman
Department of Pharmaceutical Sciences, South Dakota State University, Brookings, South Dakota, USA

Patrick J. Ronan
Department of Veterans Affairs Medical Center, Laboratory for Clinical and Translational Research in Psychiatry, Denver, Colorado, USA
Research Service, Sioux Falls VA Health Care System, Sioux Falls, South Dakota, USA
Department of Psychiatry and Division of Basic Biomedical Sciences, Sanford School of Medicine at the University of South Dakota, Sioux Falls, South Dakota, USA

Wei-Lun Sun
Department of Drug Discovery and Biomedical Sciences, South Carolina College of Pharmacy, University of South Carolina, Columbia, South Carolina, USA

Karen K. Szumlinski
Department of Psychological and Brain Sciences, University of California Santa Barbara, Santa Barbara, California, USA

Joachim D. Uys
Department of Cellular and Molecular Pharmacology and Experimental Therapeutics, Medical University of South Carolina, Charleston, South Carolina, USA

Jacqueline S. Womersley
Department of Cellular and Molecular Pharmacology and Experimental Therapeutics, Medical University of South Carolina, Charleston, South Carolina, USA

Narin Wongngamnit
Department of Veterans Affairs Medical Center, Laboratory for Clinical and Translational Research in Psychiatry, Denver, Colorado, USA
Department of Psychiatry, School of Medicine, University of Colorado, Denver, Colorado, USA
Substance Abuse Treatment Program, Department of Veterans Affairs, Denver, Colorado, USA

Nurulain T. Zaveri
Astraea Therapeutics, LLC, Mountain View, California, USA

Jun Zhu
Department of Drug Discovery and Biomedical Sciences, South Carolina College of Pharmacy, University of South Carolina, Columbia, South Carolina, USA

PREFACE

Drug addiction is the most complex and costly neuropsychiatric disorder affecting millions of people in the world. Recent surveys indicate that approximately 250 million people are illegal drug users which represent ~4% of the global population. Acute and chronic exposure to drugs of abuse produces numerous neurobiological effects, but the cellular and molecular processes involved are only partially understood. Neuroscientists around the world are searching for clues that underlie the molecular basis of drug addiction. While current scientific breakthroughs have increased the understanding on molecular determinants of drug addiction, limitations exist on effective treatment strategies for many forms of drug addiction. Thus, there is a need to translate the current knowledge regarding molecular mechanisms of drug addiction derived from neurobiological research into the discovery of new therapeutics.

This volume, *The Molecular Basis of Drug Addiction*, consists of eight chapters written by eminent experts in the field. The volume covers important aspects of neuroscience research on drug addiction associated with the neurotransmitter receptors, signaling molecules, and relevant mechanisms implicated in drug addiction. The chapters in this volume describe some of the latest concepts in emerging and innovative research, discuss new breakthrough findings, define innovative strategies, and target multiple signaling pathways and genes. The primary molecular targets discussed in this volume include extracellular signal-regulated kinase, glutamate-associated genes or proteins, S-glutathionylated proteins, cannabinoid receptor mediated signaling pathways, adenylyl cyclase/cyclic adenosine 3,5-monophosphate protein kinase A, neuronal nicotinic receptors, and nociceptin receptors involved in many forms of drug addiction. The first chapter presents and discusses the role of the extracellular signal-regulated kinase and its related intracellular signaling pathways in drug-induced neuroadaptive changes that are associated with drug-mediated psychomotor activity, rewarding properties, and relapse of drug-seeking behaviors (Zhu *et al.*). The second chapter reviews the role of glutamate neurotransmitter receptor system in mediating the development of alcohol dependence. The chapter discusses the expression levels of glutamate-associated genes and/or proteins, including metabotropic and ionotropic receptor subunits and glutamate transporters in a genetic animal model of alcoholism and highlights the changes in glutamate receptors, transporters, enzymes, and scaffolding proteins involving alcohol dependence (Bell *et al.*). The third chapter presents and highlights the evidence for S-glutathionylation as a redox-sensing mechanism and how this may be involved in the response to drug-induced oxidative stress. The function of S-glutathionylated proteins involved in neurotransmission, dendritic spine structure, and drug-induced behavioral outputs are reviewed with specific reference to alcohol, cocaine, and heroin (*Uys and Reissner*). The fourth chapter provides a comprehensive account of the state of knowledge regarding mechanisms of *Cannabis* signaling in the brain and the modulation of key brain neurotransmitter systems involved in addiction and psychiatric disorders (Ronan *et al.*). The fifth chapter reviews the existing literature on the roles of nociception receptors and associated mechanisms in the rewarding and addictive actions of cocaine (Lutfy and Zaveri). The sixth chapter presents recent insights on the rewarding effects of alcohol as they pertain to different brain nicotinic receptor subtypes and associated signaling pathways that contribute to the molecular mechanisms of alcoholism and/or comorbid brain disorders (Rahman *et al.*). The seventh chapter focuses on and reviews the adenylyl cyclase and cyclic adenosine 3,5-monophosphate/protein kinase A system as a central player in mediating the acute and chronic effects of opioids in opiate abusers (Chan and Lutfy). The eighth chapter concentrates on *Caenorhabditis elegans*, a nonvertebrate model, to study the molecular and genetic mechanisms of drug addiction and to identify potential targets for medication development (Engleman *et al.*).

Together, this body of work not only provides a deeper understanding of our current knowledge on specific neurotransmitter systems, functional proteins, signaling molecules, genes, and additional targets for drug addiction, but also indicates complex interactions between drugs of abuse, endogenous neuromodulators, signaling molecules, and the mechanisms underlying the structural and functional plasticity in the brain. I hope that the molecular basis of drug addiction research summarized in this volume will generate new ideas on diverse targets and stimulate translational research for further mechanistic understanding and insight into effective strategies for novel therapeutics in the management of drug addiction.

I would like to thank all the authors for their outstanding contributions to this volume. I am very thankful to Dr. P. Michael Conn, the Editor-in-Chief of the Book Series, for his guidance. Finally, I also thank Ms. Mary Ann

Zimmerman, the Senior Acquisitions Editor and Ms. Helene Kabes, Senior Editorial Project Manager of Elsevier, for their assistance and support in bringing this volume together. A special thanks to my wife and daughters for their understanding and love.

Shafiqur Rahman
Editor

CHAPTER ONE

Molecular Mechanism: ERK Signaling, Drug Addiction, and Behavioral Effects

Wei-Lun Sun, Pamela M. Quizon, Jun Zhu[1]

Department of Drug Discovery and Biomedical Sciences, South Carolina College of Pharmacy, University of South Carolina, Columbia, South Carolina, USA

[1] Corresponding author: e-mail address: zhuj@sccp.sc.edu.

Contents

Abstract

Addiction to psychostimulants has been considered as a chronic psychiatric disorder characterized by craving and compulsive drug seeking and use. Over the past two decades, accumulating evidence has demonstrated that repeated drug exposure causes long-lasting neurochemical and cellular changes that result in enduring neuroadaptation in brain circuitry and underlie compulsive drug consumption and relapse. Through intercellular signaling cascades, drugs of abuse induce remodeling in the rewarding circuitry that contributes to the neuroplasticity of learning and memory associated with addiction. Here, we review the role of the extracellular signal-regulated kinase (ERK), a member of the mitogen-activated protein kinase, and its related intracellular signaling pathways in drug-induced neuroadaptive changes that are associated with drug-mediated psychomotor activity, rewarding properties and relapse of drug seeking behaviors. We also discuss the neurobiological

Progress in Molecular Biology and Translational Science, Volume 137
ISSN 1877-1173
http://dx.doi.org/10.1016/bs.pmbts.2015.10.017

and behavioral effects of pharmacological and genetic interferences with ERK-associated molecular cascades in response to abused substances. Understanding the dynamic modulation of ERK signaling in response to drugs may provide novel molecular targets for therapeutic strategies to drug addiction.

ABBREVIATIONS

AC	Adenylyl cyclase
AMPH	Amphetamine
Amy	Amygdala
BDNF	Brain-derived neurotrophic factor
BNST	Bed nucleus of the striatal terminals
Ca^{2+}	Calcium
CaM	Calcium/calmodulin
CaMK	CaM kinase
CB1-R	Cannabinoid receptor 1
CB2-R	Cannabinoid receptor 2
CPP	Conditioned place preference
CPu	Caudate putamen
CREB	cAMP response element-binding protein
DA	Dopamine-regulated phosphoprotein-32
D1-R	Dopamine D1 receptor
D2-R	Dopamine D2-Receptor
ERK	Extracellular signal-regulated kinase
Glu	Glutamate
HIPP	Hippocampus
IEG	Immediate early gene
MAPK	Mitogen-activated protein kinase
MEK	MAPK kinase
METH	Methamphetamine
mGluR1/5	Metabotropic glutamate receptor-1/5
MKP-1/3	MAPK phosphatases 1 and 3
MSK	Mitogen- and stress-activated protein kinase
NAc	Nucleus accumbens
nAChRs	Nicotinic acetylcholine receptors
pCREB	Phosphorylated CREB
pERK	Phosphorylated ERK
PFC	Prefrontal cortex
pGluN2B	Phosphorylation of glutamate receptor, ionotropic, *N*-methyl D-aspartate 2B
PKA	Protein Kinase A
PKC	Protein Kinase C
pMEK	Phosphorylation of MEK
PP2A	Protein phosphatase 2A

PP2B	Protein phosphatase 2B
pSTEP	Phosphorylation of STEP
pThr75 DARPP-32	Phosphorylation of DARPP-32 at threonine 75
Ras-GRF-1	Ras-guanine nucleotide-releasing factors 1
RSK	Ribosomal S6 kinase
SA	Self-administration
STEP	Striatal-enriched protein tyrosine phosphatase
THC	Δ9-Tetrahydrocannabinol
VTA	Ventral tegmental area

1. INTRODUCTION

Drug addiction is a chronic brain disease characterized by high relapse rates and compulsive drug use despite negative consequences. To date, there is no effective treatment for drug addiction. Understanding the neurobiologic aspects underlying substance abuse provide a basis for developing potential therapeutic strategies targeting to drug addiction. Accumulating evidence demonstrates that drugs of abuse alter dopamine (DA) and glutamate (Glu) neurotransmission in the mesocorticolimbic system to exert their molecular and behavioral effects.[1–3] DA neurons in the ventral tegmental area (VTA) and their descending projections to the nucleus accumbens, prefrontal cortex (PFC) and other limbic regions, including the hippocampus (HIPP) and amygdala (Amy), comprise the mesocorticolimbic system,[4] which is crucial for reward and reinforcement processing, motivation, and goal-directed behavior.[5,6] The NAc and VTA also receive Glu output from the PFC. In addition, a reciprocal Glu connection is found between the PFC and Amy. The nigrostriatal pathway containing the DA projection from the substantia nigra to the caudate putamen (CPu/dorsal striatum) has also been implicated in molecular events, rewarding effects, and habitual behavior of drug addiction.[7,8]

The extracellular signal-regulated kinases (ERK1/2 or p44/p42 MAPK) cascade, one of the isoforms of mitogen-activated protein kinases (MAPK), is associated with the pathology of diseases due to its role in cell proliferation, differentiation, survival, and death.[9,10] Since the identification the activation of ERK by chronic morphine and cocaine administration in the VTA in 1996,[11] several lines of studies have focused ERK-mediated molecular signaling in response to various drugs of abuse during the last two decades.

Herein, we review the alterations of ERK signaling induced by abused substances including cocaine, amphetamine (AMPH), methamphetamine, marijuana, nicotine, and alcohol. In addition, most of these drugs have been shown to induce psychomotor changes, the ERK-associated molecular changes underlying drug-induced behaviors are also discussed. Further, due to the critical role of ERK in the neuroplasticity of learning and memory associated with addiction,[12] its influence on the reinforcing, rewarding, and relapse/reinstatement of drug addiction is also described.

2. ERK SIGNALING PATHWAY

Initially, intracellular ERK signaling has been characterized to respond to extracellular stimuli and regulate cell proliferation and differentiation.[13] For example, once ERK is activated by growth factors or neurotrophins, the tyrosine kinase receptors recruit Ras family G-proteins and lead to sequential activation of Raf (MAPK kinase kinase), MEK (MAPK kinase), and ERK. Once ERK is activated, the phosphorylated ERK (pERK) protein can translocate to the nucleus,[14] where they phosphorylate the ternary complex factor Elk-1.[15,16] The activated Elk-1 and other ternary complex factors associate with serum response factor, bind to the serum response element site, and promote immediate early gene (IEG) transcription related to neuroadaptation.[17–19] In addition to Elk-1, through phosphorylating ribosomal S6 kinases and mitogen- and stress-activated protein kinases (pRSKs and pMSKs, respectively), ERK has been shown to indirectly result in cAMP response element-binding protein phosphorylation (pCREB), a transcription factor that has been shown to regulate gene expression.[20–24] Increasing evidence shows a Glu linkage to ERK signaling in neurons both *in vivo* and *in vitro*. For instance, through the elevation of intracellular calcium (Ca^{2+})/calmodulin (CaM)/CaM kinases (CaMK), the activation of the Glu NMDA receptor can increase the phosphorylation of MEK (pMEK)/ERK/Elk-1 in hippocampal slices, neuronal culture,[25–27] cortical cultured neurons,[28] and striatal cultured neurons.[29–31] Inhibition of ERK activation attenuates Glu-mediated pElk-1 in the striatal slice,[32] striatum,[33–35] and in the HIPP.[17] Alternatively, in PC12 cells, Ca^{2+} may increase the intracellular cAMP through Ca^{2+}/CaM-sensitive adenylyl cyclase (AC) leading to the activation of PKA. Increase of cAMP and PKA induces pMEK via the activation of Rap1/Raf.[36,37] Consistent with these studies, pharmacologic activation of D1-R or the AC markedly stimulates ERK activity and its phosphorylation in various neuronal cells.[33,38–41] In addition, activation of group 1 metabotropic Glu receptors (mGluR1/5) has been shown to increase the intracellular Ca^{2+} and activate ERK signaling.[42–45] Although the activation of DA D2 receptor (D2-R) inhibits PKA activity, D2-R stimulation also increases ERK signaling through PKC activation.[46]

There are several families of ERK-related phosphatases. Among these, protein phosphatase 2A (PP2A) and striatal-enriched protein tyrosine phosphatase (STEP) are the best characterized. PP2A is a major serine/threonine phosphatase containing two regulatory subunits and one catalytic subunit. PP2A mediates a rapid inactivation of pERK *in vitro*. STEP is another phosphatase that regulates ERK activation. Although it is enriched in the striatum, STEP is expressed abundantly in the mesocorticolimbic system.[47,48] Through direct interaction of a kinase-interacting motif, STEP and its nonneuronal homologues have been shown to dephosphorylate pERK and prevent its nuclear translocation.[49,50] Phosphorylation of STEP (pSTEP) reduces its activity and its capacity to inhibit pERK.[49] STEP is regulated through D1-R/PKA/DARPP-32 signaling.[51] *In vitro*, D1-R activation has been shown to activate pThr34 and inhibit pThr75 DARPP-32 via PKA-activated PP2A,[52] which inhibits protein phosphatase 1 and thereby increasing pSTEP.[53] In addition, stimulation of NMDA-R has been reported to induce Ca^{2+}-activated PP2A and protein phosphatase 2B (PP2B), which inhibit DARPP-32 signaling[52,54,55] and indirectly modulate ERK activity. Therefore, the protein phosphatases of pERK are regulated by DA- and Glu-mediated transmission. Further, dual specificity MAPK phophatases 1 and 3 (MKP-1/3) are also implicated in pERK deactivation. Both *in vitro* and *in vivo* studies indicated that MKP-1/3 expression and activation is dependent on ERK signaling. Once induced and activated, MKP-1/3 reduces the ERK activation as an inhibitory feedback loop.[34,56–61] Furthermore, there is evidence demonstrating that MKP-1 is phosphorylated (pMKP-1) by pERK leading to MKP-1 protein stabilization without altering its ability to dephosphorylate pERK.[62]

3. ERK SIGNALING AND DRUG ADDICTION

ERK signaling is responsive to various abused drugs in the mesocorticolimbic system. Both acute and chronic exposure to drugs results in alteration of ERK-mediated signaling in specific brain regions underlying neuronal plasticity and drug-induced behavioral changes. Therefore, we focus on the effects of the most prevalent abused substances on ERK signaling and its relationship of drug-mediated behavioral changes across different paradigms including locomotor activity/sensitization, conditioned place preference (CPP), and self-administration (SA), if applicable. Since pharmacologic and genetic approaches have been used to interfere with the ERK signaling cascade, their effects on abused drug-mediated behaviors were summarized in Table 1 and Table 2, respectively.

3.1 Cocaine

Numerous studies have demonstrated that acute cocaine administration increases pERK in the CPu, NAc, PFC, central and basolateral Amy (CeA and BLA, respectively), HIPP, and bed nucleus of the striatal terminals (BNST).[98–112] The increased pERK and its downstream targets including pMSK-1, pElk-1, pCREB, phosphorylation of GluN2B (pGluN2B) and IEGs by acute cocaine, are dependent on the activation of MEK, D1-R/DARPP-32, and NMDA-R.[51,69,71,97–99,102,103,106,107,111] In addition to pMSK-1 induction, the pRSKs in the striatum are also increased by acute cocaine leading to the indirect activation of CREB by pERK.[97,112] In terms of protein phosphatases of pERK, acute cocaine has been shown to result in an increase of MKP-1 mRNA in the striatum and cortex.[113] In addition, depending on D1- and NMDA-Rs, the phosphorylation of MKP-1 was also enhanced in the CPu and NAc 45–60 min after acute cocaine, contributing to the transient pERK induction.[111] Further, the pSTEP was also downregulated after acute cocaine in the CPu with corresponding pERK induction.[112] Together, in a time-dependent manner, the activation and inactivation of protein phosphatases are critical for controlling the acute cocaine–augmented pERK. Behaviorally, the acute cocaine–induced locomotor activity was not affected by MEK inhibitor, SL327 (30 or 40 mg/kg), but partially inhibited or not altered with a higher dose injection (50 mg/kg), which has nonspecific sedative effect on basal locomotion.[51,69,71,75,114] Similar to acute cocaine, MEK/ERK activation is necessary for the chronic cocaine-induced IEG expression in the CPu, NAc, and Amy in a time-dependent manner.[102,103] In cocaine-sensitized animals, 7–21 days but not 1 day withdrawal resulted in increased AMPA-R subunit surface insertion and NDMA-R subunit expression with paralleled pERK induction in the NAc.[115–119] AMPA-R expression in the NAc after prolonged withdrawal from repeated cocaine injection is dependent on the activation of both GluN2B and pERK, which contributes to the development of behavioral sensitization.[117] This conclusion is further supported by a study that D1-R/Src

Table 1 Effects of MEK Inhibitors on Drug-Induced Behaviors

Drugs	MEK Inhibitors (Dose, Area)	Behavioral Effects	References
None	SL327 (50 mg/kg, i.p.)	↑ Basal locomotor activity	[63]
	SL327 (50–100 mg/kg, i.p.)	↓ Basal locomotor activity	[64–67]
	PD98059 (50 μM, continuous infusion into the PFC)	↑ Basal locomotor activity↓	[68]
Cocaine	SL327 (50 mg/kg, i.p.)	↓ Acute cocaine-induced locomotion	[69]
	SL327 (30 mg/kg, i.p.); PD98059 (10 μM, VTA)	↓ Development of locomotor sensitization (inhibitors were injected/infused before each cocaine injection)	[64,70]
	SL327 (40 mg/kg, i.p.); PD98059 (2 μg) or U0126 (1 μg, NAc)	↓ Expression of locomotor sensitization (inhibitors were injected/infused before cocaine challenge)	[71,72]
	SL327 (30 mg/kg, i.p.)	↓ Conditioned locomotor response (inhibitor was injected before each cocaine injection during conditioning)	[64]
	SL327 (50 mg/kg, i.p.); U0126 (0.1 μg, VTA)	↓ Development of CPP (inhibitors injected/infused before each cocaine injection during conditioning)	[69,73]
	U0126 (1 μg, NAc core)	↓ Expression of CPP (inhibitor was infused before CPP test)	[74]
	SL327 (30 mg/kg, i.p.); PD98059 (2 μg) or U0126 (1 μg, NAc core); U0126 (1 μg, BLA)	↓ Context- and cocaine priming-induced expression of CPP and ↓ context-induced reinstatement after SA by impairing memory reconsolidation (inhibitors were injected/infused either before or after reconsolidation phase)	[74–77]
	U0126 (1 μg, CeA)	↓ Context + cues-induced relapse after abstinence from SA (inhibitor was infused before relapse testing)	[78]
	U0126 (1 μg, VTA)	↓ BDNF/GDNF-enhanced relapse by context + cues after abstinence from SA (infusions were conducted immediately after the end of the last SA session)	[79,80]
	U0126 (0.5 μg, dmPFC)	↓ BDNF's inhibitory effect on context-, cues-, and cocaine priming-induced drug seeking after abstinence/extinction	[81]

(*Continued*)

Table 1 Effects of MEK Inhibitors on Drug-Induced Behaviors—cont'd.

Drugs	MEK Inhibitors (Dose, Area)	Behavioral Effects	References
		of SA (infusions were conducted immediately after the end of the last SA session)	
Amphetamine	SL327 (50–100 mg/kg, i.p.)	↓ Acute AMPH–induced locomotion	[69,82,83]
	PD98059 (50 μM, continuous infusion into the PFC)	↑ Acute AMPH–induced locomotor activity	[68]
	SL327 (40 mg/kg, i.p.)	↓ Expression of locomotor sensitization (inhibitors were injected/infused before AMPH challenge)	[71]
	SL327 (30 mg/kg, i.p.)	↓ Conditioned locomotor response (inhibitor was injected before each AMPH injection during conditioning)	[64]
	PD98059 (2.5 μg, NAc)	↓ Development of intra-NAc AMPH-induced CPP (inhibitor was infused before or after each intra-NAc AMPH infusion during conditioning)	[84]
	PD98059 (2 μg, NAc)	↓ Expression of AMPH-CPP (inhibitor was infused before CPP testing)	[85]
Marijuana (THC)	SL327 (50 mg/kg, i.p.)	↓ Development of THC-induced locomotion tolerance (inhibitor was injected before each THC administration)	[86]
	SL327 (50 mg/kg, i.p.)	↓ Development of THC-CPP (inhibitor was injected before each conditioning session)	[87]
Alcohol	PD98059 (30 or 90 μg, i.c.v.)	↓ Development of ACD-CPP (inhibitor was infused before each conditioning session)	[88]
	SL327 (30 mg/kg, i.p.)	↑ Ethanol SA (inhibitor was injected before SA session)	[67]
	U0216 (0.5 μg, VTA)	↓ GDNF's inhibitory effect on ethanol SA (infusions were conducted before SA session)	[89]

i.p., intraperitoneal injection; i.c.v., intracerebroventricular infusion; ↑, enhancing effect; ↓, inhibiting effect.

Table 2 Effects of Interfering ERK Signaling-Related Genes/Proteins on Drug-Induced Behaviors

Target Genes/Proteins	Behavioral Effects	References
Ca^{2+}-stimulated AC1/AC8 (KO)	↑ Acute cocaine–induced locomotion ↓ Development of cocaine locomotor sensitization	[90]
Ras-GRF-1 (KO)	↓ Development and expression of cocaine locomotor sensitization ↓ Cocaine-CPP ↓ Repeated THC-induced behavioral tolerance	[86,91,92]
Ras-GRF-1 (OE)	↑ Development and expression of cocaine locomotor sensitization ↑ cocaine-CPP	[91]
Ras-GRF-2 (KO)	↓ Ethanol intake and preference (two bottle-free choice task)	[93]
ERK1 (KO)	↑ Basal locomotor activity ↑ AMPH-induced locomotion ↑ Development of cocaine locomotor sensitization ↑ cocaine-CPP	[34,63,94,95]
ERK1 (KD in the PFC)	↑ Basal locomotor activity ↑ AMPH-induced locomotion	[68]
ERK2 (OE in the VTA)	↑ development and expression of cocaine locomotor sensitization ↑ Cocaine-CPP	[96]
ERK2 (KD in the VTA)	↓ Development and expression of cocaine locomotor sensitization ↓ Cocaine-CPP	[96]
MSK-1 (KO)	↓ Development and expression of cocaine locomotor sensitization ↑ Cocaine-CPP	[97]
Inhibition of pElk-1	↓ Development and expression of cocaine locomotor sensitization ↓ The establishment of cocaine-CPP	[98]

KO, knockout; KD, knockdown; OE, overexpression; ↑, enhancing effect; ↓, inhibiting effect.

kinase-mediated pGluN2B is necessary for the pERK induction in response to repeated cocaine administration.[106] In addition, cocaine challenge after withdrawal from repeated cocaine administration also resulted in sensitized pERK in the NAc and CPu compared to the acute cocaine effect.[108,120,121] The cocaine behavioral sensitization-induced pERK and pCREB in the NAc is dependent on ERK activation.[122] Further, the induction and expression of cocaine behavioral sensitization can be inhibited by systemic SL327 injection or intra-NAc MEK inhibitor infusion.[64,71,72] Similarly through MEK activation, the pERK induction in the VTA is required for the development of behavioral sensitization to cocaine.[11,70] Lastly, studies have indicated that, in response to D1- and NMDA-R activation, pERK induced by cocaine is responsible for the chronic cocaine-enhanced dendritic spine density and dendritic length in the CPu and NAc[123,124] providing the morphologic evidence mediated by ERK signaling after repeated cocaine administration.

Repeated pairing of a specific environment with drug administration leads to a memory association between contextual cues and the drug rewarding effect. Subsequently, the context itself directly motivates drug-seeking behavior as a measurement of the reinforcing effect of the drug,[125,126] which is associated with ERK signaling. For example, the acquisition of cocaine-CPP is accompanied by pERK induction in the NAc and PFC in a D1-R-dependent manner.[127] Systemic preadministration of SL327 (50 mg/kg) and a GluN2B antagonist inhibited the development of cocaine-CPP,[69,106] indicating the requirement of NMDA-R-mediated ERK activation in the formation of context–drug association memory. ERK activation in the VTA is necessary for the development of cocaine-CPP.[73] Cocaine challenge in the drug-paired environment resulted in pERK and pCREB induction in the subset of neurons of the NAc.[128] In animals with repeated cocaine administration, the saline challenge enhanced pERK induction in the D1-positive neurons in NAc and CPu, indicating context conditioning-induced ERK activity.[108] Similarly, after the establishment of CPP, CPP testing or re-exposure to the cocaine-associated context induced pERK, pCREB, and/or ΔFosB in the CPu, HIPP, VTA, and NAc as well as in D1-R-containing neurons of the NAc.[73,129–133] The CPP test-induced pERK expression in the VTA is dependent on mGluR1 activation and protein synthesis.[133] Further, Miller and Marshall demonstrated that CPP test-elevated pERK and drug-seeking behavior were blocked by intra-NAc core infusion of U0126 (2 μg/side).[74] In the cocaine SA paradigm, context-induced relapse is also associated with enhanced pERK in the NAc core and CPu.[134] Altogether, these results imply that, through ERK signaling, the NAc core and VTA are important for the memory formation of context–drug association. pERK in the NAc core and CPu also involve the retrieval of CPP memory and a general motor activation driven by drug-associated context, respectively.

Memory reconsolidation occurs when well-established drug-associated memories are recalled by re-exposure to drug-associated context, cues, or the drug itself during which memories can be destabilized by adding new information or subjected to manipulation.[135–137] The ability to disrupt drug-related memories provides an opportunity to promote treatment outcome and prevent relapse. The general procedure to test the memory reconsolidation on drug-seeking behavior contains two phases: re-exposing animals to drug-associated context (phase 1) followed by testing drug-seeking behavior after withdrawal (phase 2). A previous study demonstrated that, before or immediately after phase 1, intra-NAc core MEK inhibition through U0126 (1 μg/side) or PD98059 (2 μg/side) reduced cocaine-CPP during the phase 2. The protein expression of pERK, pCREB, pElk-1, and c-Fos induced by phase 2 is also attenuated with inhibiting ERK during phase 1.[74] Systemic SL327 injection after phase 1 also decreased subsequent context-induced CPP in animals conditioned by escalating doses of cocaine.[76] Similar to reactivation of CPP memory by context, the memory reconsolidation in response to cocaine is also accompanied by ERK activation in the PFC, NAc, and CPu. With or without cocaine priming, the systemic SL327 (20 mg/kg) pretreatment before phase 1 inhibits the subsequent drug-seeking behavior.[75] However, the effect of ERK on cocaine-induced memory reconsolidation is still dependent on the presence of context. Thus, the contribution of cocaine itself on memory reconsolidation is still ambiguous. After the establishment of cocaine SA, U0126 (1 μg/side) infusion into the BLA immediately after phase 1 inhibited context-induced reinstatement and the pERK induction after phase 2.[77] Taken together, these studies indicate that ERK signaling activated during memory reconsolidation is necessary for cocaine-seeking behavior. However, a critical time window, 6 h after the reactivation of memories, has been documented during which the memory is susceptible to alteration in the fear-conditioning paradigm.[138] The pretreatment before phase 1 may influence the memory retrieval instead of reconsolidation. If the ERK signaling actually involves drug-related memory reconsolidation, the difference should be found when treatment is conducted within and beyond the critical time window in terms of both behavioral and molecular aspects.

Unlike pERK sensitization in cocaine-induced behavioral sensitization, immediately after the cessation of cocaine SA, there is a dissociation between pERK induction and cocaine intake indicating the failure of developing pERK sensitization or tolerance, although with enhanced pERK expression in several brain regions.[139] However, ERK activation has been implicated in relapse after withdrawal. For example, the extinction test (conditioned cues + context) significantly increased pERK in the CeA and cocaine-seeking behavior after 30 days of withdrawal. Both enhanced pERK and relapse are dependent on MEK and NMDA-R activation.[78] Similarly, the pERK induction in the ventromedial PFC has been shown to mediate extinction-test-induced cocaine-seeking behavior after 1- or 30-day withdrawal from cocaine SA.[140] Through ERK activation, direct intra-VTA glial cell line-derived neurotrophic factor (GNDF) or brain-derived neurotrophic factor (BDNF) infusion immediately after the last session of cocaine SA induced robust drug-seeking behavior after 3 or 10 days withdrawal.[79,80] These results demonstrated that the potentiated ERK signaling underlies relapse behavior after cocaine SA. In contrast to augmented pERK induction in the PFC after 1-day abstinence of cocaine SA,[140] 2 h after the last cocaine SA session, we have demonstrated a transient reduction of pERK in the PFC.[81,141,142] The reduction of pERK is associated with an increase of STEP but not PP2A activity accompanied by decreased total GluN2B protein expression and phosphorylation, suggesting the inhibitory effect of STEP on pERK and NMDA-R.[143] Through MEK activation and normalization of pERK in the PFC, direct BDNF infusion into the dorsomedial PFC immediately after the end of the last cocaine SA session resulted in a long-term inhibition on context-, cue-, or cocaine-induced relapse.[81] Thus, it indicated that rescuing the ERK signaling or hypofunction in the PFC during early withdrawal might provide a potential therapeutic strategy for preventing cocaine relapse.

Several animal models have been used to dissect the ERK signaling cascade in cocaine-induced behavioral changes. For example, double knock-out (KO) type 1 and type 8Ca^{2+}-stimulated AC resulted in a reduction of basal pERK in medium spiny neurons in the striatum with blunted acute cocaine–induced pERK, pMSK-1, and pCREB. Behaviorally, these double KO AC mice are supersensitive to low-dose acute cocaine–induced locomotion and fail to develop behavioral sensitization in response to repeated cocaine administration.[90] Ras-guanine nucleotide-releasing factors 1 (Ras-GRF1), the upstream activator of Ras, can increase ERK signaling. In the striatum, the protein expression of Ras-GRF-1 is increased by acute psychostimulants including cocaine.[144,145] D1-R agonist and Glu-induced pERK is attenuated in the striatal slice of Ras-GRF-1 KO mice. The acute cocaine–induced pERK is downregulated and upregulated in Ras-GRF-1 KO and overexpressing (OE) mice, respectively. In addition, the development of cocaine behavioral sensitization and cocaine-CPP are attenuated in Ras-GRF-1 KO mice accompanied by a reduction of FosB/ΔFosB in the striatum. An opposite facilitation on behavior and FosB/ΔFosB was observed in Ras-GRF-1 OE mice in response to repeated cocaine.[91] ERK1 KO mice exhibit higher responsibility to morphine.[94] Similarly, in response to chronic cocaine exposure, ERK1 KO mice display enhanced behavioral sensitization and cocaine-CPP as well as *c-fos* mRNA induction in the CPu.[34] This suggests that ERK1 acts as an inhibitor on ERK2 activation and a heightened stimulus- or cocaine-induced ERK2 signaling after ERK1 KO.[146] In addition, selective ERK2 OE in the VTA resulted in an increase of sensitivity of cocaine-CPP and the repeated cocaine-mediated behavioral sensitization.[96] In contrast, inhibition of ERK2 activity in the VTA attenuated the cocaine-CPP and the development and expression of cocaine-induced locomotor sensitization. Through activating MSKs, ERK leads to the increase of CREB activity. The acute cocaine–induced pCREB and IEGs as well as histone H3 phosphorylation were attenuated in the striatum of MSK-1 KO mice, indicating the role of MSK-1 in chromatin remodeling in response to cocaine. Although showing higher sensitivity to low-dose cocaine-CPP, MSK-1 KO mice have reduced behavioral sensitization in response to repeated cocaine administration.[97] Finally, systemic injection of the peptide-inhibiting pElK-1 significantly inhibited acute cocaine–activated pElk-1, pElk-1 nuclear translocation, and histone H3 phosphorylation, as well as IEGs protein and mRNA expression in the CPu and NAc.[98,147] Further, the inhibition of pElk-1 also resulted in an attenuation of repeated cocaine-induced dendritic plasticity in the NAc shell and prevented repeated cocaine-induced behavioral sensitization and CPP.[98] Together, these studies demonstrated that ERK-associated signaling is important for the long-term cocaine-mediated behavioral alterations, rewarding effects, and neuronal plasticity. Interestingly, the acute cocaine–mediated locomotor activity was not altered in animal models with manipulation of ERK1 or downstream molecular targets of ERK (e.g., MSK-1, ElK-1), further supporting that ERK signaling is not required for the acute cocaine–induced psychomotor effect.

Since both NMDA- and D1-Rs are implicated in cocaine-induced pERK, the direct protein–protein interaction between both receptors may underlie their effects on ERK activation.[148–151] Previously, we have demonstrated the protein–protein interaction between D1-R and GluN1 of NMDA-R in the CPu. The D1-R/GluN1 complex is disrupted after acute cocaine administration which may underlie transient pERK induction by cocaine.[152] The assumption is supported by a recent finding indicating that interference of D1-R/GluN1 association *in vitro* decreases D1 agonist- and NMDA-induced pERK induction. In addition, disrupting the protein–protein interaction in the NAc also attenuates acute cocaine-induced pERK induction and repeated cocaine-induced behavioral sensitization in the two-injection protocol.[153] Further, the receptor complex of sigma-1, histamine H_3, and D1-Rs has been found in the striatum. Through binding to sigma-1-R, cocaine results in a disinhibitory effect of histamine H_3 receptor on D1-Rs leading to pERK activation after either acute cocaine injection or cocaine SA.[154] However, the impact of these receptor–receptor interactions on cocaine-induced behavioral alteration is still unknown.

3.2 Amphetamine

Acute AMPH has been shown to increase pERK in the CPu, NAc, PFC, and VTA.[51,64,71,82,83,155–157] Multiple upstream receptors and molecular activators have been implicated in acute AMPH-induced ERK signaling in a brain-region-specific manner. For instance, acute AMPH-induced pMEK and pERK in the striatum is regulated by D1-R/DARPP-32 and NMDA-R activation.[51] In contrast, pERK induction in the PFC by acute AMPH is dependent on NMDA-R, adrenoceptors, and serotonin receptors but not D1- or D2-Rs.[158] Blockade of mGluR1/5 or mGluR5 specifically in the CPu attenuates acute AMPH-induced pERK, pElk-1, pCREB, and Fos immunoreactivity.[159–161] The activation of Ca^{2+}/CaM-dependent protein kinases II (CaMK II) in the CPu is also necessary for acute AMPH-augmented pERK, pElk-1, and pCREB.[159] Direct MEK inhibition via systemic SL327 (20–100 mg/kg) administration or intra-CPu U0216 (2 μg/side) infusion attenuated acute AMPH-elevated pERK and pCREB protein expression in the CPu and NAc, and IEGs including *preproenkephalin*, *preprodynorphin*, and *c-fos* mRNA in the CPu.[71,82,83,162] However, the differential pERK induction profile in the striatum in response to acute psychostimulants is determined by the environment: acute AMPH and cocaine-induced pERK expression mainly in D1-R-expressing neurons,[51,108,163] whereas, in a novel environment, AMPH dominantly increases pERK in D2-R-containing neurons of the striatum.[162] In line with cocaine, protein phosphatases have been shown to be induced by acute AMPH administration, which may control ERK activity after AMPH stimulation. For example, in the striatum, acute AMPH significantly increases pSTEP in a DARPP-32-dependent manner.[51] In addition, acute AMPH increases the gene encoding PP2B in the striatum,[164] relevant to *MKP-1* mRNA expression and DARPP-32/STEP activity.[53,165]

Behaviorally, similar to their enhanced response to the rewarding properties of morphine and cocaine, ERK1 KO mice exhibit higher hyperlocomotion after acute AMPH injection.[34,63,94] ERK1 KO mice display increased basal locomotor activity accompanied by a reduction of pRSK expression in the PFC and striatum,[63,94,95] indicating a blunted ERK-mediated signaling after ERK1 ablation. The increased basal and acute AMPH–induced locomotion as well as the reduction of pRSK can be replicated by chronic and continuous infusion of MEK inhibitor, PD98059 (50 μM), and selective knockdown (KD) of ERK1 in the PFC.[68] Although the predominant hypothesis indicates that enhanced stimuli-activated ERK2 signaling in the striatum in ERK1 KO mice is responsible for increased behavioral responses to abused drugs,[34,94] the reduction of ERK-mediated molecular cascade, at least in the PFC, may also contribute to both basal and drug-induced behavioral phenotype due to a general inhibition of ERK1 and ERK2 activity by MEK inhibitor. The latter assumption is supported by our recent finding demonstrating that rats raised in enriched environment have an augmented basal pERK induction in the PFC associated with lower basal and repeated nicotine-induced locomotion compared to control animals.[166] The acute AMPH–induced hyperactivity was not altered by SL327 (30–40 mg/kg) but attenuated by high doses of SL327 (50–100 mg/kg) with a potentially inhibitory effect on basal locomotion.[64,65,71,82,83] Although inhibiting acute AMPH–induced locomotor activity, acute systemic MEK inhibition by SL327 (50 mg/kg) resulted an enhancement to the basal locomotion.[167] The discrepancy may be accounted for experimental procedure, since a potentiated acute AMPH–activated locomotor activity was documented after pERK suppression in the CPu of rats without habituating to the behavioral apparatus.[161]

In a D1- and D2-Rs dependent manner, AMPH challenge after withdrawal from repeated AMPH exposure resulted in behavioral sensitization, which is associated with pERK and pCREB sensitization in the CPu.[168,169] The chronic AMPH-augmented pERK and pCREB induction is attenuated by D1- but not D2-Rs antagonist. Thus, although antagonism of both D1- and D2-Rs can inhibit the expression of behavioral sensitization, only D1-R-mediated ERK and CREB activation is critical for the expression of behavioral sensitization of the AMPH challenge. In contrast to the CPu, the expression of AMPH-induced behavioral sensitization is required for ERK's inhibitory effect on CREB activity modulated by Ca^{2+} voltage-gated channels in the NAc.[71] However, in the VTA, withdrawal from repeated AMPH exposure results in elevated MKP-1 and PP2B protein expression to downregulate the AMPH-mediated pERK induction.[157] Systemic administration of SL327 (30 or 40 mg/kg) dose-dependently prevents the development and expression of behavioral sensitization as well as the acquisition of conditioned locomotor response to AMPH administration.[64,71] A previous study demonstrated that intra-NAc AMPH infusion led to pERK and the establishment of CPP.[84] The AMPH-CPP was prevented by direct intra-NAc PD98059 (2.5 μg/side) infusion either before or after each conditioning session, suggesting the role of ERK on memory acquisition and consolidation of association of contextual rewarding effect of AMPH. However, the enhanced locomotor response by intra-NAc AMPH infusion is not affected by MEK inhibition. Altogether, it seems that ERK plays an important role in chronic AMPH-induced behavioral alterations ranging from behavioral sensitization, conditioned locomotor response to CPP. However, dynamic molecular mechanisms underlying behaviors including ERK-mediated downstream targets and the modulatory effect of ERK-related protein phosphatases should be further elucidated in specific brain regions associated to AMPH.

3.3 Methamphetamine

Methamphetamine (METH) is a highly addictive psychostimulant causing a serious and growing worldwide problem associated with medical, socioeconomic, and legal domains.[170,171] Although accumulating evidence has implicated the Glu and DA neurotransmission in METH-induced behavioral changes,[172–176] a direct exploration of their downstream target, ERK signaling, is limited. Acute METH (3 mg/kg) injection significantly increases pERK in the striatum, which is attenuated in serine racemase KO mice.[177] Serine racemase is an enzyme synthesizing D-serine, an endogenous coagonist of NMDA-R, thereby, partially supporting the requirement of NMDA-R for acute METH-induced pERK. In contrast, a recent study demonstrated that acute METH (2 mg/kg) did not affect pERK in either CPu or NAc.[178] The dose of METH, routes of administration, or the timing of collecting tissue may contribute to the discrepancy.

METH challenge after withdrawal from repeated METH exposure has been shown to induce behavioral sensitization related to pERK induction in both CPu and NAc as well as ΔFosB expression in the CPu.[142,178,179] The development and expression of METH behavioral sensitization and challenge-augmented pERK induction were inhibited by levo-tetrahydropalmatine, an antagonist of D1- and D2-Rs,[178,180,181] suggesting the involvement of DA receptors in chronic METH-induced pERK and behavioral sensitization. However, the METH challenge-elevated pERK is associated to the consequences of acute stimulation, since the pERK protein expression in the striatum is transiently increased during early withdrawal or not altered after long-term abstinence.[142,182] In agreement with the increase of pERK induction in the NAc shell after 1-day withdrawal from METH sensitization,[142] 2 h withdrawal from METH SA resulted in elevated D1-R, pCREB, and ΔFosB protein expression as well as transcriptional regulating genes including CREB, Elk-1, and Fos family in the striatum.[183,184] Genes associated with dual-specificity phosphatases 12 and protein tyrosine phosphatase were also upregulated, implying an inhibitory mechanism to dampen ERK signaling during the early phase of withdrawal from METH SA.[184–186] In both D1- and NMDA-Rs dependent manners, acute or chronic METH administration results in increases of *MKP-1* and *MKP-3* mRNA in several brain regions including the PFC, orbital cortex, CPu, NAc, and HIPP.[187,188] Therefore, the ERK-driven MKPs expression and other phosphatases represent a positive feedback to gate the transient ERK activation in response to acute or chronic METH exposure.

The increase of pERK, pElk-1, pCREB, and/or ΔFosB protein expression in the CPu, NAc, or PFC is related to METH-induced CPP.[85,189] Specifically, the acquisition of CPP and pERK induction in the NAc by METH-CPP require D1-R but not NMDA-R activation. Intra-NAc infusion of MEK inhibitor, PD98059 (2 μg/side), also prevents the expression of METH-CPP and pERK induction.[85] Therefore, this demonstrates the importance of the activation of D1-R/MEK/ERK/pElk-1 in the NAc on the development and expression of METH-CPP. In contrast, the METH-CPP testing reduced pERK and pCREB in the NAc after a single pairing session with 2 days withdrawal,[190] suggesting either a compensatory reduction in response to overactivation of ERK signaling during conditioning and withdrawal or other molecular cascades are required for the initial acquisition of METH-CPP. Both assumptions should be further deciphered to identify molecular mechanisms underlying the difference between single and multiple condition session-mediated METH-CPP.

Chronic METH use causes cognitive deficits associated with altered neurotransmission.[191–194] In animal studies, repeated METH administration leads to spatial learning and memory impairment, which is associated with reduced total ERK in the PFC.[195] In addition, deficits in spatial working memory and novel object recognition (NOR) are accompanied by an inability of pERK induction in the HIPP and PFC by the learning process or stimuli.[196–199] Interestingly, intra-PFC infusion of PD98509 (2 μg/side) mimics the METH-induced cognitive impairment in NOR,[196] indicating that a reduced pERK signaling is responsible for the cognitive dysfunction after long-term METH exposure. Several drugs have been demonstrated to ameliorate the cognitive deficit by METH through ERK signaling. For example, depending on nicotinic acetylcholine receptors (nAChRs), D1-R and MEK activation, galantamine, a drug used to treat Alzheimer's disease by inhibiting acetylcholinesterase and allosterically modulating nAChRs, alleviates NOR impairment through pERK induction in the PFC.[198] Similarly, modafinil, a cognitive enhancer with a weak DA-transporter-inhibiting effect, also activates pERK in the PFC to rescue the NOR deficit[199,200] probably through increasing extracellular DA levels. Finally, clozapine, an atypical antipsychotic medication, reverses dysfunctional pERK signaling in the HIPP with an attenuating effect on spatial working memory impairment induced by chronic METH.[197] Taken together, these results demonstrate that the cognitive impairment induced by chronic METH is attributed to the downregulation of ERK signaling during learning and memory-a potential therapeutic molecular biomarker for future drug development.

3.4 Marijuana

Δ9-Tetrahydrocannabinol (THC) is the main psychoactive component of marijuana, which is one of the most used illicit drugs.[201] Cannabinoid receptors 1 and 2(CB1-R and CB2-R) have been identified and located mainly in neuronal and peripheral tissues, respectively. Activation of the CB1-R leads to the closing of calcium and the opening of potassium channel, subsequently inhibiting AC and activating protein kinase including ERK.[202] Acute low-dose THC injection (1 mg/kg) has been demonstrated to increase pERK expression in the mesocorticolimbic system.[99,203] Specifically, in the striatum, the THC-activated pERK is mediated by CB1-, D1-, D2-, and NMDA-Rs indicating a synergistic action among cannabinoid, DA and Glu neurotransmission. Acute THC–induced ERK downstream targets, pElk-1 and *zif268* mRNA, were inhibited by D1-R antagonist and MEK inhibitor, SL327 (100 mg/kg). Further, in response to repeated low dose of THC injection, the development of THC-CPP was attenuated by SL327 (50 mg/kg), suggesting that ERK-regulated signaling is involved in THC-rewarding effect.[87]

Similarly, acute THC (1 mg/kg) induced transient pERK induction in the HIPP was dependent on the activation of CB1- and NMDA-Rs. SL327 (100 mg/kg) pretreatment also inhibited the acute THC–induced IEG expression (c-Fos protein, *Zif268*, and *BDNF* mRNAs) in the HIPP.[204] However, the relevance between acute THC–induced behavioral changes and ERK signaling should be further elucidated, since there is no significant locomotor activity alteration by low-dose THC.[205]

In contrast to the low dose of THC, high dose of THC (≥10 mg/kg) acutely resulted in hypolocomotor activity in the striatum and cerebellum with pERK, pCREB, and *c-fos* induction depending on CB1-R and Ras-GRF1.[86,92,203] Although the ERK signaling in both brain regions is distinct from the acute THC–induced hypolocomotor, it is necessary for the development of behavioral tolerance, a gradual behavioral recovery from the initial hypolocomotor by acute THC, after repeated THC injection.[86,92,206] In an ERK-dependent manner, the behavioral tolerance is mediated by recruiting G-protein-coupled receptor kinases and β-arrestins to desensitize and internalize CB1-R in the striatum and cerebellum. The chronic THC-mediated cerebellar synaptic transmission and plasticity as well as reduced sensitivity of CB1-R activation were also prevented in Ras-GRF1 KO mice.[207] In addition, chronic THC-exposure-induced pCREB and FosB protein expression in the PFC and HIPP is inhibited by either SL327 (50 mg/kg) or in Ras-GRF1null mice.[206] Taken together, the results demonstrated that, in response to a high dose of THC, the activation of pERK-mediated signaling in the striatum and cerebellum is critical for the development of behavioral tolerance. In the PFC and HIPP, the ERK-associated molecular cascade may underlie the addicted state for THC. However, the latter assumption needs to be examined due to a similar analgesic tolerance effect after a chronic high dose of THC. Since THC-induced behavioral sensitization and SA have been documented,[208–210] it will be worthwhile to determine the role of ERK on reinforcing/rewarding effects of THC in specific brain region(s).

In addition to the THC action on CB-1R, DA agonists and psychostimulants have been shown to increase endocannabinoid release.[211–213] A previous study has indicated that acute cocaine–induced pERK and c-Fos in the CPu and NAc was inhibited by CB1-R antagonist pretreatment, and mice with CB1-R KO or conditional deletion in the forebrain neurons.[104] In addition, the elevated pERK protein expression-induced by chronic cocaine in the VTA is dependent on CB1-R activation. The development of cocaine-CPP and underlying pERK induction were also inhibited by intra-VTA CB1-R antagonist infusion,[73] implicating the role of CB1-R and endocannabinoids in regulating the rewarding effect of cocaine mediated by ERK signaling activation.

3.5 Nicotine

Cigarette smoking is the largest preventable cause of death and diseases worldwide with an estimated 6 million deaths each year.[214] It has been shown that nicotine, through activation of the DA- and Glu-related signaling in the mesocorticolimbic system, exerts its reinforcing effects.[215–217] Several *in vitro* studies have demonstrated that activation of ERK and CREB by acute and chronic administration of nicotine depends on nAChRs, CaMKs, PKA, and MEK activity.[218–222] A genome-wide expression analysis revealed acute nicotine exposure, through activation of ERK signaling, induced alterations of gene expression.[223] Similarly, acute nicotine–induced transient ERK activation through nAChRs, Ca^{2+} voltage-gated channels, CaMKs, and MEK in primary cortical and hippocampal neurons;[224,225] however, only PKA is required for pERK induction by nicotine in the hippocampal neurons, suggesting differential upstream activators for ERK activity in distinct neuronal types. Chronic nicotine exposure in mesencephalic dopaminergic neuronal culture resulted in increases of dendritic length and soma size through nAChRs- and D3-R-recrutied ERK signaling,[226] demonstrating that ERK involves nicotine-mediated structural neuronal plasticity. *In vivo*, acute nicotine administration increases pERK levels in the NAc, CPu, PFC, Amy, and BNST.[51,99,227,228] In the striatum, acute nicotine–induced pERK is mediated by D1-R/PKA/DARPP-32 signaling pathways,[99,100] indicating the relevance of dopaminergic neurotransmission in response to nicotine. After chronic oral consumption of nicotine,the levels of pERK and pCREB were increased in the PFC, but pCREB was decreased in the NAc,[227] suggesting an increase of PFC excitatory output into the NAc. Indeed, pERK was increased in the NAc of nicotine-induced CPP animals,[228] supporting the role of PFC-NAc projection in the conditioned rewarding effect of nicotine. Interestingly, a direct protein–protein interaction between α7nAChR and GluN2A has been identified in the HIPP, which can be upregulated by chronic nicotine exposure.[229] After nicotine SA, disruption of the α7nAChR–NMDAR complex decreased ERK activity and blocked cue-induced reinstatement of nicotine-seeking behavior.[229] Taken together, these results demonstrate that the ERK signaling pathway is a key integrator of the DA/D1 receptor and

Glu/NMDA receptor signaling that induces long-term cellular alterations and behavioral adaptation in response to nicotine exposure. However, a direct manipulation on ERK is warranted to examine its effect on nicotine-induced behavioral changes.

Environment is an important factor affecting the vulnerability for drug abuse.[230–232] Exposure to an environmental enrichment paradigm results in neurobiologic adaptations, particularly in the PFC of the mesocorticolimbic dopaminergic system.[166,233–235] Our recent study has demonstrated that the basal level of pERK was higher in animals housing in an enriched environmental condition compared to animals housing in an impoverished condition, which was negatively correlated with their respective baseline locomotor activities.[166] After nicotine sensitization or nicotine SA, the pERK induction was significantly increased in the PFC of rats raised in either impoverished or standard condition. In contrast, due to their higher basal ERK activity in the PFC, nicotine did not alter pERK protein levels in animals raised in an enriched environmental condition with a decreased sensitivity in response to chronic nicotine.[166] Regardless of raising conditions, the pERK induction is positively correlated to the amount of nicotine intake during nicotine SA. Thus, these results suggest that pERK induction in the PFC may underlie the rewarding effect of nicotine, which is consistent with a previous study demonstrating a preference for ERK-mediated signaling pathway activation in the PFC after nicotine SA.[236]

3.6 Alcohol (Ethanol)

In a time-dependent manner, acute injection with higher doses of ethanol (EtOH 2.5–4.7 g/kg) reduced pERK and pCREB in the PFC, NAc, CPu, Amy, HIPP, cerebellum, and BNST in various ages of rodents.[237–240] In contrast, acute administration of a lower dose of EtOH (1 g/kg) significantly increased pERK in various regions including the NAc and CeA in theD1-R and neuropeptide S receptor dependent manner.[241,242] The acute EtOH-induced *c-fos* induction in the medial Amy was inhibited by the MEK inhibitor, U0126.[243] Similarly, acute acetaldehyde (ACD), the first and main metabolite of EtOH, enhanced pERK in the NAc, CeA, and BNST through activation of D1-R and opioid receptors.[244,245] Behaviorally, the low dose of acute EtOH (1 mg/kg) is associated to an anxiolytic response accompanied by the rapid increase of spine density in the CeA and MeA through the BDNF-mediated TrkB phosphorylation and pERK/pElk-1/pCREB and *Arc* induction.[246] In addition to the acute EtOH-mediated pERK signaling, its intrinsic activation state may contribute to the alcohol intake or preference. For example, despite mixed results of acute EtOH-mediated pERK levels in alcohol-preferring animals, they have higher basal levels of pERK in the PFC and NAc as well as higher expression levels of Ras-GRF2, an upstream activator of MEK, in the brain, compared to their alcohol non-preferring counterparts.[93,247–249] The Ras-GRF2 KO mice exhibited lower EtOH intake associated with an aberrant DA transmission in the VTA-NAc projection mediated by ERK activation,[93] revealing a functional role of ERK on acute EtOH–mediated DA signaling underlying the preference of alcohol.

The effect of chronic EtOH exposure on pERK is heavily dependent on administering paradigms, time of withdrawal, and brain regions. Immediate cessation of repeated EtOH oral consumption and vaporized EtOH exposure has been demonstrated to decrease pERK in the PFC, NAc, CPu, Amy, and HIPP, although with an enhanced pERK induction after 7–11 h withdrawal.[240,250,251] In contrast, the chronic EtOH-attenuated neuronal plasticity during early withdrawal (e.g., within 1-day withdrawal) is associated with the downregulation of pERK and the inability of pERK induction in response to stimulus in the HIPP.[252] Similarly, a desensitization/tolerance of pERK or *c-fos* expression in response to EtOH re-exposure or challenge after withdrawal from repeated EtOH has been found in the PFC and HIPP.[243,251] Further, a paralleled attenuated phosphorylation of GluN1and CaMKII is also documented immediately after chronic EtOH exposure.[240,251] Taken together, these results demonstrate that a reduction of Glu-receptor-mediated ERK activity during early withdrawal leads to the desensitization of subsequent EtOH-induced pERK signaling. In contrast, Pandey and coworkers demonstrated that 24-h withdrawal from repeated EtOH consumption produced anxiety-like behavior followed by blunted BDNF/TrkB/pERK/pElk-1 and Arc protein expression with reduced spine density in the CeA and MeA.[246] Intra-CeA BDNF infusion restored the early withdrawal-induced ERK signaling dysfunction and inhibited the anxiety-like behavior. Similarly, KD of the BDNF-mediated ERK signaling in the CeA and MeA induces anxiety and promotes EtOH intake.[253] This suggests that withdrawal syndrome after chronic EtOH consumption accompanied by physical signs and negative emotional state (e.g., anxiety, depression, and irritability)[254,255] may precipitate the relapse of EtOH intake. In addition, withdrawal from repeated EtOH has been demonstrated to result in an enhancement of fear conditioning depending on pERK activation in the BLA by NMDA-R and MEK activation.[256] Thus, the ERK signaling in the Amy complex is important for the development and further acquisition of the negative-affective state underlying the vulnerability for subsequent alcohol-seeking behavior after withdrawal.

The role of ERK signaling pathway in EtOH-mediated rewarding effect has been documented. In the CPP model, the alcohol metabolite ACD-CPP is dependent on D1-R activation and the development of ACD-CPP can be attenuated by MEK inhibitor, PD98059.[88] In contrast, a previous study has indicated that the systemic SL327 administration in the ERK-independent learning mechanism in EtOH-CPP did not affect the acquisition, expression, and extinction of EtOH-CPP (2 g/kg) as well as pERK after acute EtOH administration (2.5 g/kg).[66] However, EtOH-CPP has been shown to be established by the lower dose of EtOH (1 g/kg) in D1-R-dependent manner,[88] which is sufficient to induce pERK induction as described earlier. Based on the D1-R-activated pERK and significant pERK induction after acute EtOH, it is required to further test the effect of D1-R/ERK signaling on EtOH-CPP by the lower dose of EtOH. The ERK signaling is also implicated in the operant rewarding effect of EtOH. For instance, systemic MEK inhibition resulted in an increase of EtOH SA,[67] indicating the antagonism of acute pharmacologic effect of alcohol promoting the drug-taking behavior. After abstinence from EtOH SA, re-exposure to conditioned cues induced alcohol-seeking behavior accompanied by pERK and c-Fos expression in the BLA.[257] The ERK signaling is also critical for the Glu-transmission-mediated alcohol seeking. After extinction from EtOH SA, systemic mGluR5 inhibition attenuated cue-induced reinstatement and the cue-induced pERK expression in the BLA and NAc shell in alcohol-preferring rats.[258] Probably through restoring the Glu transmission, L-cysteine prevents EtOH SA and EtOH-primed-induced drug seeking.[259] In addition, the reinstatement-induced pERK in the NAc shell is also inhibited by the L-cysteine pretreatment. Finally, ERK activation in the VTA has been demonstrated to mediate the inhibitory effect of GDNF in preventing EtOH intake and reacquisition of EtOH SA after extinction.[89]

4. CONCLUSIONS AND FUTURE DIRECTIONS

Drug addiction is a significant public health problem and has been considered as a chronic psychiatric disorder, characterized by craving and compulsive drug-seeking and use. The main obstacle in drug-addiction treatment is the cycle for relapse/reinstatement from drugs of abuse. This review summarizes the current understanding of the role of ERK signaling and its associated intracellular signaling pathways in drug-induced neuroadaptive changes underlying the rewarding and reinforcing mechanisms in response to abused drugs. Despite the differential regulatory pathways in which all drugs of abuse can affect ERK signaling, one evolving theme in all cases is the regulation of the ERK molecules at the phosphorylation level. It is therefore important to understand the precise mechanisms that underlie the regulation of ERK phosphorylation by different drugs of abuse. The ERK signaling pathway may play a critical role in the early intervention during withdrawal from chronic drug administration. For example, recent studies demonstrated that normalizing the prefrontal ERK signaling pathway during the early withdrawal from repeated cocaine exposure leads to a long-term inhibitory effect on cocaine relapse[81,141] and restores the extracellular Glu dysregulation in the NAc.[260] In contrast, after prolonged withdrawal from cocaine, an increase of PKA-mediated signaling is dominant in the PFC and NAc responsible for cocaine seeking,[81,116,261–263] implicating that ERK activity in the PFC-NAc projection is dynamically regulated by multiple intracellular pathways. Future studies will evaluate the ability of novel therapeutic interventions to restore normal ERK signaling activity in the brain for inhibiting addictive drug-seeking behavior. On the other hand, environmental factors can also influence vulnerability to drug addiction. We demonstrated that environmental enrichment induces compensatory alterations of D1R/DARPP-32 and ERK1/2 signaling pathways in the PFC, which may contribute to environmental enrichment-dependent reduction of susceptibility to nicotine.[166,264] Although current knowledge of multiple factors regulating ERK activity has greatly expanded, many aspects of this regulation remain to be elucidated. For example, OE of microRNA-221 attenuates nicotine-induced pERK,[265] whereas activation of ERK can regulate microRNA-221 expression.[266] Furthermore, ERK signaling has been associated with epigenetic mechanisms including chromatin remodeling through histone methylation and DNA methylation associated with drug addiction,[267–269] which are critical for the regulation of gene expression, neuronal plasticity, and drug-induced behavioral alteration. In summary, several molecular signaling pathways are involved in the complexity of drug addiction; with ERK being the most highly characterized during the past two decades. Herein, we provide the general role of ERK-mediated molecular cascade in response to various abused drugs, but it is by no means exhaustive. Future studies are warranted to dissect the ERK signaling pathway, providing a better understanding for the development of feasible and potential therapeutic strategies for drug addiction and related disorders.

ACKNOWLEDGMENT

This research was supported by a grant from the National Institute on Drug Abuse to Jun Zhu (DA035714). We acknowledge Dr. Jacqueline F. McGinty (Medical University of South Carolina, Charleston, SC) for commenting on the manuscript.

REFERENCES

1. Nestler EJ. Molecular basis of long-term plasticity underlying addiction. *Nat Rev Neurosci.* 2001;2(2):119–128.
2. Kalivas PW. Glutamate systems in cocaine addiction. *Curr Opin Pharmacol.* 2004;4(1): 23–29.
3. Feltenstein MW, See RE. The neurocircuitry of addiction: an overview. *Br J Pharmacol.* 2008;154(2):261–274.
4. Swanson LW. The projections of the ventral tegmental area and adjacent regions: a combined fluorescent retrograde tracer and immunofluorescence study in the rat. *Brain Res Bull.* 1982;9(1–6):321–353.
5. Schultz W. Predictive reward signal of dopamine neurons. *J Neurophysiol.* 1998;80(1): 1–27.
6. Wise RA. Dopamine, learning and motivation. *Nat Rev Neurosci.* 2004;5(6):483–494.
7. Canales JJ. Stimulant-induced adaptations in neostriatal matrix and striosome systems: transiting from instrumental responding to habitual behavior in drug addiction. *Neurobiol Learn Mem.* 2005;83(2):93–103.
8. Wise RA. Roles for nigrostriatal—not just mesocorticolimbic—dopamine in reward and addiction. *Trends Neurosci.* 2009;32(10):517–524.
9. Kyosseva SV. Mitogen-activated protein kinase signaling. *Int Rev Neurobiol.* 2004;59:201–220.
10. Kim EK, Choi EJ. Pathological roles of MAPK signaling pathways in human diseases. *Biochim Biophys Acta.* 2010;1802(4):396–405.
11. Berhow MT, Hiroi N, Nestler EJ. Regulation of ERK (extracellular signal regulated kinase), part of the neurotrophin signal transduction cascade, in the rat mesolimbic dopamine system by chronic exposure to morphine or cocaine. *J Neurosci.* 1996;16(15):4707–4715.
12. Atkins CM, Selcher JC, Petraitis JJ, Trzaskos JM, Sweatt JD. The MAPK cascade is required for mammalian associative learning. *Nat Neurosci.* 1998;1(7):602–609.
13. Seger R, Krebs EG. The MAPK signaling cascade. *FASEB J.* 1995;9(9):726–735.
14. Chen RH, Sarnecki C, Blenis J. Nuclear localization and regulation of erk- and rsk-encoded protein kinases. *Mol Cell Biol.* 1992;12(3):915–927.
15. Gille H, Strahl T, Shaw PE. Activation of ternary complex factor Elk-1 by stress-activated protein kinases. *Curr Biol.* 1995;5(10):1191–1200.
16. Gille H, Sharrocks AD, Shaw PE. Phosphorylation of transcription factor p62TCF by MAP kinase stimulates ternary complex formation at c-fos promoter. *Nature.* 1992;358 (6385):414–417.
17. Davis S, Vanhoutte P, Pages C, Caboche J, Laroche S. The MAPK/ERK cascade targets both Elk-1 and cAMP response element-binding protein to control long-term potentiation-dependent gene expression in the dentate gyrus *in vivo*. *J Neurosci.* 2000;20(12): 4563–4572.
18. Hill CS, Marais R, John S, Wynne J, Dalton S, Treisman R. Functional analysis of a growth factor-responsive transcription factor complex. *Cell.* 1993;73(2):395–406.

19. Treisman R. Regulation of transcription by MAP kinase cascades. *Curr Opin Cell Biol.* 1996;8(2):205–215.
20. Pende M, Fisher TL, Simpson PB, Russell JT, Blenis J, Gallo V. Neurotransmitter- and growth factor-induced cAMP response element binding protein phosphorylation in glial cell progenitors: role of calcium ions, protein kinase C, and mitogen-activated protein kinase/ribosomal S6 kinase pathway. *J Neurosci.* 1997;17(4):1291–1301.
21. Frodin M, Gammeltoft S. Role and regulation of 90 kDa ribosomal S6 kinase (RSK) in signal transduction. *Mol Cell Endocrinol.* 1999;151(1–2):65–77.
22. Xing J, Kornhauser JM, Xia Z, Thiele EA, Greenberg ME. Nerve growth factor activates extracellular signal-regulated kinase and p38 mitogen-activated protein kinase pathways to stimulate CREB serine 133 phosphorylation. *Mol Cell Biol.* 1998;18(4): 1946–1955.
23. Merienne K, Pannetier S, Harel-Bellan A, Sassone-Corsi P. Mitogen-regulated RSK2–CBP interaction controls their kinase and acetylase activities. *Mol Cell Biol.* 2001;21(20):7089–7096.
24. Arthur JS, Cohen P. MSK1 is required for CREB phosphorylation in response to mitogens in mouse embryonic stem cells. *FEBS Lett.* 2000;482(1–2):44–48.
25. Kurino M, Fukunaga K, Ushio Y, Miyamoto E. Activation of mitogen-activated protein kinase in cultured rat hippocampal neurons by stimulation of glutamate receptors. *J Neurochem.* 1995;65(3):1282–1289.
26. Robinson TE, Kolb B. Alterations in the morphology of dendrites and dendritic spines in the nucleus accumbens and prefrontal cortex following repeated treatment with amphetamine or cocaine. *Eur J Neurosci.* 1999;11(5):1598–1604.
27. Chandler LJ, Sutton G, Dorairaj NR, Norwood D. *N*-methyl D-aspartate receptor-mediated bidirectional control of extracellular signal-regulated kinase activity in cortical neuronal cultures. *J Biol Chem.* 2001;276(4):2627–2636.
28. Xia Z, Dudek H, Miranti CK, Greenberg ME. Calcium influx via the NMDA receptor induces immediate early gene transcription by a MAP kinase/ERK-dependent mechanism. *J Neurosci.* 1996;16(17):5425–5436.
29. Mao L, Tang Q, Samdani S, Liu Z, Wang JQ. Regulation of MAPK/ERK phosphorylation via ionotropic glutamate receptors in cultured rat striatal neurons. *Eur J Neurosci.* 2004;19(5):1207–1216.
30. Perkinton MS, Ip JK, Wood GL, Crossthwaite AJ, Williams RJ. Phosphatidylinositol 3-kinase is a central mediator of NMDA receptor signalling to MAP kinase (Erk1/2), Akt/PKB and CREB in striatal neurones. *J Neurochem.* 2002;80(2):239–254.
31. Vincent SR, Sebben M, Dumuis A, Bockaert J. Neurotransmitter regulation of MAP kinase signaling in striatal neurons in primary culture. *Synapse.* 1998;29(1): 29–36.
32. Vanhoutte P, Barnier JV, Guibert B, et al. Glutamate induces phosphorylation of Elk-1 and CREB, along with c-fos activation, via an extracellular signal-regulated kinase-dependent pathway in brain slices. *Mol Cell Biol.* 1999;19(1):136–146.
33. Choe ES, McGinty JF. Cyclic AMP and mitogen-activated protein kinases are required for glutamate-dependent cyclic AMP response element binding protein and Elk-1 phosphorylation in the dorsal striatum *in vivo*. *J Neurochem.* 2001;76(2):401–412.
34. Ferguson SM, Fasano S, Yang P, Brambilla R, Robinson TE. Knockout of ERK1 enhances cocaine-evoked immediate early gene expression and behavioral plasticity. *Neuropsychopharmacology.* 2006;31(12):2660–2668.
35. Seyler CE. Nonlinear 3-D evolution of bounded kinetic Alfven waves due to shear flow and collisionless tearing instability. *Geophys Res Lett.* 1998;15(8):756.
36. Grewal SS, Horgan AM, York RD, Withers GS, Banker GA, Stork PJ. Neuronal calcium activates a Rap1 and B-Raf signaling pathway via the cyclic adenosine monophosphate-dependent protein kinase. *J Biol Chem.* 2000;275(5):3722–3728.

37. Vossler MR, Yao H, York RD, Pan MG, Rim CS, Stork PJ. cAMP activates MAP kinase and Elk-1 through a B-Raf- and Rap1-dependent pathway. *Cell.* 1997;89(1):73–82.
38. Gerfen CR, Miyachi S, Paletzki R, Brown P. D1 dopamine receptor supersensitivity in the dopamine-depleted striatum results from a switch in the regulation of ERK1/2/MAP kinase. *J Neurosci.* 2002;22(12):5042–5054.
39. Impey S, Obrietan K, Wong ST, et al. Cross talk between ERK and PKA is required for Ca^{2+} stimulation of CREB-dependent transcription and ERK nuclear translocation. *Neuron.* 1998;21(4):869–883.
40. Roberson ED, English JD, Adams JP, Selcher JC, Kondratick C, Sweatt JD. The mitogen-activated protein kinase cascade couples PKA and PKC to cAMP response element binding protein phosphorylation in area CA1 of hippocampus. *J Neurosci.* 1999;19(11):4337–4348.
41. Zanassi P, Paolillo M, Feliciello A, Avvedimento EV, Gallo V, Schinelli S. cAMP-dependent protein kinase induces cAMP-response element-binding protein phosphorylation via an intracellular calcium release/ERK-dependent pathway in striatal neurons. *J Biol Chem.* 2001;276(15):11487–11495.
42. Dale LB, Babwah AV, Bhattacharya M, Kelvin DJ, Ferguson SS. Spatial-temporal patterning of metabotropic glutamate receptor-mediated inositol 1,4,5-triphosphate, calcium, and protein kinase C oscillations: protein kinase C-dependent receptor phosphorylation is not required. *J Biol Chem.* 2001;276(38):35900–35908.
43. Kawabata S, Kohara A, Tsutsumi R, et al. Diversity of calcium signaling by metabotropic glutamate receptors. *J Biol Chem.* 1998;273(28):17381–17385.
44. Schnabel R, Palmer MJ, Kilpatrick IC, Collingridge GL. A CaMKII inhibitor, KN-62, facilitates DHPG-induced LTD in the CA1 region of the hippocampus. *Neuropharmacology.* 1999;38(4):605–608.
45. Choe ES, Wang JQ. Group I metabotropic glutamate receptor activation increases phosphorylation of cAMP response element-binding protein, Elk-1, and extracellular signal-regulated kinases in rat dorsal striatum. *Brain Res Mol Brain Res.* 2001;94(1–2):75–84.
46. Yan Z, Feng J, Fienberg AA, Greengard P. D(2) dopamine receptors induce mitogen-activated protein kinase and cAMP response element-binding protein phosphorylation in neurons. *Proc Natl Acad Sci USA.* 1999;96(20):11607–11612.
47. Boulanger LM, Lombroso PJ, Raghunathan A, During MJ, Wahle P, Naegele JR. Cellular and molecular characterization of a brain-enriched protein tyrosine phosphatase. *J Neurosci.* 1995;15(2):1532–1544.
48. Lombroso PJ, Naegele JR, Sharma E, Lerner M. A protein tyrosine phosphatase expressed within dopaminoceptive neurons of the basal ganglia and related structures. *J Neurosci.* 1993;13(7):3064–3074.
49. Nika K, Hyunh H, Williams S, et al. Haematopoietic protein tyrosine phosphatase (HePTP) phosphorylation by cAMP-dependent protein kinase in T-cells: dynamics and subcellular location. *Biochem J.* 2004;378(Pt 2):335–342.
50. Zuniga A, Torres J, Ubeda J, Pulido R. Interaction of mitogen-activated protein kinases with the kinase interaction motif of the tyrosine phosphatase PTP-SL provides substrate specificity and retains ERK2 in the cytoplasm. *J Biol Chem.* 1999;274(31):21900–21907.
51. Valjent E, Pascoli V, Svenningsson P, et al. Regulation of a protein phosphatase cascade allows convergent dopamine and glutamate signals to activate ERK in the striatum. *Proc Natl Acad Sci USA.* 2005;102(2):491–496.
52. Nishi A, Bibb JA, Snyder GL, Higashi H, Nairn AC, Greengard P. Amplification of dopaminergic signaling by a positive feedback loop. *Proc Natl Acad Sci USA.* 2000;97(23):12840–12845.
53. Greengard P, Allen PB, Nairn AC. Beyond the dopamine receptor: the DARPP-32/protein phosphatase-1 cascade. *Neuron.* 1999;23(3):435–447.

54. Halpain S, Greengard P. Activation of NMDA receptors induces rapid dephosphorylation of the cytoskeletal protein MAP2. *Neuron.* 1990;5(3):237–246.
55. Nishi A, Bibb JA, Matsuyama S, et al. Regulation of DARPP-32 dephosphorylation at PKA- and Cdk5-sites by NMDA and AMPA receptors: distinct roles of calcineurin and protein phosphatase-2A. *J Neurochem.* 2002;81(4):832–841.
56. Bokemeyer D, Sorokin A, Yan M, Ahn NG, Templeton DJ, Dunn MJ. Induction of mitogen-activated protein kinase phosphatase 1 by the stress-activated protein kinase signaling pathway but not by extracellular signal-regulated kinase in fibroblasts. *J Biol Chem.* 1996;271(2):639–642.
57. Brondello JM, Brunet A, Pouyssegur J, McKenzie FR. The dual specificity mitogen-activated protein kinase phosphatase-1 and -2 are induced by the p42/p44MAPK cascade. *J Biol Chem.* 1997;272(2):1368–1376.
58. Muda M, Theodosiou A, Rodrigues N, et al. The dual specificity phosphatases M3/6 and MKP-3 are highly selective for inactivation of distinct mitogen-activated protein kinases. *J Biol Chem.* 1996;271(44):27205–27208.
59. Camps M, Nichols A, Gillieron C, et al. Catalytic activation of the phosphatase MKP-3 by ERK2 mitogen-activated protein kinase. *Science.* 1998;280(5367): 1262–1265.
60. Hafen E. Kinases and phosphatases—a marriage is consummated. *Science.* 1998;280 (5367):1212–1213.
61. Zhou B, Wang ZX, Zhao Y, Brautigan DL, Zhang ZY. The specificity of extracellular signal-regulated kinase 2 dephosphorylation by protein phosphatases. *J Biol Chem.* 2002;277(35):31818–31825.
62. Brondello JM, Pouyssegur J, McKenzie FR. Reduced MAP kinase phosphatase-1 degradation after p42/p44MAPK-dependent phosphorylation. *Science.* 1999;286 (5449):2514–2517.
63. Engel SR, Creson TK, Hao Y, et al. The extracellular signal-regulated kinase pathway contributes to the control of behavioral excitement. *Mol Psychiatry.* 2009;14(4):448–461.
64. Valjent E, Corvol JC, Trzaskos JM, Girault JA, Herve D. Role of the ERK pathway in psychostimulant-induced locomotor sensitization. *BMC Neurosci.* 2006;7:20.
65. Carr KD, de Vaca SC, Sun Y, Chau LS, Pan Y, Dela Cruz J. Effects of the MEK inhibitor, SL-327, on rewarding, motor- and cellular-activating effects of D-amphetamine and SKF-82958, and their augmentation by food restriction in rat. *Psychopharmacology (Berl).* 2009;201(4):495–506.
66. Groblewski PA, Franken FH, Cunningham CL. Inhibition of extracellular signal-regulated kinase (ERK) activity with SL327 does not prevent acquisition, expression, and extinction of ethanol-seeking behavior in mice. *Behav Brain Res.* 2011;217(2):399–407.
67. Faccidomo S, Besheer J, Stanford PC, Hodge CW. Increased operant responding for ethanol in male C57BL/6J mice: specific regulation by the ERK1/2, but not JNK, MAP kinase pathway. *Psychopharmacology (Berl).* 2009;204(1):135–147.
68. Creson TK, Hao Y, Engel S, et al. The anterior cingulate ERK pathway contributes to regulation of behavioral excitement and hedonic activity. *Bipolar Disord.* 2009;11 (4):339–350.
69. Valjent E, Corvol JC, Pages C, Besson MJ, Maldonado R, Caboche J. Involvement of the extracellular signal-regulated kinase cascade for cocaine-rewarding properties. *J Neurosci.* 2000;20(23):8701–8709.
70. Pierce RC, Pierce-Bancroft AF, Prasad BM. Neurotrophin-3 contributes to the initiation of behavioral sensitization to cocaine by activating the Ras/mitogen-activated protein kinase signal transduction cascade. *J Neurosci.* 1999;19(19):8685–8695.
71. Giordano TP, Tropea TF, Satpute SS, et al. Molecular switch from L-type Ca v 1.3 to Ca v 1.2 Ca^{2+} channel signaling underlies long-term psychostimulant-induced behavioral and molecular plasticity. *J Neurosci.* 2010;30(50):17051–17062.

72. Kim S, Shin JK, Yoon HS, Kim JH. Blockade of ERK phosphorylation in the nucleus accumbens inhibits the expression of cocaine-induced behavioral sensitization in rats. *Korean J Physiol Pharmacol.* 2011;15(6):389–395.
73. Pan B, Zhong P, Sun D, Liu QS. Extracellular signal-regulated kinase signaling in the ventral tegmental area mediates cocaine-induced synaptic plasticity and rewarding effects. *J Neurosci.* 2011;31(31):11244–11255.
74. Miller CA, Marshall JF. Molecular substrates for retrieval and reconsolidation of cocaine-associated contextual memory. *Neuron.* 2005;47(6):873–884.
75. Valjent E, Corbille AG, Bertran-Gonzalez J, Herve D, Girault JA. Inhibition of ERK pathway or protein synthesis during reexposure to drugs of abuse erases previously learned place preference. *Proc Natl Acad Sci USA.* 2006;103(8):2932–2937.
76. Liddie S, Itzhak Y. Variations in the stimulus salience of cocaine reward influences drug-associated contextual memory. *Addict Biol.* 2014 Oct 28. doi: 10.1111/adb.12191. [Epub ahead of print].
77. Wells AM, Arguello AA, Xie X, et al. Extracellular signal-regulated kinase in the basolateral amygdala, but not the nucleus accumbens core, is critical for context-response-cocaine memory reconsolidation in rats. *Neuropsychopharmacology.* 2013;38(5):753–762.
78. Lu L, Hope BT, Dempsey J, Liu SY, Bossert JM, Shaham Y. Central amygdala ERK signaling pathway is critical to incubation of cocaine craving. *Nat Neurosci.* 2005;8(2): 212–219.
79. Lu L, Wang X, Wu P, et al. Role of ventral tegmental area glial cell line-derived neurotrophic factor in incubation of cocaine craving. *Biol Psychiatry.* 2009;66(2):137–145.
80. Lu L, Dempsey J, Liu SY, Bossert JM, Shaham Y. A single infusion of brain-derived neurotrophic factor into the ventral tegmental area induces long-lasting potentiation of cocaine seeking after withdrawal. *J Neurosci.* 2004;24(7):1604–1611.
81. Whitfield Jr TW, Shi X, Sun WL, McGinty JF. The suppressive effect of an intra-prefrontal cortical infusion of BDNF on cocaine-seeking is Trk receptor and extracellular signal-regulated protein kinase mitogen-activated protein kinase dependent. *J Neurosci.* 2011;31(3):834–842.
82. Beaulieu JM, Sotnikova TD, Gainetdinov RR, Caron MG. Paradoxical striatal cellular signaling responses to psychostimulants in hyperactive mice. *J Biol Chem.* 2006;281 (43):32072–32088.
83. Shi X, McGinty JF. Extracellular signal-regulated mitogen-activated protein kinase inhibitors decrease amphetamine-induced behavior and neuropeptide gene expression in the striatum. *Neuroscience.* 2006;138(4):1289–1298.
84. Gerdjikov TV, Ross GM, Beninger RJ. Place preference induced by nucleus accumbens amphetamine is impaired by antagonists of ERK or p38 MAP kinases in rats. *Behav Neurosci.* 2004;118(4):740–750.
85. Mizoguchi H, Yamada K, Mizuno M, et al. Regulations of methamphetamine reward by extracellular signal-regulated kinase 1/2/ets-like gene-1 signaling pathway via the activation of dopamine receptors. *Mol Pharmacol.* 2004;65(5):1293–1301.
86. Rubino T, Forlani G, Vigano D, Zippel R, Parolaro D. Ras/ERK signalling in cannabinoid tolerance: from behaviour to cellular aspects. *J Neurochem.* 2005;93(4):984–991.
87. Valjent E, Pages C, Rogard M, Besson MJ, Maldonado R, Caboche J. Delta 9-tetrahydrocannabinol-induced MAPK/ERK and Elk-1 activation *in vivo* depends on dopaminergic transmission. *Eur J Neurosci.* 2001;14(2):342–352.
88. Spina L, Longoni R, Vinci S, et al. Role of dopamine D1 receptors and extracellular signal regulated kinase in the motivational properties of acetaldehyde as assessed by place preference conditioning. *Alcohol Clin Exp Res.* 2010;34(4):607–616.
89. Carnicella S, Kharazia V, Jeanblanc J, Janak PH, Ron D. GDNF is a fast-acting potent inhibitor of alcohol consumption and relapse. *Proc Natl Acad Sci USA.* 2008;105 (23):8114–8119.

90. DiRocco DP, Scheiner ZS, Sindreu CB, Chan GC, Storm DR. A role for calmodulin-stimulated adenylyl cyclases in cocaine sensitization. *J Neurosci.* 2009;29(8): 2393–2403.
91. Fasano S, D'Antoni A, Orban PC, et al. Ras-guanine nucleotide-releasing factor 1 (Ras-GRF1) controls activation of extracellular signal-regulated kinase (ERK) signaling in the striatum and long-term behavioral responses to cocaine. *Biol Psychiatry.* 2009;66 (8):758–768.
92. Rubino T, Forlani G, Vigano D, Zippel R, Parolaro D. Modulation of extracellular signal-regulated kinases cascade by chronic delta 9-tetrahydrocannabinol treatment. *Mol Cell Neurosci.* 2004;25(3):355–362.
93. Stacey D, Bilbao A, Maroteaux M, et al. RASGRF2 regulates alcohol-induced reinforcement by influencing mesolimbic dopamine neuron activity and dopamine release. *Proc Natl Acad Sci USA.* 2012;109(51):21128–21133.
94. Mazzucchelli C, Vantaggiato C, Ciamei A, et al. Knockout of ERK1 MAP kinase enhances synaptic plasticity in the striatum and facilitates striatal-mediated learning and memory. *Neuron.* 2002;34(5):807–820.
95. Selcher JC, Nekrasova T, Paylor R, Landreth GE, Sweatt JD. Mice lacking the ERK1 isoform of MAP kinase are unimpaired in emotional learning. *Learn Mem.* 2001;8 (1):11–19.
96. Iniguez SD, Warren BL, Neve RL, Russo SJ, Nestler EJ, Bolanos-Guzman CA. Viral-mediated expression of extracellular signal-regulated kinase-2 in the ventral tegmental area modulates behavioral responses to cocaine. *Behav Brain Res.* 2010;214(2): 460–464.
97. Brami-Cherrier K, Valjent E, Herve D, et al. Parsing molecular and behavioral effects of cocaine in mitogen- and stress-activated protein kinase-1-deficient mice. *J Neurosci.* 2005;25(49):11444–11454.
98. Besnard A, Bouveyron N, Kappes V, et al. Alterations of molecular and behavioral responses to cocaine by selective inhibition of Elk-1 phosphorylation. *J Neurosci.* 2011; 31(40):14296–14307.
99. Valjent E, Pages C, Herve D, Girault JA, Caboche J. Addictive and non-addictive drugs induce distinct and specific patterns of ERK activation in mouse brain. *Eur J Neurosci.* 2004;19(7):1826–1836.
100. Valjent E, Herve D, Girault JA. Drugs of abuse, protein phosphatases, and ERK pathway. *Med Sci (Paris).* 2005;21(5):453–454.
101. Valjent E, Bertran-Gonzalez J, Aubier B, Greengard P, Herve D, Girault JA. Mechanisms of locomotor sensitization to drugs of abuse in a two-injection protocol. *Neuropsychopharmacology.* 2010;35(2):401–415.
102. Radwanska K, Valjent E, Trzaskos J, Caboche J, Kaczmarek L. Regulation of cocaine-induced activator protein 1 transcription factors by the extracellular signal-regulated kinase pathway. *Neuroscience.* 2006;137(1):253–264.
103. Radwanska K, Caboche J, Kaczmarek L. Extracellular signal-regulated kinases (ERKs) modulate cocaine-induced gene expression in the mouse amygdala. *Eur J Neurosci.* 2005; 22(4):939–948.
104. Corbille AG, Valjent E, Marsicano G, et al. Role of cannabinoid type 1 receptors in locomotor activity and striatal signaling in response to psychostimulants. *J Neurosci.* 2007;27(26):6937–6947.
105. Zhang L, Lou D, Jiao H, et al. Cocaine-induced intracellular signaling and gene expression are oppositely regulated by the dopamine D1 and D3 receptors. *J Neurosci.* 2004;24(13):3344–3354.
106. Pascoli V, Besnard A, Herve D, et al. Cyclic adenosine monophosphate-independent tyrosine phosphorylation of NR2B mediates cocaine-induced extracellular signal-regulated kinase activation. *Biol Psychiatry.* 2011;69(3):218–227.

107. Jiao H, Zhang L, Gao F, Lou D, Zhang J, Xu M. Dopamine D(1) and D(3) receptors oppositely regulate NMDA- and cocaine-induced MAPK signaling via NMDA receptor phosphorylation. *J Neurochem.* 2007;103(2):840–848.
108. Bertran-Gonzalez J, Bosch C, Maroteaux M, et al. Opposing patterns of signaling activation in dopamine D1 and D2 receptor-expressing striatal neurons in response to cocaine and haloperidol. *J Neurosci.* 2008;28(22):5671–5685.
109. Xu S, Kang UG. Cocaine induces ubiquitination of Egr-1 in the rat dorsal striatum. *Neuroreport.* 2014;25(17):1362–1367.
110. Jenab S, Festa ED, Nazarian A, et al. Cocaine induction of ERK proteins in dorsal striatum of Fischer rats. *Brain Res Mol Brain Res.* 2005;142(2):134–138.
111. Sun WL, Zhou L, Hazim R, Quinones-Jenab V, Jenab S. Effects of dopamine and NMDA receptors on cocaine-induced Fos expression in the striatum of Fischer rats. *Brain Res.* 2008;1243:1–9.
112. Sun WL, Zhou L, Hazim R, Quinones-Jenab V, Jenab S. Effects of acute cocaine on ERK and DARPP-32 phosphorylation pathways in the caudate-putamen of Fischer rats. *Brain Res.* 2007;1178:12–19.
113. Berke JD, Paletzki RF, Aronson GJ, Hyman SE, Gerfen CR. A complex program of striatal gene expression induced by dopaminergic stimulation. *J Neurosci.* 1998;18(14):5301–5310.
114. Pascoli V, Turiault M, Luscher C. Reversal of cocaine-evoked synaptic potentiation resets drug-induced adaptive behaviour. *Nature.* 2011;481(7379):71–75.
115. Boudreau AC, Reimers JM, Milovanovic M, Wolf ME. Cell surface AMPA receptors in the rat nucleus accumbens increase during cocaine withdrawal but internalize after cocaine challenge in association with altered activation of mitogen-activated protein kinases. *J Neurosci.* 2007;27(39):10621–10635.
116. Boudreau AC, Ferrario CR, Glucksman MJ, Wolf ME. Signaling pathway adaptations and novel protein kinase A substrates related to behavioral sensitization to cocaine. *J Neurochem.* 2009;110(1):363–377.
117. Schumann J, Yaka R. Prolonged withdrawal from repeated noncontingent cocaine exposure increases NMDA receptor expression and ERK activity in the nucleus accumbens. *J Neurosci.* 2009;29(21):6955–6963.
118. Kim S, Kim JH. Time-dependent change of ERK phosphorylation levels in the nucleus accumbens during withdrawals from repeated cocaine. *Neurosci Lett.* 2008;436(2):107–110.
119. Peng Q, Sun X, Liu Z, Yang J, Oh KW, Hu Z. Microinjection of CART (cocaine- and amphetamine-regulated transcript) peptide into the nucleus accumbens inhibits the cocaine-induced upregulation of dopamine receptors and locomotor sensitization. *Neurochem Int.* 2014;75:105–111.
120. Yoon HS, Kim S, Park HK, Kim JH. Microinjection of CART peptide 55–102 into the nucleus accumbens blocks both the expression of behavioral sensitization and ERK phosphorylation by cocaine. *Neuropharmacology.* 2007;53(2):344–351.
121. Janes AC, Kantak KM, Cherry JA. The involvement of type IV phosphodiesterases in cocaine-induced sensitization and subsequent pERK expression in the mouse nucleus accumbens. *Psychopharmacology (Berl).* 2009;206(2):177–185.
122. Mattson BJ, Bossert JM, Simmons DE, et al. Cocaine-induced CREB phosphorylation in nucleus accumbens of cocaine-sensitized rats is enabled by enhanced activation of extracellular signal-related kinase, but not protein kinase A. *J Neurochem.* 2005;95(5):1481–1494.
123. Ren Z, Sun WL, Jiao H, et al. Dopamine D1 and *N*-methyl-D-aspartate receptors and extracellular signal-regulated kinase mediate neuronal morphological changes induced by repeated cocaine administration. *Neuroscience.* 2010;168(1):48–60.
124. Zhang L, Li J, Liu N, et al. Signaling via dopamine D1 and D3 receptors oppositely regulates cocaine-induced structural remodeling of dendrites and spines. *Neurosignals.* 2012;20(1):15–34.

125. Cardinal RN, Everitt BJ. Neural and psychological mechanisms underlying appetitive learning: links to drug addiction. *Curr Opin Neurobiol.* 2004;14(2):156–162.
126. Kelley AE. Memory and addiction: shared neural circuitry and molecular mechanisms. *Neuron.* 2004;44(1):161–179.
127. Chen L, Xu M. Dopamine D1 and D3 receptors are differentially involved in cue-elicited cocaine seeking. *J Neurochem.* 2010;114(2):530–541.
128. Marin MT, Berkow A, Golden SA, Koya E, Planeta CS, Hope BT. Context-specific modulation of cocaine-induced locomotor sensitization and ERK and CREB phosphorylation in the rat nucleus accumbens. *Eur J Neurosci.* 2009;30(10):1931–1940.
129. Fricks-Gleason AN, Marshall JF. Role of dopamine D1 receptors in the activation of nucleus accumbens extracellular signal-regulated kinase (ERK) by cocaine-paired contextual cues. *Neuropsychopharmacology.* 2011;36(2):434–444.
130. Nygard SK, Klambatsen A, Hazim R, et al. Sexually dimorphic intracellular responses after cocaine-induced conditioned place preference expression. *Brain Res.* 2013;1520:121–133.
131. Tropea TF, Guerriero RM, Willuhn I, et al. Augmented D1 dopamine receptor signaling and immediate-early gene induction in adult striatum after prenatal cocaine. *Biol Psychiatry.* 2008;63(11):1066–1074.
132. Tropea TF, Kosofsky BE, Rajadhyaksha AM. Enhanced CREB and DARPP-32 phosphorylation in the nucleus accumbens and CREB, ERK, and GluR1 phosphorylation in the dorsal hippocampus is associated with cocaine-conditioned place preference behavior. *J Neurochem.* 2008;106(4):1780–1790.
133. Yu F, Zhong P, Liu X, Sun D, Gao HQ, Liu QS. Metabotropic glutamate receptor I (mGluR1) antagonism impairs cocaine-induced conditioned place preference via inhibition of protein synthesis. *Neuropsychopharmacology.* 2013;38(7):1308–1321.
134. Edwards S, Bachtell RK, Guzman D, Whisler KN, Self DW. Emergence of context-associated GluR(1) and ERK phosphorylation in the nucleus accumbens core during withdrawal from cocaine self-administration. *Addict Biol.* 2011;16(3):450–457.
135. Suzuki A, Josselyn SA, Frankland PW, Masushige S, Silva AJ, Kida S. Memory reconsolidation and extinction have distinct temporal and biochemical signatures. *J Neurosci.* 2004;24(20):4787–4795.
136. Reichelt AC, Lee JL. Memory reconsolidation in aversive and appetitive settings. *Front Behav Neurosci.* 2013;7:118.
137. Tronson NC, Taylor JR. Addiction: a drug-induced disorder of memory reconsolidation. *Curr Opin Neurobiol.* 2013;23(4):573–580.
138. Nader K, Schafe GE, Le Doux JE. Fear memories require protein synthesis in the amygdala for reconsolidation after retrieval. *Nature.* 2000;406(6797):722–726.
139. Edwards S, Graham DL, Bachtell RK, Self DW. Region-specific tolerance to cocaine-regulated cAMP-dependent protein phosphorylation following chronic self-administration. *Eur J Neurosci.* 2007;25(7):2201–2213.
140. Koya E, Uejima JL, Wihbey KA, Bossert JM, Hope BT, Shaham Y. Role of ventral medial prefrontal cortex in incubation of cocaine craving. *Neuropharmacology.* 2009;56(suppl 1):177–185.
141. Berglind WJ, See RE, Fuchs RA, et al. A BDNF infusion into the medial prefrontal cortex suppresses cocaine seeking in rats. *Eur J Neurosci.* 2007;26(3):757–766.
142. Sun LL, Zhang Y, Liu JF, et al. Role of melanin-concentrating hormone in the nucleus accumbens shell in rats behaviourally sensitized to methamphetamine. *Int J Neuropsychopharmacol.* 2013;16(8):1767–1780.
143. Sun WL, Zelek-Molik A, McGinty JF. Short and long access to cocaine self-administration activates tyrosine phosphatase STEP and attenuates GluN expression but

differentially regulates GluA expression in the prefrontal cortex. *Psychopharmacology (Berl).* 2013;229(4):603–613.
144. Parelkar NK, Jiang Q, Chu XP, Guo ML, Mao LM, Wang JQ. Amphetamine alters Ras-guanine nucleotide-releasing factor expression in the rat striatum *in vivo*. *Eur J Pharmacol.* 2009;619(1–3):50–56.
145. Zhang GC, Hoffmann J, Parelkar NK, et al. Cocaine increases Ras-guanine nucleotide-releasing factor 1 protein expression in the rat striatum *in vivo*. *Neurosci Lett.* 2007;427(2):117–121.
146. Brambilla R. Targeting Ras/ERK signaling in the striatum: will it help? *Mol Psychiatry.* 2003;8(4):366–368.
147. Lavaur J, Bernard F, Trifilieff P, et al. A TAT-DEF-Elk-1 peptide regulates the cytonuclear trafficking of Elk-1 and controls cytoskeleton dynamics. *J Neurosci.* 2007;27(52):14448–14458.
148. Fiorentini C, Gardoni F, Spano P, Di Luca M, Missale C. Regulation of dopamine D1 receptor trafficking and desensitization by oligomerization with glutamate *N*-methyl-D-aspartate receptors. *J Biol Chem.* 2003;278(22):20196–20202.
149. Lee FJ, Liu F. Direct interactions between NMDA and D1 receptors: a tale of tails. *Biochem Soc Trans.* 2004;32(Pt 6):1032–1036.
150. Lee FJ, Xue S, Pei L, et al. Dual regulation of NMDA receptor functions by direct protein–protein interactions with the dopamine D1 receptor. *Cell.* 2002;111(2):219–230.
151. Pei L, Lee FJ, Moszczynska A, Vukusic B, Liu F. Regulation of dopamine D1 receptor function by physical interaction with the NMDA receptors. *J Neurosci.* 2004;24(5):1149–1158.
152. Sun WL, Zhou L, Quinones-Jenab V, Jenab S. Cocaine effects on dopamine and NMDA receptors interactions in the striatum of Fischer rats. *Brain Res Bull.* 2009;80(6):377–381.
153. Cahill E, Pascoli V, Trifilieff P, et al. D1R/GluN1 complexes in the striatum integrate dopamine and glutamate signalling to control synaptic plasticity and cocaine-induced responses. *Mol Psychiatry.* 2014;19(12):1295–1304.
154. Moreno E, Moreno-Delgado D, Navarro G, et al. Cocaine disrupts histamine H3 receptor modulation of dopamine D1 receptor signaling: sigma1-D1-H3 receptor complexes as key targets for reducing cocaine's effects. *J Neurosci.* 2014;34(10):3545–3558.
155. Corvol JC, Valjent E, Pascoli V, et al. Quantitative changes in Galphaolf protein levels, but not D1 receptor, alter specifically acute responses to psychostimulants. *Neuropsychopharmacology.* 2007;32(5):1109–1121.
156. Mao LM, Reusch JM, Fibuch EE, Liu Z, Wang JQ. Amphetamine increases phosphorylation of MAPK/ERK at synaptic sites in the rat striatum and medial prefrontal cortex. *Brain Res.* 2013;1494:101–108.
157. Rajadhyaksha A, Husson I, Satpute SS, et al. L-type Ca^{2+} channels mediate adaptation of extracellular signal-regulated kinase 1/2 phosphorylation in the ventral tegmental area after chronic amphetamine treatment. *J Neurosci.* 2004;24(34):7464–7476.
158. Pascoli V, Valjent E, Corbille AG, et al. cAMP and extracellular signal-regulated kinase signaling in response to d-amphetamine and methylphenidate in the prefrontal cortex *in vivo*: role of beta 1-adrenoceptors. *Mol Pharmacol.* 2005;68(2):421–429.
159. Choe ES, Wang JQ. CaMKII regulates amphetamine-induced ERK1/2 phosphorylation in striatal neurons. *Neuroreport.* 2002;13(8):1013–1016.
160. Choe ES, Chung KT, Mao L, Wang JQ. Amphetamine increases phosphorylation of extracellular signal-regulated kinase and transcription factors in the rat striatum via group I metabotropic glutamate receptors. *Neuropsychopharmacology.* 2002;27(4):565–575.

161. Schwendt M, Sigmon SA, McGinty JF. RGS4 overexpression in the rat dorsal striatum modulates mGluR5- and amphetamine-mediated behavior and signaling. *Psychopharmacology (Berl).* 2012;221(4):621–635.
162. Ferguson SM, Robinson TE. Amphetamine-evoked gene expression in striatopallidal neurons: regulation by corticostriatal afferents and the ERK/MAPK signaling cascade. *J Neurochem.* 2004;91(2):337–348.
163. Gerfen CR, Paletzki R, Worley P. Differences between dorsal and ventral striatum in Drd1a dopamine receptor coupling of dopamine- and cAMP-regulated phosphoprotein-32 to activation of extracellular signal-regulated kinase. *J Neurosci.* 2008;28(28):7113–7120.
164. Wang XB, Uhl GR. Subtracted differential display: genes with amphetamine-altered expression patterns include calcineurin. *Brain Res Mol Brain Res.* 1998;53(1–2):344–347.
165. Lim HW, New L, Han J, Molkentin JD. Calcineurin enhances MAPK phosphatase-1 expression and p38 MAPK inactivation in cardiac myocytes. *J Biol Chem.* 2001;276(19):15913–15919.
166. Gomez AM, Sun WL, Midde NM, Harrod SB, Zhu J. Effects of environmental enrichment on ERK1/2 phosphorylation in the rat prefrontal cortex following nicotine-induced sensitization or nicotine self-administration. *Eur J Neurosci.* 2015;41(1):109–119.
167. Einat H, Yuan P, Gould TD, et al. The role of the extracellular signal-regulated kinase signaling pathway in mood modulation. *J Neurosci.* 2003;23(19):7311–7316.
168. Shi X, McGinty JF. Repeated amphetamine treatment increases phosphorylation of extracellular signal-regulated kinase, protein kinase B, and cyclase response element-binding protein in the rat striatum. *J Neurochem.* 2007;103(2):706–713.
169. Shi X, McGinty JF. D1 and D2 dopamine receptors differentially mediate the activation of phosphoproteins in the striatum of amphetamine-sensitized rats. *Psychopharmacology (Berl).* 2011;214(3):653–663.
170. Gonzales R, Mooney L, Rawson RA. The methamphetamine problem in the United States. *Annu Rev Public Health.* 2010;31:385–398.
171. Lu L, Wang X. Drug addiction in China. *Ann NY Acad Sci.* 2008;1141:304–317.
172. Brennan KA, Carati C, Lea RA, Fitzmaurice PS, Schenk S. Effect of D1-like and D2-like receptor antagonists on methamphetamine and 3,4-methylenedioxymethamphetamine self-administration in rats. *Behav Pharmacol.* 2009;20(8):688–694.
173. Zhang Y, Loonam TM, Noailles PA, Angulo JA. Comparison of cocaine- and methamphetamine-evoked dopamine and glutamate overflow in somatodendritic and terminal field regions of the rat brain during acute, chronic, and early withdrawal conditions. *Ann NY Acad Sci.* 2001;937:93–120.
174. Ohmori T, Abekawa T, Muraki A, Koyama T. Competitive and noncompetitive NMDA antagonists block sensitization to methamphetamine. *Pharmacol Biochem Behav.* 1994;48(3):587–591.
175. Kuribara H, Asami T, Ida I, Iijima Y, Tadokoro S. Effects of repeated MK-801 on ambulation in mice and in sensitization following methamphetamine. *Psychopharmacology (Berl).* 1992;108(3):271–275.
176. Kim HS, Jang CG. MK-801 inhibits methamphetamine-induced conditioned place preference and behavioral sensitization to apomorphine in mice. *Brain Res Bull.* 1997;44(3):221–227.
177. Horio M, Kohno M, Fujita Y, et al. Role of serine racemase in behavioral sensitization in mice after repeated administration of methamphetamine. *PLoS One.* 2012;7(4):e35494.
178. Zhao N, Chen Y, Zhu J, et al. Levo-tetrahydropalmatine attenuates the development and expression of methamphetamine-induced locomotor sensitization and the

accompanying activation of ERK in the nucleus accumbens and caudate putamen in mice. *Neuroscience.* 2014;258:101–110.
179. Yan T, Li L, Sun B, et al. Luteolin inhibits behavioral sensitization by blocking methamphetamine-induced MAPK pathway activation in the caudate putamen in mice. *PLoS One.* 2014;9(6):e98981.
180. Mantsch JR, Li SJ, Risinger R, et al. Levo-tetrahydropalmatine attenuates cocaine self-administration and cocaine-induced reinstatement in rats. *Psychopharmacology (Berl).* 2007;192(4):581–591.
181. Wang JB, Mantsch JR. 1-Tetrahydropalamatine: a potential new medication for the treatment of cocaine addiction. *Future Med Chem.* 2012;4(2):177–186.
182. McDaid J, Graham MP, Napier TC. Methamphetamine-induced sensitization differentially alters pCREB and DeltaFosB throughout the limbic circuit of the mammalian brain. *Mol Pharmacol.* 2006;70(6):2064–2074.
183. Krasnova IN, Chiflikyan M, Justinova Z, et al. CREB phosphorylation regulates striatal transcriptional responses in the self-administration model of methamphetamine addiction in the rat. *Neurobiol Dis.* 2013;58:132–143.
184. Cadet JL, Brannock C, Jayanthi S, Krasnova IN. Transcriptional and epigenetic substrates of methamphetamine addiction and withdrawal: evidence from a long-access self-administration model in the rat. *Mol Neurobiol.* 2015;51(2):696–717.
185. Caunt CJ, Keyse SM. Dual-specificity MAP kinase phosphatases (MKPs): shaping the outcome of MAP kinase signalling. *FEBS J.* 2013;280(2):489–504.
186. Toledano-Katchalski H, Kraut J, Sines T, et al. Protein tyrosine phosphatase epsilon inhibits signaling by mitogen-activated protein kinases. *Mol Cancer Res.* 2003;1(7):541–550.
187. Takaki M, Ujike H, Kodama M, Takehisa Y, Nakata K, Kuroda S. Two kinds of mitogen-activated protein kinase phosphatases, MKP-1 and MKP-3, are differentially activated by acute and chronic methamphetamine treatment in the rat brain. *J Neurochem.* 2001;79(3):679–688.
188. Ujike H, Takaki M, Kodama M, Kuroda S. Gene expression related to synaptogenesis, neuritogenesis, and MAP kinase in behavioral sensitization to psychostimulants. *Ann NY Acad Sci.* 2002;965:55–67.
189. Son JS, Jeong YC, Kwon YB. Regulatory effect of bee venom on methamphetamine-induced cellular activities in prefrontal cortex and nucleus accumbens in mice. *Biol Pharm Bull.* 2015;38(1):48–52.
190. Herrold AA, Shen F, Graham MP, et al. Mirtazapine treatment after conditioning with methamphetamine alters subsequent expression of place preference. *Drug Alcohol Depend.* 2009;99(1–3):231–239.
191. Nordahl TE, Salo R, Leamon M. Neuropsychological effects of chronic methamphetamine use on neurotransmitters and cognition: a review. *J Neuropsychiatry Clin Neurosci.* 2003;15(3):317–325.
192. Chang L, Ernst T, Speck O, et al. Perfusion MRI and computerized cognitive test abnormalities in abstinent methamphetamine users. *Psychiatry Res.* 2002;114(2):65–79.
193. Sekine Y, Iyo M, Ouchi Y, et al. Methamphetamine-related psychiatric symptoms and reduced brain dopamine transporters studied with PET. *Am J Psychiatry.* 2001;158(8):1206–1214.
194. Volkow ND, Chang L, Wang GJ, et al. Association of dopamine transporter reduction with psychomotor impairment in methamphetamine abusers. *Am J Psychiatry.* 2001;158(3):377–382.
195. Chen YJ, Liu YL, Zhong Q, et al. Tetrahydropalmatine protects against methamphetamine-induced spatial learning and memory impairment in mice. *Neurosci Bull.* 2012;28(3):222–232.

196. Kamei H, Nagai T, Nakano H, et al. Repeated methamphetamine treatment impairs recognition memory through a failure of novelty-induced ERK1/2 activation in the prefrontal cortex of mice. *Biol Psychiatry.* 2006;59(1):75–84.
197. Nagai T, Takuma K, Dohniwa M, et al. Repeated methamphetamine treatment impairs spatial working memory in rats: reversal by clozapine but not haloperidol. *Psychopharmacology (Berl).* 2007;194(1):21–32.
198. Noda Y, Mouri A, Ando Y, et al. Galantamine ameliorates the impairment of recognition memory in mice repeatedly treated with methamphetamine: involvement of allosteric potentiation of nicotinic acetylcholine receptors and dopaminergic-ERK1/2 systems. *Int J Neuropsychopharmacol.* 2010;13(10):1343–1354.
199. Gonzalez B, Raineri M, Cadet JL, Garcia-Rill E, Urbano FJ, Bisagno V. Modafinil improves methamphetamine-induced object recognition deficits and restores prefrontal cortex ERK signaling in mice. *Neuropharmacology.* 2014;87:188–197.
200. Mereu M, Bonci A, Newman AH, Tanda G. The neurobiology of modafinil as an enhancer of cognitive performance and a potential treatment for substance use disorders. *Psychopharmacology (Berl).* 2013;229(3):415–434.
201. Adams IB, Martin BR. Cannabis: pharmacology and toxicology in animals and humans. *Addiction.* 1996;91(11):1585–1614.
202. Howlett AC, Barth F, Bonner TI, et al. International Union of Pharmacology. XXVII. Classification of cannabinoid receptors. *Pharmacol Rev.* 2002;54(2):161–202.
203. Daigle TL, Wetsel WC, Caron MG. Opposite function of dopamine D1 and *N*-methyl-D-aspartate receptors in striatal cannabinoid-mediated signaling. *Eur J Neurosci.* 2011;34(9):1378–1389.
204. Derkinderen P, Valjent E, Toutant M, et al. Regulation of extracellular signal-regulated kinase by cannabinoids in hippocampus. *J Neurosci.* 2003;23(6):2371–2382.
205. Long LE, Chesworth R, Huang XF, McGregor IS, Arnold JC, Karl T. A behavioural comparison of acute and chronic Delta9-tetrahydrocannabinol and cannabidiol in C57BL/6JArc mice. *Int J Neuropsychopharmacol.* 2010;13(7):861–876.
206. Rubino T, Vigano D, Premoli F, et al. Changes in the expression of G protein-coupled receptor kinases and beta-arrestins in mouse brain during cannabinoid tolerance: a role for RAS-ERK cascade. *Mol Neurobiol.* 2006;33(3):199–213.
207. Tonini R, Ciardo S, Cerovic M, et al. ERK-dependent modulation of cerebellar synaptic plasticity after chronic Delta9-tetrahydrocannabinol exposure. *J Neurosci.* 2006;26(21):5810–5818.
208. Rubino T, Vigano D, Massi P, Parolaro D. The psychoactive ingredient of marijuana induces behavioural sensitization. *Eur J Neurosci.* 2001;14(5):884–886.
209. Tanda G, Munzar P, Goldberg SR. Self-administration behavior is maintained by the psychoactive ingredient of marijuana in squirrel monkeys. *Nat Neurosci.* 2000;3(11):1073–1074.
210. Cadoni C, Pisanu A, Solinas M, Acquas E, Di Chiara G. Behavioural sensitization after repeated exposure to Delta 9-tetrahydrocannabinol and cross-sensitization with morphine. *Psychopharmacology (Berl).* 2001;158(3):259–266.
211. Giuffrida A, Parsons LH, Kerr TM, Rodriguez de Fonseca F, Navarro M, Piomelli D. Dopamine activation of endogenous cannabinoid signaling in dorsal striatum. *Nat Neurosci.* 1999;2(4):358–363.
212. Patel S, Rademacher DJ, Hillard CJ. Differential regulation of the endocannabinoids anandamide and 2-arachidonylglycerol within the limbic forebrain by dopamine receptor activity. *J Pharmacol Exp Ther.* 2003;306(3):880–888.
213. Centonze D, Battista N, Rossi S, et al. A critical interaction between dopamine D2 receptors and endocannabinoids mediates the effects of cocaine on striatal gabaergic transmission. *Neuropsychopharmacology.* 2004;29(8):1488–1497.

214. Nguyen K, Marshall L, Hu S, Neff L. State-specific prevalence of current cigarette smoking and smokeless tobacco use among adults aged ≥18 years—United States, 2011–2013. *MMWR Morb Mortal Wkly Rep.* 2015;64(19):532–536.
215. Pontieri FE, Tanda G, Orzi F, Di Chiara G. Effects of nicotine on the nucleus accumbens and similarity to those of addictive drugs. *Nature.* 1996;382(6588):255–257.
216. Dani JA, Jenson D, Broussard JI, De Biasi M. Neurophysiology of nicotine addiction. *J Addict Res Ther.* 2011;S1(1). pii: 001.
217. Lenoir M, Kiyatkin EA. Intravenous nicotine injection induces rapid, experience-dependent sensitization of glutamate release in the ventral tegmental area and nucleus accumbens. *J Neurochem.* 2013;127(4):541–551.
218. Dajas-Bailador F, Wonnacott S. Nicotinic acetylcholine receptors and the regulation of neuronal signalling. *Trends Pharmacol Sci.* 2004;25(6):317–324.
219. Nakayama H, Shimoke K, Isosaki M, Satoh H, Yoshizumi M, Ikeuchi T. Subtypes of neuronal nicotinic acetylcholine receptors involved in nicotine-induced phosphorylation of extracellular signal-regulated protein kinase in PC12 h cells. *Neurosci Lett.* 2006; 392(1–2):101–104.
220. Nakayama H, Numakawa T, Ikeuchi T, Hatanaka H. Nicotine-induced phosphorylation of extracellular signal-regulated protein kinase and CREB in PC12 h cells. *J Neurochem.* 2001;79(3):489–498.
221. Nakayama H, Numakawa T, Ikeuchi T. Nicotine-induced phosphorylation of Akt through epidermal growth factor receptor and Src in PC12h cells. *J Neurochem.* 2002; 83(6):1372–1379.
222. Utsugisawa K, Nagane Y, Obara D, Tohgi H. Over-expression of alpha7 nicotinic acetylcholine receptor induces sustained ERK phosphorylation and N-cadherin expression in PC12 cells. *Brain Res Mol Brain Res.* 2002;106(1–2):88–93.
223. Wang J, Cui W, Wei J, et al. Genome-wide expression analysis reveals diverse effects of acute nicotine exposure on neuronal function-related genes and pathways. *Front Psychiatry.* 2011;2:5.
224. Dajas-Bailador FA, Soliakov L, Wonnacott S. Nicotine activates the extracellular signal-regulated kinase 1/2 via the alpha7 nicotinic acetylcholine receptor and protein kinase A, in SH-SY5Y cells and hippocampal neurones. *J Neurochem.* 2002;80(3):520–530.
225. Steiner RC, Heath CJ, Picciotto MR. Nicotine-induced phosphorylation of ERK in mouse primary cortical neurons: evidence for involvement of glutamatergic signaling and CaMKII. *J Neurochem.* 2007;103(2):666–678.
226. Collo G, Bono F, Cavalleri L, et al. Nicotine-induced structural plasticity in mesencephalic dopaminergic neurons is mediated by dopamine D3 receptors and Akt-mTORC1 signaling. *Mol Pharmacol.* 2013;83(6):1176–1189.
227. Brunzell DH, Russell DS, Picciotto MR. *In vivo* nicotine treatment regulates mesocorticolimbic CREB and ERK signaling in C57Bl/6J mice. *J Neurochem.* 2003;84(6): 1431–1441.
228. Neugebauer NM, Henehan RM, Hales CA, Picciotto MR. Mice lacking the galanin gene show decreased sensitivity to nicotine conditioned place preference. *Pharmacol Biochem Behav.* 2011;98(1):87–93.
229. Li S, Li Z, Pei L, Le AD, Liu F. The α7nACh–NMDA receptor complex is involved in cue-induced reinstatement of nicotine seeking. *Diff id721[721.TDDIFF]/Diff] Exp Med.* 2012;209(12):2141–2147.
230. Leshner AI. Vulnerability to addiction: new research opportunities. *Am J Med Genet.* 2000;96(5):590–591.
231. Rhee SH, Hewitt JK, Young SE, Corley RP, Crowley TJ, Stallings MC. Genetic and environmental influences on substance initiation, use, and problem use in adolescents. *Arch Gen Psychiatry.* 2003;60(12):1256–1264.

232. Stairs DJ, Bardo MT. Neurobehavioral effects of environmental enrichment and drug abuse vulnerability. *Pharmacol Biochem Behav.* 2009;92(3):377–382.
233. Bowling SL, Rowlett JK, Bardo MT. The effect of environmental enrichment on amphetamine-stimulated locomotor activity, dopamine synthesis and dopamine release. *Neuropharmacology.* 1993;32(9):885–893.
234. Zhu J, Green T, Bardo MT, Dwoskin LP. Environmental enrichment enhances sensitization to GBR 12935-induced activity and decreases dopamine transporter function in the medial prefrontal cortex. *Behav Brain Res.* 2004;148(1–2): 107–117.
235. Del Arco A, Mora F, Mohammed AH, Fuxe K. Stimulation of D2 receptors in the prefrontal cortex reduces PCP-induced hyperactivity, acetylcholine release and dopamine metabolism in the nucleus accumbens. *J Neural Transm.* 2007;114(2): 185–193.
236. Konu O, Kane JK, Barrett T, et al. Region-specific transcriptional response to chronic nicotine in rat brain. *Brain Res.* 2001;909(1–2):194–203.
237. Kalluri HS, Ticku MK. Ethanol-mediated inhibition of mitogen-activated protein kinase phosphorylation in mouse brain. *Eur J Pharmacol.* 2002;439(1–3):53–58.
238. Chandler LJ, Sutton G. Acute ethanol inhibits extracellular signal-regulated kinase, protein kinase B, and adenosine 3':5'-cyclic monophosphate response element binding protein activity in an age- and brain region-specific manner. *Alcohol Clin Exp Res.* 2005; 29(4):672–682.
239. Davis MI, Szarowski D, Turner JN, Morrisett RA, Shain W. *In vivo* activation and *in situ* BDNF-stimulated nuclear translocation of mitogen-activated/extracellular signal-regulated protein kinase is inhibited by ethanol in the developing rat hippocampus. *Neurosci Lett.* 1999;272(2):95–98.
240. Zhu Y, Wang Y, Zhao B, et al. Differential phosphorylation of GluN1-MAPKs in rat brain reward circuits following long-term alcohol exposure. *PLoS One.* 2013;8(1): e54930.
241. Ibba F, Vinci S, Spiga S, et al. Ethanol-induced extracellular signal regulated kinase: role of dopamine D1 receptors. *Alcohol Clin Exp Res.* 2009;33(5):858–867.
242. Thorsell A, Tapocik JD, Liu K, et al. A novel brain penetrant NPS receptor antagonist, NCGC00185684, blocks alcohol-induced ERK-phosphorylation in the central amygdala and decreases operant alcohol self-administration in rats. *J Neurosci.* 2013;33 (24):10132–10142.
243. Hansson AC, Rimondini R, Neznanova O, Sommer WH, Heilig M. Neuroplasticity in brain reward circuitry following a history of ethanol dependence. *Eur J Neurosci.* 2008;27 (8):1912–1922.
244. Vinci S, Ibba F, Longoni R, Spina L, Spiga S, Acquas E. Acetaldehyde elicits ERK phosphorylation in the rat nucleus accumbens and extended amygdala. *Synapse.* 2010; 64(12):916–927.
245. Peana AT, Muggironi G, Fois GR, Zinellu M, Vinci S, Acquas E. Effect of opioid receptor blockade on acetaldehyde self-administration and ERK phosphorylation in the rat nucleus accumbens. *Alcohol.* 2011;45(8):773–783.
246. Pandey SC, Zhang H, Ugale R, Prakash A, Xu T, Misra K. Effector immediate-early gene arc in the amygdala plays a critical role in alcoholism. *J Neurosci.* 2008;28(10): 2589–2600.
247. Neznanova O, Bjork K, Rimondini R, et al. Acute ethanol challenge inhibits glycogen synthase kinase-3beta in the rat prefrontal cortex. *Int J Neuropsychopharmacol.* 2009;12 (2):275–280.
248. Nuutinen S, Kiianmaa K, Panula P. DARPP-32 and Akt regulation in ethanol-preferring AA and ethanol-avoiding ANA rats. *Neurosci Lett.* 2011;503(1):31–36.

249. Rosas M, Zaru A, Sabariego M, et al. Differential sensitivity of ethanol-elicited ERK phosphorylation in nucleus accumbens of Sardinian alcohol-preferring and -non preferring rats. *Alcohol.* 2014;48(5):471–476.
250. Sanna PP, Simpson C, Lutjens R, Koob G. ERK regulation in chronic ethanol exposure and withdrawal. *Brain Res.* 2002;948(1–2):186–191.
251. Wang Y, Cui H, Wang W, Zhao B, Lai J. The region-specific activation of Ca^{2+}/calmodulin dependent protein kinase II and extracellular signal-regulated kinases in hippocampus following chronic alcohol exposure. *Brain Res Bull.* 2012;89(5–6): 191–196.
252. Roberto M, Nelson TE, Ur CL, Brunelli M, Sanna PP, Gruol DL. The transient depression of hippocampal CA1 LTP induced by chronic intermittent ethanol exposure is associated with an inhibition of the MAP kinase pathway. *Eur J Neurosci.* 2003;17(8): 1646–1654.
253. Pandey SC, Zhang H, Roy A, Misra K. Central and medial amygdaloid brain-derived neurotrophic factor signaling plays a critical role in alcohol-drinking and anxiety-like behaviors. *J Neurosci.* 2006;26(32):8320–8331.
254. De Witte P, Pinto E, Ansseau M, Verbanck P. Alcohol and withdrawal: from animal research to clinical issues. *Neurosci Biobehav Rev.* 2003;27(3):189–197.
255. Markou A, Kosten TR, Koob GF. Neurobiological similarities in depression and drug dependence: a self-medication hypothesis. *Neuropsychopharmacology.* 1998;18(3): 135–174.
256. Bertotto ME, Maldonado NM, Bignante EA, et al. ERK activation in the amygdala and hippocampus induced by fear conditioning in ethanol withdrawn rats: modulation by MK-801. *Eur Neuropsychopharmacol.* 2011;21(12):892–904.
257. Radwanska K, Wrobel E, Korkosz A, et al. Alcohol relapse induced by discrete cues activates components of AP-1 transcription factor and ERK pathway in the rat basolateral and central amygdala. *Neuropsychopharmacology.* 2008;33(8):1835–1846.
258. Schroeder JP, Spanos M, Stevenson JR, Besheer J, Salling M, Hodge CW. Cue-induced reinstatement of alcohol-seeking behavior is associated with increased ERK1/2 phosphorylation in specific limbic brain regions: blockade by the mGluR5 antagonist MPEP. *Neuropharmacology.* 2008;55(4):546–554.
259. Peana AT, Giugliano V, Rosas M, Sabariego M, Acquas E. Effects of L-cysteine on reinstatement of ethanol-seeking behavior and on reinstatement-elicited extracellular signal-regulated kinase phosphorylation in the rat nucleus accumbens shell. *Alcohol Clin Exp Res.* 2013;37(suppl 1):E329–E337.
260. Berglind WJ, Whitfield Jr TW, LaLumiere RT, Kalivas PW, McGinty JF. A single intra-PFC infusion of BDNF prevents cocaine-induced alterations in extracellular glutamate within the nucleus accumbens. *J Neurosci.* 2009;29(12):3715–3719.
261. Hope BT, Crombag HS, Jedynak JP, Wise RA. Neuroadaptations of total levels of adenylate cyclase, protein kinase A, tyrosine hydroxylase, cdk5 and neurofilaments in the nucleus accumbens and ventral tegmental area do not correlate with expression of sensitized or tolerant locomotor responses to cocaine. *J Neurochem.* 2005;92(3):536–545.
262. Lu L, Grimm JW, Shaham Y, Hope BT. Molecular neuroadaptations in the accumbens and ventral tegmental area during the first 90 days of forced abstinence from cocaine self-administration in rats. *J Neurochem.* 2003;85(6):1604–1613.
263. Sun WL, Coleman NT, Zelek-Molik A, Barry SM, Whitfield Jr TW, McGinty JF. Relapse to cocaine-seeking after abstinence is regulated by cAMP-dependent protein kinase A in the prefrontal cortex. *Addict Biol.* 2014;19(1):77–86.
264. Gomez AM, Midde NM, Mactutus CF, Booze RM, Zhu J. Environmental enrichment alters nicotine-mediated locomotor sensitization and phosphorylation of DARPP-32 and CREB in rat prefrontal cortex. *PLoS One.* 2012;7(8):e44149.

265. Gomez AM, Altomare D, Sun WL, Midde NM, Ji H, Shtutman M, Turner JR, Creek KE, Zhu J. Prefrontal microRNA-221 Mediates Environmental Enrichment-Induced Increase of Locomotor Sensitivity to Nicotine. *Int J Neuropsychopharmacol.* 2015, doi: 10.1093/ijnp/pyv090.
266. Terasawa K, Ichimura A, Sato F, Shimizu K, Tsujimoto G. Sustained activation of ERK1/2 by NGF induces microRNA-221 and 222 in PC12 cells. *FEBS J.* 2009;276 (12):3269–3276.
267. Moonat S, Starkman BG, Sakharkar A, Pandey SC. Neuroscience of alcoholism: molecular and cellular mechanisms. *Cell Mol Life Sci.* 2010;67(1):73–88.
268. Schmidt HD, McGinty JF, West AE, Sadri-Vakili G. Epigenetics and psychostimulant addiction. *Cold Spring Harb Perspect Med.* 2013;3(3):a012047.
269. Ciccarelli A, Giustetto M. Role of ERK signaling in activity-dependent modifications of histone proteins. *Neuropharmacology.* 2014;80:34–44.

CHAPTER TWO

Ethanol-Associated Changes in Glutamate Reward Neurocircuitry: A Minireview of Clinical and Preclinical Genetic Findings

Richard L. Bell*[,1], Sheketha R. Hauser*, Jeanette McClintick[†], Shafiqur Rahman[‡], Howard J. Edenberg[†], Karen K. Szumlinski[§], William J. McBride*

[*] Department of Psychiatry, Indiana University School of Medicine, Indianapolis, Indiana, USA
[†] Departments of Biochemistry and Molecular Biology and Medical and Molecular Genetics, Indiana University School of Medicine, Indianapolis, Indiana , USA
[‡] Department of Pharmaceutical Sciences, South Dakota State University, Brookings, South Dakota, USA
[§] Department of Psychological and Brain Sciences, University of California Santa Barbara, Santa Barbara, California, USA
[1] Corresponding author: e-mail address: ribell@iupui.edu.

Contents

Progress in Molecular Biology and Translational Science, Volume 137
ISSN 1877-1173
http://dx.doi.org/10.1016/bs.pmbts.2015.10.018

Abstract

Herein, we have reviewed the role of glutamate, the major excitatory neurotransmitter in the brain, in a number of neurochemical, -physiological, and -behavioral processes mediating the development of alcohol dependence. The findings discussed include results from both preclinical as well as neuroimaging and postmortem clinical studies. Expression levels for a number of glutamate-associated genes and/or proteins are modulated by alcohol abuse and dependence. These changes in expression include metabotropic receptors and ionotropic receptor subunits as well as different glutamate transporters. Moreover, these changes in gene expression parallel the pharmacologic manipulation of these same receptors and transporters. Some of these gene expression changes may have predated alcohol abuse and dependence because a number of glutamate-associated polymorphisms are related to a genetic predisposition to develop alcohol dependence. Other glutamate-associated polymorphisms are linked to age at the onset of alcohol-dependence and initial level of response/sensitivity to alcohol. Finally, findings of innate and/or ethanol-induced glutamate-associated gene expression differences/changes observed in a genetic animal model of alcoholism, the P rat, are summarized. Overall, the existing literature indicates that changes in glutamate receptors, transporters, enzymes, and scaffolding proteins are crucial for the development of alcohol dependence and there is a substantial genetic component to these effects. This indicates that continued research into the genetic underpinnings of these glutamate-associated effects will provide important novel molecular targets for treating alcohol abuse and dependence.

1. ALCOHOLISM AND GENETICS

Over half of adult Americans have a family history of alcoholism or alcohol (ethanol) abuse and one in four Americans have had an alcohol use disorder (AUD) during their lifetime, costing the US economy an estimated $225 billion per year (Research Society on Alcoholism).[1,2] AUDs continue to be ranked as the third leading cause of preventable death by the Centers for Disease Control and Prevention.[3] Moreover, research supports a causal relationship between AUDs and at least 50 different medical conditions.[4–6]

The well-documented familial incidence of alcoholism indicates that heredity contributes significantly to a predisposition toward, and the development of, AUDs.[7–9] In fact, family history positive (FHP) individuals are at a three- to sevenfold increased risk to develop alcoholism, relative to those who are family history negative (FHN).[10] This genetic proposal has been confirmed by multiple gene studies (e.g., the Collaborative Study On the Genetics of Alcoholism (COGA), the Study of Addiction: Genes and Environment (SAGE), and the European research project on risk-taking behavior in teenagers (IMAGEN)) examining the association between diagnostic criteria for alcohol dependence/addiction and the presence of single nucleotide polymorphisms (SNPs) in alcohol-dependent individuals.[11–17]

Similar to humans, different lines of heterogeneous stock rats display a wide-range of ethanol-consumption levels.[18] The very early work by Williams and coworkers,[19] as well as Mardones and coworkers,[20] resulted in the hypothesis that ethanol intake in rodents is also under substantial genetic control. From their early work and that of four other international sites, bidirectional selective breeding has resulted in at least five high alcohol–consuming versus their respective low alcohol–consuming rat lines.[21] One of these selectively bred high alcohol–consuming rat lines is the alcohol-preferring P rat. Essentially, starting from a closed colony of Wistar rats, the highest alcohol drinkers were mated together and the lowest alcohol drinkers were mated together, which resulted in the P and NP lines, respectively.[21,22] The selectively bred alcohol-preferring P rat meets all of the criteria put forth for a valid animal model[23,24] of alcoholism.[21,22] It also meets the more recently proposed criteria including relapse-like,[25] binge-like,[21,22,26,27] and early/adolescent-onset of excessive drinking, which results in blood alcohol concentrations (BACs) greater than the threshold (80 mg%) of NIAAA's[28] definition for binge drinking.[26,29] By the nature of selective breeding, P rats represent multigenerational FHP subjects and their counterparts, the selectively bred alcohol-nonpreferring NP rats, represent multigenerational-FHN individuals. Regarding the point about representing FHP individuals, the P rat has some neurochemical, neurophysiological, and behavioral characteristics similar to those seen clinically in FHP individuals.[21,26,30]

Some of these neurochemical characteristics of P rats involve the glutamatergic system and these findings parallel clinical findings in both FHP individuals and chronic alcoholics. The present minireview will focus on the central glutamatergic system and its role in alcohol dependence. The basic structure and function of the glutamatergic system will be outlined; as it pertains to its activity in the brain, its receptors, and transporters as well as the excitatory synapse. A summary of the existing literature on how ethanol affects the structure and function of the central glutamatergic system will be presented in the context of both preclinical and clinical research. A synopsis of the genetic influence on the development and/or expression of alcohol dependence will be described. An overview of the current knowledge regarding how the ethanol–glutamate interaction affects gene and/or protein expression will be presented, again from both preclinical and clinical perspectives. This will be followed by a compilation of our findings with the P rat as it pertains to innate differences in gene and/or protein expression, relative to its NP counterpart, as well as ethanol's modulation of gene and/or protein expression in subregions of the mesocorticolimbic reward circuitry. This minireview will then be concluded with some closing thoughts on some limitations observed in the existing literature.

2. CENTRAL GLUTAMATERGIC ACTIVITY

The amino acid glutamate is the primary excitatory neurotransmitter in the central nervous system (CNS). Therefore, it is not surprising that glutamate receptors are located throughout the brain (see Fig. 1 for glutamatergic projections). In addition, given the ubiquitous distribution of glutamate and its receptors, its function as the primary excitatory neurotransmitter is crucial for many processes, especially those mediating neuroplasticity, learning and memory.[31–33] Glutamate interacts with both metabotropic mGlu1–mGlu8 (Grm1–Grm8 = mGluR1–mGluR8) and ionotropic receptors, which include those that can bind to *N*-methyl-D-aspartate (NMDA) (subunits: GluN1 (Grin1 = NR1); GluN2a–GluN2d (Grin2a–Grin2d = Nr2a–Nr2d); GluN3a–GluN3b (Grin3a–Grin3b = NR3a–NR3b), those that can bind to α-amino-3-hydroxy-5-methyl-4-isoxazole propionic acid (AMPA) subunits: GluA1–GluA4 (Gria1–Gria4 = GluR1–GluR4) or kainite subunits GluK1–GluK4 (Grik1–Grik4 = GluR5–GluR7 + KA1–KA2); these nomenclatures reflect IUPHAR, HUGO, and "old" symbols respectively.[34–36] Due to glutamate's role in excitotoxicity, extracellular glutamate must be tightly controlled.[34,37,38] This is accomplished, for the most part, by multiple glutamate transporters.[34] The human excitatory amino acid transporter 2 (EAAT2) and its rodent analog glutamate transporter 1 (GLT1) appear to be the main transporters performing this function centrally.[34,37]

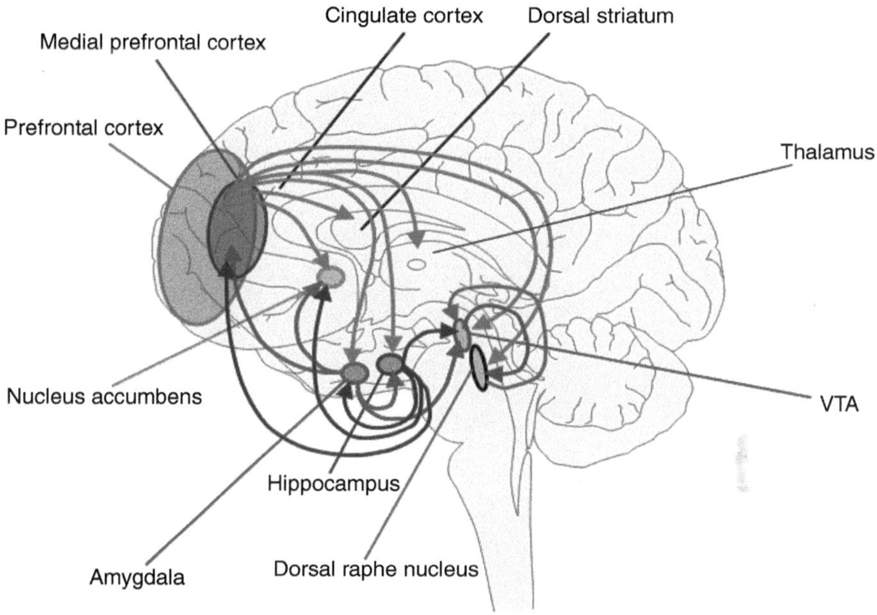

Figure 1 A simplified diagram depicting glutamatergic projections of the mesocorticolimbic, extended amygdala, and brainstem reward neurocircuitry. The online version includes color coding of these projections; with green representing the PFC/mPFC, red representing the amygdala, purple representing the hippocampus, blue representing the VTA, and gray representing the DRN.

2.1 Metabotropic Glutamate Receptors (Fig. 2)

The mGlu receptors are G-protein-coupled protein receptors (GPCRs) located at the neuronal synapse, extrasynaptically as well as on glial cells (Fig. 2). These receptors are divided into three groups. Group I mGluRs (mGlu1 and mGlu5) are predominately postsynaptic and engage in slow excitatory neurotransmission; Group II mGluRs (mGlu2 and mGlu3) are predominately presynaptic, with some postsynaptic and glial localization (Fig. 2) where they engage in slow inhibitory neurotransmission; and Group III mGluRs (mGlu4, mGlu6, mGlu7, and mGlu8) are generally restricted to the presynaptic terminal and, like Group II mGluRs, engage in slow inhibitory neurotransmission. The mesocorticolimbic and associated reward circuitry express high levels of mGlu1, mGlu2, mGlu3, and/or mGlu5, notably in the Acb, caudate nucleus, cortex, lateral septum, dorsal striatum, amygdala, and hippocampus.[39–43]

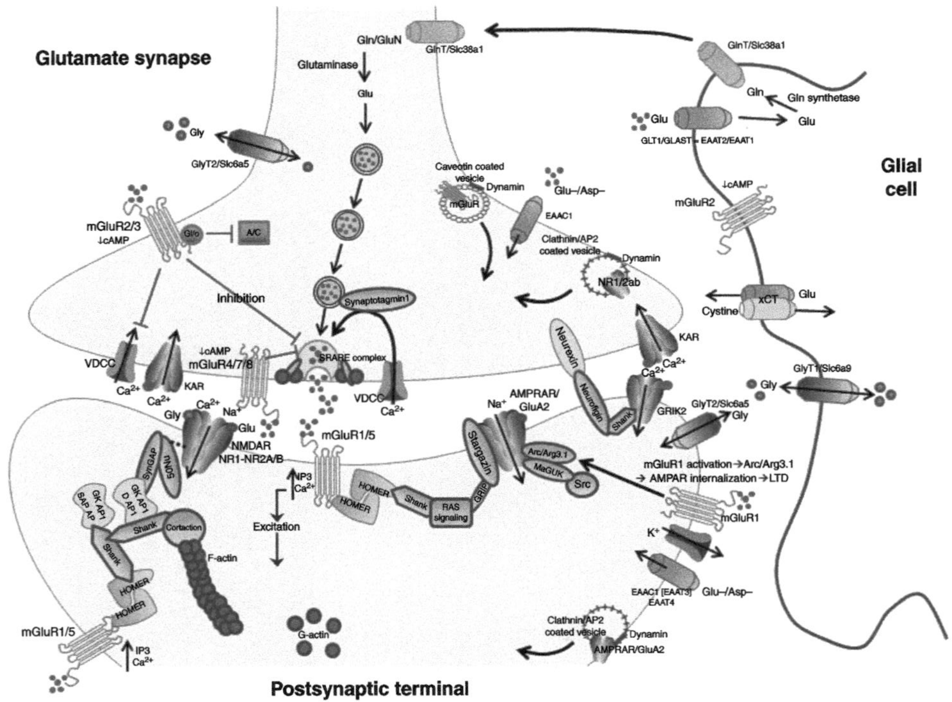

Figure 2 A simplified diagram of a prototypic glutamatergic synapse in the brain. The figure depicts some of the intra-, inter-, and extracellular activities of glutamate-associated plasticity. Abbreviations: AC, adenylate cyclase; AP2, adaptor protein 2; Arc, activity-regulated cytoskeleton-association protein; Asp, aspartate; GKAP/DAP1, disks large associated protein 1 (part of PSD); Gln or GluN, glutamine; GlnT, glutamine transporter; GLT, glutamate transporter; Glu, glutamate; Gly, glycine; GlyT, glycine transporter; LTD, long-term depression; nNOS, neuronal nitric oxide synthase; Src, a tyrosine kinase; VDCC, voltage dependent calcium channel. Other abbreviations are in the text and tables.

2.2 Ionotropic Glutamate Receptors (Fig. 2)

Ionotropic glutamate receptors are ligand-gated ion channels involved in fast excitatory transmission in the CNS. There are several types of glutamate ionotropic receptors including AMPA, kainate, and NMDA-receptor subunits. Most ionotropic glutamate receptors are located postsynaptically, although some are located presynaptically and on glia cells as well (Fig. 2). Similar to the mGlu receptors, they are found throughout the brain including cortical regions, hippocampus, amygdala, basal ganglia, midbrain, hindbrain, and brainstem nuclei.[44–49] A recent functional addition to the ionotropic glutamate receptors are the delta1 and delta2 subunits symbolized by *Glud1* and *Glud2*. In the past these subunits were considered orphans and not functional but more recent findings indicate that they are indeed functional, by modulating LTD and prepulse inhibition of the acoustic startle response (sensorimotor gating), and their localization is not restricted to the cerebellum but includes cortical and limbic regions as well.[50–54]

2.3 Glutamate Transporters and Carriers (Fig. 2)

As discussed earlier, excessive glutamate in the synapse (e.g., the addiction-related hyperglutamatergic state) can lead to excitotoxicity and neuronal death. Thus, glutamate uptake/transport from the synapse and perisynapse is required to prevent the plasticity associated with addiction to become excitotoxicity.[34,38] There are several transporters that regulate extracellular glutamate levels including the EAAT1 (glutamate aspartate transporter: GLAST or *Slc1a3*), EAAT2 (GLT1), EAAT3 (excitatory amino acid carrier 1: EAAC1 or *Slc1a1*), EAAT4 (*Slc1a6*), and EAAT5; where EAAT indicates the human homolog.[55–57] There are also intracellular vesicular glutamate transporters (VGLUT1–3 of the *Slc17* family of genes) that mediate the uptake of glutamate into synaptic vesicles. Intracellular glutamate carriers, other than the vGLUTs, include the *Slc25a* family of genes. Since the NMDAR NR1 subunit contains the glycine-binding site, it is important to recognize the role of bidirectional glycine transporters (GlyT1 which is primarily glial and GlyT2 which is primarily neuronal) at the excitatory synapse (GlyTs are represented by gene families *Slc6a* and *Slc7a*). Given the increased interest in *N*-acetyl-cysteine's role in mental health, it is equally important to recognize the cystine–glutamate exchanger's (xCT = *Slc7a11*) role in reversing damage induced by excessive extracellular glutamate. The xCT, generally located on glial cells, takes up cystine and releases glutamate molecules. Cystine is then converted to cysteine, which is used to synthesize glutathione and other proteins. Glutathione is a key antioxidant and important in reversing neuronal damage induced by excitotoxicity and oxidative stress.[58]

2.4 Glutamate Synthesis and Metabolism (Fig. 2)

Another biologic method for preventing excessive glutamate in the synapse is glutamate's synthesis from glutamine intracellularly.[34] Glutamine can be transported into and out of the synapse without inducing neurotoxicity. The metabolism of glutamate to glutamine occurs primarily in glial cells via the enzyme glutamine synthetase. Glutamine is then transported (glutamine transporter (GlnT), which is common to both glia and neurons and encoded by the *Slc38a* gene family) out of the glial cell. In turn, glutamine is transported out of the perisynapse and into the excitatory presynaptic compartment. Glutamine is then converted to glutamate by glutaminase and transported to the synaptic membrane, or shunted into GABA synthesis via glutamate decarboxylase (GAD). Glutamate is also metabolized by glutamate dehydrogenase–yielding alpha-ketoglutarate, which enters the TCA/Krebs cycle.

2.5 Glutamate and the Postsynaptic Density (Fig. 2)

The postsynaptic density (PSD) is a cytoskeletal specialization that is located beneath the postsynaptic membrane and directly contiguous with the presynaptic "active zone" of excitatory synapses. Glutamate receptors, synaptic proteins, scaffolding proteins, kinases, and other downstream-signaling proteins are located within this PSD. There are several scaffolding proteins within the PSD including membrane-associated guanylate kinases (MAGUKs), Shanks, and Homers.[59–61] Scaffolding proteins can be defined as molecules binding at least two other signaling proteins together. These scaffolding proteins are crucial for synaptic plasticity (e.g., learning and memory) by (1) acting as platforms where signaling molecules can assemble; (2) localizing signaling molecules at specific intracellular sites; (3) coordinating positive and negative feedback signals to modify intra- and extra-cellular signaling pathways; as well as (4) protecting these signaling pathways from inactivation, generally by preventing and/or disrupting phosphorylation.[62–64] In general, scaffolding proteins act as signaling proteins for neuromodulator receptors and anchor these receptors (e.g., glutamate receptors) to the synaptic membrane.[65] Given their role in receptor anchoring, dysregulation of these scaffolding proteins can lead to a number of neurological diseases.[66]

MAGUKs are expressed widely throughout the central nervous system. They are the scaffolding proteins closest to the surface of the postsynaptic membrane and they contain multiple PDZ domains.[67] The PSD-95 protein is one of the most studied MAGUK scaffolding proteins and is involved in postsynaptic stability as well as excitatory receptor insertion.[68] PSD-95 binds to numerous proteins associated with AMPAR and NMDAR complexes. Schnell *et al.*[69] found that interaction between PSD-95 and the AMPA receptor-interacting protein, Stargazin, determines the density of AMPARs at the synapse and through this interaction can regulate synaptic maturation.[70] The PSD-95 anchors NMDARs to the postsynaptic membrane and it acts as a signaling scaffold mediating the activation of neuronal nitric oxide synthase (nNOS) by calcium–calmodulin activity following, for example, entry of calcium through NMDAR channels.[71] Differences between AMPA and NMDA receptors influence on synaptic events is due, in part, to their respective cytosolic C-terminal binding sites to the PSD-95/discs large/zona occludens-1 (PDZ) domain-containing scaffolding proteins.[72–74] The PDZ domain's function is to regulate protein–protein interactions by binding to the C-terminus of each respective target protein; thus, highlighting its crucial role in neuroplasticity, dendritic growth, and dendritic arborization.[68,75]

The ionotropic NMDA and AMPA receptors are primarily concentrated in the PSD but there are distinct differences in synaptic regulation of these receptors.[76,77] Additionally, NMDARs initiate synaptic plasticity by interacting with other components of the PSD.[78–82] Similarly, AMPARs are involved in rapid synaptic transmission and these receptors cycle on and off in a manner tightly controlled by neuronal activity/plasticity in the PSD.[78–82] The recycling (i.e., insertion and removal) of these AMPARs at the synapse modulates synaptic efficiency and thus, like NMDARs, influence learning and memory.[76,77]

Considerable evidence indicates the crucial role of Shank and Homer proteins in neuroplasticity, as well as alcohol and drug dependence.[83–89] Shank and the Homer scaffolding proteins are located deeper within the PSD than the MAGUKs (i.e., toward the cytoplasm).[70] Three genes encode for Shank proteins (Shank1, Shank2, and Shank3) and Shank scaffolding proteins bind to neuroligins and neurexins, as well as NMDAR complexes in the PSD.[70] Shanks are capable of binding other scaffolding molecules within the PSD (e.g., Homers) and thus are capable of linking mGlu receptors, as well as AMPA and kainate ionotropic receptors to NMDARs. Therefore, scaffolding proteins present in the PSD mediate a number of plasticity-associated events through reorganization of PSD-associated proteins, regulation of membrane protein trafficking and activity, as well as the maintenance of associated epithelial cell polarity and morphology.[90–92] Moreover, the connection between Shank proteins and neuroligins/neurexins indicates that the PSD may bridge with the "active zone" of the presynaptic terminal. The Homer family of scaffolding proteins is encoded by 3 genes (*Homer1*, *Homer2*, and *Homer3*). Homer scaffolding proteins interact with the C-terminus of Group I mGluRs, bind to Shank/PSD-95/NMDA-receptor complexes, and can also interact with a number of downstream effectors of mGlu1/5 including: IP3 receptors, diacylglycerol lipase-2, and PI3K enhancer (PIKE). Homer proteins are best characterized for their role in regulating mGlu1/5 trafficking, PSD localization, and signaling of mGlu1/5 and NMDA receptors, but are also critical in the regulation of actin and dendritic morphology.[93] Furthermore, through their ability to associate with Shank, Homers facilitate cross-talk between mGlu1/5 and NMDA receptors and the integration of their calcium-dependent intracellular events underpinning synaptic plasticity.[94]

3. CENTRAL GLUTAMATE ACTIVITY AND ALCOHOL DEPENDENCE

3.1 Alcohol's Effects on Glutamate Activity and Extracellular Levels

Substantial preclinical evidence indicates that glutamatergic activity mediates natural as well as alcohol- and drug-associated reward through direct and indirect interactions with other neurotransmitter/neuromodulatory systems within the mesocorticolimbic, extended amygdala, and associated reward neurocircuitry (Fig. 1).[95–102] A key hypothesis that has received considerable attention postulates that the mesocorticolimbic and extended amygdala reward circuits, in the presence of alcohol, lose homeostasis between excitatory and inhibitory transmission and revert to a hyperglutamatergic/hyperexcitatory state resulting in the development and expression of alcohol/drug dependence.[37,98,102–107]

In general, ethanol consumption and/or exposure to low or moderate doses of ethanol elevate glutamatergic transmission and/or extracellular levels of glutamate in the nucleus accumbens (Acb)[108]; Acb shell (AcbSh)[88,89,109–111]; basolateral amygdala (BLA)[112]; cortex[113]; Hippocampus[114,115]; ventral tegmental area (VTA)[116]; and posterior VTA (pVTA).[110,117] It has also been shown that genetics influence ethanol-induced increases in extracellular glutamate within the Acb and/or PFC, such that rats with a predisposition for higher ethanol intake (P and Lewis rats) display greater elevations in glutamate relative to rats with a predisposition for lower ethanol intake (NP and F344 rats).[118,119] Interestingly, similar relations have been observed between alcohol-preferring versus nonpreferring inbred strains of mice.[86,120] In addition, these elevations in glutamatergic activity can be conditioned to the environment in which the animal had access to ethanol, with glutamatergic increases seen in the Acb core (AcbCo) or basolateral Amyg (BLA).[103,121,122]

A recent comprehensive preclinical study provides an excellent example of glutamate's role in the development and expression of alcohol dependence.[123] Griffin and coworkers[123] evaluated whether free-choice ethanol access would increase Acb extracellular glutamate levels and found that dependence-induced (chronic intermittent access via ethanol-vapor chambers) ethanol-drinking doubled Acb glutamate levels over those seen in nondependent mice. Moreover, these authors reported that this doubling of glutamate was observed a week later indicating that these increases in glutamate activity were not dependent upon ethanol withdrawal itself. Consistent with prior neuropharmacologic results in studies of low versus high alcohol–consuming inbred mice,[120] Griffin *et al.*[123] showed that pharmacologic elevation of glutamate in the Acb, with a pan-glutamate-reuptake inhibitor (Threo-beta-benzyloxyaspartate, TBOA), increased the ethanol intake of nondependent mice to the levels observed in dependent animals, with TBOA also increasing ethanol intake further in the dependent mice. Also consistent with the results of Kapasova and Szumlinski,[120] when the mGlu2/3 autoreceptor agonist LY379268 was microinjected into the Acb to lower glutamate levels, ethanol intake was decreased in both the dependent and nondependent mice.[123] These latter results parallel earlier work indicating that manipulations of extracellular glutamate, including mGlu2/3 activation, actively regulate ethanol intake in multiple animal models of alcoholism (discussed later in the chapter).

Finally, evidence for altered glutamate neurotransmission within the previously mentioned brain regions, as well as the anterior cingulate cortex (ACC), has been reported in clinical studies of alcohol-dependent individuals as well.[124–131] For example, a proton magnetic resonance spectroscopy (MRS) study examining the role of hippocampal glutamate in major depression and risky alcohol drinking revealed that elevated glutamate levels in the hippocampus were directly associated with both the presence of major depression and self-reported risky drinking.[132] These authors noted that the major depression and risky drinking group did not differ from the control group in age-of-first alcohol use, Alcohol Use Disorders Identification Test (AUDIT) survey scores or smoking behavior; but, this group did have significantly more FHP individuals (approximately six-fold) indicating a possible confound. Another recent MRS study provided support for differences in glutamate activity of FHP versus FHN individuals.[133] These authors reported that glutamate/glutamine ratios increased significantly between adolescence and emerging adulthood in FHN, but not FHP, individuals. This suggests that having a familial history of AUDs may genetically predispose an individual for abnormal developmental changes in glutamatergic neurotransmission across periadolescence.[133]

3.2 Metabotropic Glutamate Receptors and Alcohol

It has been shown that ethanol-binge drinking by mice upregulates mGlu1/5-Homer2 signaling in several mesocorticolimbic structures including the Acb[83,84] and the central amygdala (CeA)[85] and activation of either or both of these Group I mGluRs is required for mice to manifest binge-drinking behavior.[83–85]

Several studies have examined the effects of mGlu5 antagonists on operant ethanol self-administration behaviors in P rats as well. Systemic administration of the mGlu5 antagonist 2-methyl-6-(phenylethynyl)pyridine (MPEP) can reduce operant ethanol self-administration,[134] reduce ethanol breakpoint without affecting sucrose breakpoint or locomotor activity,[135] and block the repeated alcohol-deprivation effect (ADE).[134] The effects of MPEP in rats were extended to mice by demonstrating that this mGlu5 antagonist interfered with the acquisition and maintenance of ethanol drinking by C57BL/6J mice as well,[136,137] which appears to depend upon a protein kinase C-epsilon (PKC-epsilon) pathway.[138] A subsequent study using P rats[139] examined the effects of systemic MPEP on the extracellular signal-regulated kinase (ERK1/2) pathway,[140] which is downstream of mGlu5 and implicated in addiction. MPEP attenuated cue-induced reinstatement of ethanol-seeking behavior, which was associated with decreased phosphorylated (p)ERK1/2 immunoreactivity (IR) in the BLA, but not CeA, and AcbSh, but not AcbCo.[139] These findings support a role for ERK1/2 phosphorylation in the BLA and AcbSh in mediating cue-induced reinstatement of ethanol-seeking behavior. A third study from this laboratory[141] confirmed a role for mGlu5 within the AcbCo in ethanol self-administration; such that local application of MPEP into the AcbCo reduced ethanol operant responding without affecting locomotor activity, sucrose or water responding. In contrast, MPEP infused into the dorsomedial caudate nucleus or mPFC did not alter operant ethanol self-administration.[141] In a study that examined the effects of mGlu5 antagonist 3-[(2-methyl-1,3-tiazol-4-yl)ethynyl]-pyridine (MTEP) on ethanol self-administration by two high alcohol–consuming rat lines (inbred P (iP) and Fawn-Hooded (FH)) reported that MTEP significantly decreased intake.[142] However, these authors also reported that MTEP induced mild sedative effects in iP but not FH rats. Together, these findings indicate that mGlu5 receptors play an important role in regulating different aspects of alcoholism-related behavior in both rat and mouse models.

Other mGluR ligands have also been tested for their effects on excessive ethanol drinking or ethanol reinforcement. Earlier studies revealed that systemic pretreatment with the selective mGlu1 antagonist, (−)-ethyl (7E)-7-hydroxyimino-1,7a-dihydrocyclopropa[b]chromene-1a-carboxylate (CPCCOEt), produced inconsistent effects on operant ethanol self-administration in P rats[134] versus C57BL/6J mice,[137] such that intra-Acb infusions of CPCCOEt were unable to alter binge drinking in mice.[83] These inconsistent effects of CPCCOEt on measures of alcohol intake likely reflect its relative insolubility. For instance, the more soluble, highly selective, mGlu1 antagonist JNJ 16259685 lowers operant ethanol self-administration and ethanol breakpoint in P rats when administered systemically[135,143] and reduces binge drinking when infused into the AcbSh of mice.[144] However, systemic JNJ 16259685 pretreatment has nonselective effects in that it also reduces locomotor activity,[135,143] which may reflect the high abundance of mGlu1 receptors in the cerebellum and their effects on its control of motor movement.[134,135,143]

Studies have also examined the effects of targeting mGlu2/3 receptors within the contexts of operant ethanol self-administration. When given systemically, the mGlu2/3 antagonist LY341495 did not alter operant responding for ethanol by P rats,[134] although systemic pretreatment with the mGlu2/3 receptor agonist LY404039 reduced ethanol-seeking and -relapse-like behavior.[145] However, these latter authors reported that LY404039 did not alter the maintenance of operant ethanol self-administration in these animals.[145] Interestingly, the local application of the mGlu2/3 agonist LY379268 in the AcbCo was sufficient to reduce operant ethanol self-administration by P rats,[141] in a manner akin to the aforementioned studies of mice drinking under free-access conditions in the home cage.[120,123] While understudied, the effects of mGlu2/3 agonists appear to depend upon the route of administration or the experimental procedures employed. However, as observed with mGlu1 antagonists,[134,135,143] the effect of intra-AcbCo mGlu2/3 agonism on operant ethanol drinking by P rats is also not specific due to effects on locomotor activity.[141] Other lines of evidence support an important role for mGlu2 receptors in drug addiction as well.[97,102,146]

3.3 Ionotropic Glutamate Receptors and Alcohol

As noted earlier, ionotropic glutamate receptors play an important part in the development of alcohol abuse and dependence, with many of ethanol's effects mediated by these receptors. For instance, Enoch and coworkers[147] as well as Jin and coworkers[148] have reported that numerous ionotropic glutamate receptor subunit expression levels are significantly altered in chronic alcoholics and cocaine addicts, relative to control samples. Nevertheless, there has been limited behavioral–pharmacology research on the involvement of ionotropic receptors in excessive ethanol intake by P rats. Pretreatment with aniracetam, a selective positive allosteric modulator of AMPA receptors, increased operant responding for ethanol and cue-induced reinstatement of operant responding for ethanol by P rats, while not altering locomotor activity or operant responding for sucrose.[149] These authors confirmed the involvement of AMPA receptors by demonstrating that the AMPA receptor antagonist 6,7-dinitroquinoxaline-2,3-dione blocked aniracetam-induced increases in ethanol self-administration. Besides the P rat, the selectively bred ALKO alcohol-accepting (high ethanol-consuming rat line from Finland) has also been used to assess the role of AMPA/kainate receptors in excessive ethanol intake, with systemic administration of the antagonist CNQX significantly reducing operant ethanol-seeking behavior by these rats.[150] In addition, systemic administration of the mixed NMDA/glycine receptor antagonist L-701,324 also significantly reduces operant ethanol-seeking behavior in AA rats.[150] While MTEP is an mGlu5 receptor antagonist and decreases ethanol self-administration, it also decreases mRNA expression for both *Glua2* and *Glun1* in the cingulate cortex of iP and FH rats.[142] Thus, the effects of mGlu-receptor activity on ethanol intake are paralleled by its regulation of ionotropic glutamate receptor subunit gene expression underscoring the interaction of these two classes of glutamate receptors. Although there has been limited research on ionotropic glutamate receptors regarding alcohol- and/or drug-intake, -seeking, etc.[151]; there is clear evidence that NMDA and AMPA receptors are affected by ethanol, which in turn affects neuroplasticity, learning and memory.[31,32,95,151,152]

3.4 Glutamate Transporters/Carriers and Alcohol

Chronic ethanol-drinking or ethanol exposure significantly reduces glutamate uptake in the brain through downregulation of glutamate transporter, or antiporter, expression[153–156] and reductions in glutamate transporter expression have been confirmed in postmortem evaluation of brains from alcoholics.[128] Recent studies have examined the effects of modulating GLT-1, glial EAAT2, and the cystine–glutamate antiporter (xCT) on ethanol intake. Ceftriaxone, a beta-lactam antibiotic, increases glutamate reuptake by upregulating GLT1 expression.[157] Given the general hyperglutamatergic state in alcohol/drug dependence, it is not surprising that this compound also decreases ethanol consumption, ethanol dependence (ED)-associated withdrawal signs, and withdrawal-associated escalation of ethanol intake in P rats.[158–162] These authors reported a consistent upregulation of GLT1 expression in the Acb and mPFC, which was negatively associated with the observed reductions in ethanol consumption. Moreover, these authors have shown that ceftriaxone, or its analogs, significantly reduces ethanol intake and reverses chronic ethanol-induced downregulation of GLT1 expression in the mPFC, Acb, Amyg, and hippocampus, as well as reversing downregulation of xCT in some brain regions.[153,163–166]

Given that chronic ethanol induces excitotoxicity and oxidative stress, it is noteworthy that ethanol-induced increases in *Slc7a11* (the protein product being xCT, the cystine–glutamate antiporter) are not dependent upon excitotoxicity or oxidative stress.[167] These authors reported that ethanol itself can inhibit octamer-binding transcription factor 1's (OCT-1) repression of the *Slc7a11* promoter *in vitro*, which in turn elevates *Slc7a11* transcription. This increase in xCT would putatively increase the import of cystine into glial cells, where it would be converted to cysteine and subsequently into glutathione resulting in increased glutathione and decreased neurotoxicity.[58] Regarding intracellular glutamate transport, an early study[168] examined the effects of continuous ethanol exposure versus exposure interspersed with repeated deprivations on these vesicular glutamate transporters. These authors reported that repeated deprivations increased vGLUT2-immunostained terminals in the AcbSh compared to the water control group. However, ethanol exposure did not alter the level of vGLUT1-immunostained terminals in this brain region. These results suggest that the presence of multiple withdrawal episodes preferentially increases vGLUT2 expression in glutamate terminals in the AcbSh of P rats.[168]

3.5 Glutamate-Associated Enzyme Activity and Alcohol

Early work revealed that chronic ethanol exposure decreased glutamine synthetase while increasing glutamate and GAD in the cortex of rats.[169] A more recent preclinical study found decreased GAD-67 expression levels, in the BLA, 2 months after 3 weeks of ethanol-diet initiated in adulthood, but not adolescence.[170] Another study reported that chronic ethanol consumption decreased glutamine synthetase in the striatum (dorsal vs. ventral was not delineated) but not cortex of rats.[171] A contemporary study also reported reduced glutamine synthetase in the brain, although the area of the brain was not identified, after chronic ethanol consumption, which started at the beginning of adolescence.[172] A postmortem study indicated that glutamine synthetase was downregulated in the hippocampus of alcoholics without hepatic pathology.[173] These consistent reductions in glutamine synthetase following ethanol exposure or consumption suggest the presence of astrocytic pathology and, by extension, increased neurotoxicity. Regarding glutamate dehydrogenase, which metabolizes glutamate; adolescent binge-like drinking by rats resulted in a 40% decrease in hippocampal glutamate dehydrogenase 1, which was not seen in rats that received the same protocol during adulthood.[174] Given this finding, it is noteworthy that ethanol inhibits NMDA excitation and LTP to a greater extent in hippocampal slices from adolescent versus adult rats.[175,176] Therefore, significant differences in ethanol's effects on glutamatergic activity occur across periadolescence and adulthood (see later in the chapter).

3.6 The Postsynaptic Density and Alcohol

Recent research indicates that alcohol and/or drugs of abuse have a profound developmental effect on the PSD as well[174,177,178]; such that, similar to earlier reports on the vulnerability of the adolescent hippocampus to alcohol and/or drug exposure,[175,176] there is a differential effect of binge-like ethanol exposure between adolescent and adult rats. Risher *et al.*[177] reported that adolescent intermittent ethanol (AIE) exposure in rats reduced PSD-95 expression levels in the hippocampus, leading to the retention of immature-like dendritic spine phenotypes into adulthood. There was also a reduction in the number of VGlut1/PSD-95 and VGlut1/SAP102 (another MAGUK) colocalized synaptic puncta and these effects were driven by decreases in PSD-95 and SAP102 density with no effect on presynaptic VGlut1 expression levels.[177] In contrast, chronic intermittent ethanol (CIE) during adulthood[178] did not alter PSD-95 expression in the hippocampus as a whole. However, these authors indicated that adult CIE could alter dendritic complexity in a subregion-specific manner, with a partial return to basal levels after protracted abstinence.[178] Taken together these studies suggest that the PSD-95, and glutamate activity, may be more vulnerable to ethanol-induced changes during adolescence than during adulthood and that adolescent ethanol-induced changes in PSD-95 may interfere with the maturation of dendritic spines. Similarly, considerable evidence indicates a crucial role for Shank and Homer proteins in neuroplasticity as well as alcohol and drug dependence.[86–89,179] For example, significant increases in both AcbCo and CeA Homer2a/b expression levels were seen 24 h after the removal of ethanol from chronically drinking P rats.[180]

4. GLUTAMATE-ASSOCIATED GENETIC VARIATIONS AND ALCOHOLISM

An early Genome Wide Association Study (GWAS) followed by a Gene Set Enrichment Analysis (GSEA) found that when gene variations were analyzed for grouping, neuronal signaling genes dominated other associations with an individual's level of response to alcohol and glutamate was the primary neurotransmitter system implicated.[181] These authors also noted that FHP individuals show an altered level of response to alcohol and ketamine (an NMDA antagonist), thus confirming a genetic risk for alcoholism and altered glutamatergic function.[181,182] Similarly, a pathway analysis of variants in 130 addiction-related candidate genes confirmed a significant role for glutamate signaling in alcohol dependence, with the odds ratio of *mGlu1*-rs2300620 (>1.6) exceeding that of any other significant gene variant.[183] A contemporary study using pathway analysis revealed the NMDA-dependent AMPA-trafficking cascade centered on the gene encoding the multiple PDZ domain protein (*Mpdz*) was significantly associated with alcohol dependence in a subset of the SAGE study.[184]

Regarding particular gene variants, polymorphisms of *Glun2a*, NR2a subunit of the NMDA receptor, are significantly associated with being an FHP individual, an early onset of risky drinking during adolescence as well as the maximum number of drinks in adulthood, with this association having been replicated in a second large sample.[185] A subsequent study also found that a polymorphism in the promoter region of *Glun2a* is significantly associated with alcohol dependence, with this finding replicated in a second sample.[186] A variant of another ionotropic glutamate receptor subunit, *Gluk3*, was found to be associated with alcohol withdrawal–associated delirium tremens, but not seizures, in a German cohort.[187] Another study reported that a polymorphism in *Gluk1* is significantly associated with alcohol dependence in Caucasians, although many of the alcohol-dependent, but none of the control, subjects were polysubstance users, which may represent a confound in this study.[188]

Following earlier work indicating theta event-related oscillations (EROs) are associated with the P300 event-related potential (ERP), whose amplitude is negatively associated with FHP,[189,190] a Family Based Association Test (FBAT) was carried out on a subset of the COGA sample.[191] These authors reported that multiple polymorphisms of the *mGlu8* gene are significantly associated with theta EROs and a genetic risk to develop alcoholism.[191] A subsequent study by this group confirmed that polymorphisms of the *mGlu8* gene are associated with a genetic risk to develop alcohol dependence.[192] In another recent study, the *mGlu3*-rs6465084 polymorphism was found to be significantly associated with alcohol dependence in a male Han Chinese cohort.[193]

A recent study examining the association between polymorphisms in vesicular glutamate transporters and severe alcoholism in a female Swedish cohort indicated a nominally significant association with *Slc17a6*-rs2290045.[194] A study evaluating associations between alcohol dependence and GAD genes (*Gad1* and *Gad2*), the enzyme responsible for conversion of glutamate to GABA, in a subset of the Irish Affected Sib Pair Study of Alcohol Dependence sample reported that two polymorphisms in *Gad1* were associated with initial sensitivity to alcohol and a different polymorphism in *Gad1* was associated with age at alcohol dependence onset.[195] An analysis of the relationship between *Gad1* variants and alcohol dependence in an Italian cohort revealed that the *Gad1*-rs11542313 polymorphism was significant.[196] An earlier study evaluating polymorphisms of *Gad2* indicated that the functional promoter *Gad2*-243A>G variant was significantly associated with alcohol dependence in Russian but not European American males.[197] A contemporary study found that three polymorphisms of *Gad1*, but no polymorphisms of *Gad2*, were significantly associated with alcohol dependence in Han Taiwanese men.[198] In a recent genome-wide DNA methylation study, it was reported that methylation of *Gad1* is significantly associated with alcohol dependence in a cohort of Han Chinese men, suggesting epigenetic effects of chronic alcohol abuse.[199] These results indicate that polymorphisms in *Gad1* are significantly associated with alcohol dependence, or related phenotypes, in multiple populations around the world.

5. ALCOHOL AND GLUTAMATE GENE EXPRESSION

Extending the previously mentioned behavioral pharmacologic and neuropharmacologic studies are the results of a very comprehensive study by Meinhardt and coworkers,[200] demonstrating that: (1) in ethanol-dependent rats, glutamate-related gene changes were primarily seen in the mPFC, rather than the Acb or amygdala; (2) within the mPFC of ethanol-dependent rats, downregulation of the NMDA-receptor subunits *Glun2a* and *Glun2b*, as well as *mGlu2*, but not *mGlu3*, and *egr1* (early growth response protein 1: *Zif268*, a transcription factor involved in neuroplasticity and vesicular exocytosis at excitatory synapses) were significant in the infralimbic mPFC only, with *mGlu2* displaying the greatest reduction; (3) the AcbSh receives its glutamatergic projections primarily from the infralimbic mPFC, according to retrograde tracing, and these projection neurons displayed significant ethanol-dependence-associated downregulation of *mGlu2*, *Egr2*, and *Egr4*; (4) while basal glutamate in the AcbSh did not differ between ethanol-dependent and control rats, peripheral administration of the mGlu2/3 agonist LY379268 significantly reduced extracellular glutamate in the AcbSh of control but not ethanol-dependent rats, suggesting a downregulation of mGlu2/3 function; (5) ethanol-dependent rats displayed greater relapse (almost twice as many responses) and progressive ratio than controls; (6) lentivial knockdown of *mGlu2* in the infralimbic mPFC significantly attenuated cue-induced reinstatement of responding in ethanol-dependent but not nondependent rats; (7) *mGlu2* knockdown rats did not differ from control rats in operant relapse responding, operant responding for sweetened condensed milk, or locomotor activity in an open-field test; and (8) a RT-qPCR analysis of postmortem ACC samples revealed that alcoholics had significantly less *mGlu2* mRNA than their respective controls. An early postmortem study revealed that individuals who had abused nicotine, but not alcohol, displayed greater expression of *Slc17a6* and *Slc17a7* vesicular glutamate transporters in the VTA.[201] However, in individuals who abused both nicotine and alcohol, these increases in *Slc17a6* and *Slc17a7* were significantly reduced.[201]

In another study, Enoch and coworkers[147] reported that hippocampal glutamate receptor subunit gene expression levels differed between alcoholics, cocaine addicts, and healthy controls. Specifically, *Glua4*, *Gluk3*, and *mGlu4* expression was significantly higher in alcoholics, relative to both controls and cocaine addicts; *Glun2b* expression was higher in both alcoholics and cocaine addicts, relative to control levels; and *Glun2d* and *mGlu3* was upregulated in alcoholics while being downregulated in cocaine addicts, relative to control levels.[147] Substantial changes of glutamate receptor subunit gene expression in the hippocampal dentate gyrus and to a lesser extent orbital frontal cortex, but not the dorsal-lateral prefrontal cortex, of alcoholics versus healthy controls was confirmed in another recent postmortem study.[148] In particular, the data from the dentate gyrus revealed that alcoholics had higher expression levels of *Glua2, Glua3, Gluk2, Gluk3, Gluk5, Glun1, Glun2a, Glun2c, Glun2d*, and *Glun3a*; whereas only *Glun3a* expression was higher than controls in the orbital frontal cortex.[148]

6. ALCOHOL-ASSOCIATED CHANGES IN GENE/PROTEIN EXPRESSION OF P RATS

An examination of protein expression changes in subregions of the Acb and Amyg of chronic ethanol-drinking P rats revealed at 24-h withdrawal that GluN2a expression levels were increased, whereas GluN2b expression levels were decreased in the AcbSh.[180] These authors also reported that Homer2a/b, mGlu1, mGlu5, GluN2a, and GluN2b expression levels were all consistently increased in the AcbCo and CeA.[180] To test the hypothesized genetic role for the mGlu2 receptor in alcohol dependence, a recent RNA and exome sequencing study revealed that a SNP which creates a stop codon in the *mGlu2* gene is present in P, but not NP, rats.[202] This stop codon results in the absence of mGlu2 receptors, impaired glutamatergic synaptic transmission, and altered levels of multiple genes associated with synaptic function. These authors also examined F2 rats from a PxNP–NPxP cross and found that *mGlu2* expression levels were significantly and inversely related to ethanol consumption, with decreases in *mGlu2* and mGlu2 receptor expression associated with significant increases in ethanol-drinking behavior.[202] Similarly, mGlu2 knockout mice display significantly greater ethanol consumption and preference than their wild-type counterparts.[202] As mGlu2 receptors function as autoreceptors, the results indicate an inverse relationship between mGlu2 gene/protein expression and ethanol intake, which is consistent with the aforementioned hypothesis that excessive drinking phenotypes are associated with a hyperglutamatergic state.

Our laboratories have published a number studies that examined central gene and/or protein expression profiles of P rats.[180,203–215] Here we highlight alcohol-associated glutamate-related genes whose expression levels have been identified as significantly changed by ethanol self-administration in P rats, or innately different between P and NP rats, in brain regions implicated in alcohol reward and reinforcement (Fig. 1).

6.1 Metabotropic Glutamate Receptor Expression Differences (Table 1)

A recent study revealed that periadolescent binge-ethanol consumption, using our drinking-in-the-dark—multiple-scheduled-access model,[27,29] resulted in a general elevation of *mGlu1* and *mGlu4* (twice or three times that of controls, respectively) in the dorsal raphe nucleus (DRN) and *mGlu7* (~1.4-fold increase) in the CeA.[213] However, there was a significant decrease (~1.7-fold decrease) in *mGlu6* expression within the DRN as well. Given that *mGlu1*, for the most part, is excitatory and the others inhibitory, this suggests a balance between ethanol-induced up- and downregulation of gene expression for these metabotropic receptors.

mGlu2 and *mGlu3* expression levels were significantly lower in the AcbSh and CeA of adult ethanol-naive P versus NP rats.[209,211] As discussed earlier, Zhou and coworkers[202] reported a stop-codon polymorphism of *mGlu2* in P, but not present in NP rats that appear to predispose them to high ethanol consumption. And, multiple studies[134,135,141,143,145] have shown that mGlu2/3 agonists, presumably acting at the presynaptic autoreceptor, block ethanol-self-administration, -seeking, and -relapse behavior in P rats. Noteworthy is the fact that *mGlu4*, *mGlu5*, *mGlu7*, and *mGlu8* (all are involved with inhibitory activity, generally at the presynaptic terminal) are also significantly lower in the AcbSh of P versus NP rats. Gene expression for only one mGluR, *mGlu1*, was ~1.4-fold higher in the AcbSh of P versus NP rats, which is a finding consistent with prior results from inbred C57BL/6J versus DBA2/J mice.[86] Regarding the effects of ethanol, operant ethanol self-administration by adult P rats significantly reduced *mGlu1* (~20% decrease) expression levels in the AcbSh, relative to ethanol-naive P rats.[214] Given *mGlu1*'s basal elevation, relative to NP rats, it is possible that ethanol self-administration could reverse these levels; although, it is also possible for mGlu1 protein levels to be elevated by ethanol with mRNA levels responding in the opposite direction. Regarding this hypothesis, our laboratory has shown that extended (6 months) ethanol drinking by P rats does indeed increase mGlu1, and mGlu5, receptor protein expression in the AcbCo and CeA.[180] Again, previous work with P rats found that an mGlu1 antagonist significantly reduced operant ethanol self-administration and

Table 1 Metabotropic Glutamate Receptor (mGluR) Gene (*mGlu*) Expression Differences Between P and NP Rats (i.e., Fold-Change (F-C) P vs. NP) or Changes Induced by Ethanol Consumption in P Rats, Relative to Ethanol-Naive P Rats (i.e., F-C Ethanol (E) vs. Control (C))

Gene ID	Gene Names	F-C P vs. NP	F-C E vs. C	Age of Rat	Brain Region
mGlu7	Glutamate receptor, metabotropic 7 (*Grm7:* mGluR7)		1.37	Adolescent	CeA
mGlu1	Glutamate receptor, metabotropic 1 (*Grm1:* mGluR1)		2.06	Adolescent	DRN
mGlu4	Glutamate receptor, metabotropic 4 (*Grm4:* mGluR4)		2.98	Adolescent	DRN
mGlu6	Glutamate receptor, metabotropic 6 (*Grm6:* mGluR6)		*−1.70*	Adolescent	DRN
mGlu1	Glutamate receptor, metabotropic 1 (*Grm1:* mGluR1)	1.39		Adult	AcbSh
mGlu1	Glutamate receptor, metabotropic 1 (*Grm1:* mGluR1)		*−1.19*	Adult	AcbSh
mGlu2	Glutamate receptor, metabotropic 2 (*Grm2:* mGluR2)	*−1.46*		Adult	AcbSh
mGlu3	Glutamate receptor, metabotropic 3 (*Grm3:* mGluR3)	*−1.22*		Adult	AcbSh
mGlu4	Glutamate receptor, metabotropic 4 (*Grm4:* mGluR4)	*−1.39*		Adult	AcbSh
mGlu5	Glutamate receptor, metabotropic 5 (*Grm5:* mGluR5)	*−1.30*		Adult	AcbSh
mGlu7	Glutamate receptor, metabotropic 7 (*Grm7:* mGluR7)	*−1.38*		Adult	AcbSh
mGlu8	Glutamate receptor, metabotropic 8 (*Grm8:* mGluR8)	*−1.33*		Adult	AcbSh
mGlu2	Glutamate receptor, metabotropic 2 (*Grm2:* mGluR2)	*−1.84*		Adult	CeA
mGlu3	Glutamate receptor, metabotropic 3 (*Grm3:* mGluR3)	*−1.24*		Adult	CeA

Age of rat refers to whether the P rats had access to ethanol during (peri-)adolescence (postnatal days (PNDs) 30–50) or during adulthood>PND75. Brain regions published thus far: CeA, DRN, and AcbSh. To facilitate distinction between directions of expression (e.g., up- vs. downregulation), downregulation, or lower level F-Cs are in italics and are below the findings of upregulation or higher-level F-Cs for that age and brain region.

breakpoint, although motor activity appeared to be affected as well.[135,143] In general, these findings indicate that the effects of ethanol self-administration on Group1 *mGluR* mRNA expression can be distinguished from effects upon protein expression and/or that different ethanol-drinking experiences/procedures elicit distinct changes in receptor mRNA/protein expression within Acb subregions, with more protracted drinking regimens eliciting changes within the more dorsal AcbCo.

Overall, adolescent binge-like ethanol drinking upregulated two of the three mGlu receptors identified as significantly changed. In the AcbSh of adult animals, P rats had only one of the seven mGlu genes identified as significantly greater than that seen in NP rats. It is noteworthy that the one gene that had higher expression levels in the AcbSh of adult P rats was downregulated following ethanol self-administration. Both genes identified as significantly different in the CeA of adult rats were Group II (mGlu2 and mGlu3) metabotropic receptors and were lower in P versus NP rats.

6.2 Ionotropic Glutamate Receptor Expression Differences (Table 2)

Ionotropic glutamate receptor subunits *Glua3*, *Glua4*, and *Gluk3* were all elevated in the DRN following periadolescent binge-ethanol drinking, but not to the same extent as *mGlu1* and *mGlu4*.[213] Gene expression changes were also observed for the "orphan" ionotropic glutamate delta receptor subunits *Glud1* and *Glud2*. Regarding *Glud2*, there was a 3.5-fold increase in *Glud2ip* (delphilin) after periadolescent binge drinking as well;[213] Glud2ip is a scaffolding protein for Glud2. To some extent, this parallels increases in expression levels of *mGlu7*, in the CeA, induced by periadolescent binge-ethanol drinking. In the periadolescent binge-drinking P rats, the *GluN* subunits were generally downregulated, with only the *Glun2c* subunit being upregulated.[213]

In adult P rats, whereas metabotropic glutamate receptor gene expression levels were consistently lower than those of NP rats; this was not true for *Glua* and *Gluk* gene expression,[209,211] which was mixed. *Glua1*, *Glua2*, *Glua3*, and *Gluk3* expression levels were all lower in the AcbSh of P versus NP rats; whereas *Glua4* and *Gluk1* expression levels were higher in the AcbSh of P versus NP rats.[211] Ethanol-binge drinking by adult P rats reversed the ~30% deficit of *Glua1* in the AcbSh through a 15% increase, relative to ethanol-naive controls.[208] However, operant ethanol self-administration by adult P rats significantly reduced *Glua2* and *Glua3* expression levels, versus ethanol-naive controls, in the AcbSh.[214] Thus, in spite of AcbSh deficits in

Table 2 Ionotropic Glutamate Receptor Subunit Gene Expression Differences Between P and NP Rats (i.e., Fold-Change (F-C) P vs. NP) or Changes Induced by Ethanol Consumption in P Rats, Relative to Ethanol-Naive P Rats (i.e., F-C Ethanol (E) vs. Control (C))

Gene ID	Gene Names	F-C P vs. NP	F-C E vs. C	Age of Rat	Brain Region
Glua3	Glutamate receptor, ionotropic (*Gria3*: AMPA3)		1.29	Adolescent	DRN
Glua4	Glutamate receptor, ionotropic (*Gria4*: AMPA4)		1.31	Adolescent	DRN
Glud1	Glutamate receptor, ionotropic (*Grid1*: delta1)		1.20	Adolescent	DRN
Glud2	Glutamate receptor, ionotropic (*Grid2*: delta2)		2.31	Adolescent	DRN
Glua3	Glutamate receptor, ionotropic (*Grik3*: kainate3)		1.35	Adolescent	DRN
Glun2c	Glutamate receptor, ionotropic NMDA2c (*Grin2c*: NR2c)		2.61	Adolescent	DRN
Glun2b	Glutamate receptor, ionotropic, NMDA2b (*Grin2b*: NR2b)		*−1.56*	Adolescent	DRN
Glun2d	Glutamate receptor, ionotropic, NMDA2d (*Grin2d*: NR2d)		*−1.40*	Adolescent	DRN
Glun3a	Glutamate receptor, ionotropic, NMDA3a (*Grin3a*: NR3a)		*−1.39*	Adolescent	DRN
Glunl1a	Glutamate receptor, ionotropic, NMDA1a like (*Grinl1a*: NR1a like)		*−1.32*	Adolescent	DRN
Glua1	Glutamate receptor, ionotropic (*Gria1*: AMPA1)		1.15	Adult	AcbSh
Glua4	Glutamate receptor, ionotropic (*Gria4*: AMPA4)	1.26		Adult	AcbSh
Glud2ip	Glutamate receptor, ionotropic (*Grid2ip*: delta2 interacting protein)		3.49	Adult	AcbSh
Gluk1	Glutamate receptor, ionotropic (*Grik1*: kainate1)	1.73		Adult	AcbSh
Glun1	Glutamate receptor, ionotropic, NMDA1 (*Grin1*: NR1)		1.09	Adult	AcbSh
Glua1	Glutamate receptor, ionotropic (*Gria1*: AMPA1)	*−1.32*		Adult	AcbSh
Glua2	Glutamate receptor, ionotropic (*Gria2*: AMPA2)	*−1.21*		Adult	AcbSh
Glua2	Glutamate receptor, ionotropic (*Gria2*: AMPA2)		*−1.16*	Adult	AcbSh
Glua3	Glutamate receptor, ionotropic (*Gria3*: AMPA3)	*−1.52*		Adult	AcbSh
Glua3	Glutamate receptor, ionotropic (*Gria3*: AMPA3)		*−1.19*	Adult	AcbSh

Gluk2	Glutamate receptor, ionotropic (*Grik2*: kainate2)	*−1.33*		Adult	AcbSh
Glua1	Glutamate receptor, ionotropic (*Gria1*: AMPA1)		1.36	Adult	CeA
Glua2	Glutamate receptor, ionotropic (*Gria2*: AMPA2)		1.34	Adult	CeA
Glua3	Glutamate receptor, ionotropic (*Gria3*: AMPA3)		1.57	Adult	CeA
Gluk2	Glutamate receptor, ionotropic (*Grik2*: kainate2)		1.14	Adult	CeA
Gluk5	Glutamate receptor, ionotropic (*Grik5*: kainate5)		1.30	Adult	CeA
Glun3a	Glutamate receptor, ionotropic, NMDA3a (*Grin3a*: NR3a)		1.23	Adult	CeA

Refer to Table 1 for details.

Glua2 and *Glua3* of adult P versus NP rats,[211] operant ethanol self-administration appears to exacerbate this condition with further decreases in expression levels.

Surprisingly, despite no adult CeA *Glua, Gluk*, or *Glun* gene expression differences between P and NP rats; ethanol-binge drinking by adult P rats elevated *Glua1*, *Glua2*, *Glua3*, *Gluk2*, *Gluk5*, and *Glun3a* expression levels, with no significant downregulation of ionotropic glutamate receptor genes in this region.[208]

Overall, periadolescent binge-like drinking decreased gene expression for certain NMDA receptor subunits in the DRN, but increased particular AMPA, delta, and kainate receptor subunits within this region as well. Ethanol self-administration by adult P rats did not downregulate any NMDA receptor subunits in the AcbSh but did upregulate an AMPA, delta, and NMDA-receptor subunit.

6.3 Expression Differences for Glutamate Transporters, Enzymes, and Postsynaptic Density (Table 3)

One of the most striking findings is that periadolescent binge-like drinking significantly downregulated gene expression for only three ancillary proteins (*Mpp5*, a MAGUK subfamily member, by ~30%; as well as *Vglut2* and *Vglut3* by ~2-fold in the DRN); while upregulating expression levels for *Homer3* and *Slc1a1* in the CeA along with nine genes from the PSD (e.g., *Homer3* ~4-fold and *Tjp3* ~2.5-fold), as well as all three Shanks, two glutamate transporters (e.g., *Slc1a6* ~3-fold), and two glutamate carriers in the DRN.[213]

In the adult AcbSh, P rats had lower gene expression of *Homer1* than NP rats.[211] However, operant self-administration of ethanol by adult P rats upregulated *Homer1* by 3.5-fold in the AcbSh.[214] These data are interesting as studies of rodents with free-access to ethanol[83–85,89,180] or rodents injected repeatedly with alcohol[86,88] have consistently detected increases in Homer2 protein expression within Acb and amygdala structures, without detecting significant changes in Homer1 protein expression. This raises the possibility that ethanol-induced changes in Homer1 mRNA/protein expression may depend upon nonpharmacologic factors associated with the act of ethanol taking, which has recently been demonstrated to be the case with respect to intravenous cocaine taking.[216] Again in the AcbSh, P rats had higher expression levels of *Shank1*, whereas NP rats had higher expression levels of *Shank2* and *Shank3*.[211] Our data suggest a dissociation between the effects of home-cage ethanol drinking which decreased expression of *Shank1* by ~15%[204] as opposed to operant ethanol self-administration which

Table 3 Gene Expression, for Ancillary Proteins of the Excitatory Synapse, Differences Between P and NP Rats (i.e., Fold-Change (F-C) P vs. NP) or Changes Induced by Ethanol Consumption in P Rats, Relative to Ethanol-Naive P Rats (i.e., F-C Ethanol (E) vs. Control (C))

Gene ID	Gene Names	F-C P vs. NP	F-C E vs. C	Age of Rat	Brain Region
Homer3	Homer homolog 3		1.15	Adolescent	CeA
Slc1a1	Solute carrier family 1 (neuronal high affinity glutamate transporter), member 1		1.14	Adolescent	CeA
Dlg1/Sap97	Discs, large homolog 1 (Sap97, AMPAR trafficking)		1.28	Adolescent	DRN
Homer3	Homer homolog 3		3.75	Adolescent	DRN
Mpp4	Membrane protein, palmitoylated 4 (MAGUK p55 subfamily member 4)		1.68	Adolescent	DRN
Mpp6	Membrane protein, palmitoylated 6 (MAGUK p55 subfamily member 6)		1.23	Adolescent	DRN
Pdlim5	PDZ and LIM domain 5		1.28	Adolescent	DRN
Shank1	SH3 and multiple ankyrin repeat domains 1		1.41	Adolescent	DRN
Shank2	SH3 and multiple ankyrin repeat domains 2		1.79	Adolescent	DRN
Shank3	SH3 and multiple ankyrin repeat domains 3		1.82	Adolescent	DRN
Slc1a3	Solute carrier family 1 (glial glutamate transporter), member 3		1.70	Adolescent	DRN
Slc1a6	Solute carrier family 1 (glial glutamate transporter), member 6		2.79	Adolescent	DRN
Slc17a7/Vglut1	Solute carrier family 17, member 7 (vesicular glutamate transporter)		5.06	Adolescent	DRN
Slc25a18	Solute carrier family 25 (mitochondrial glutamate carrier), member 18		1.45	Adolescent	DRN
Slc25a22	Solute carrier family 25 (mitochondrial glutamate carrier), member 22		1.30	Adolescent	DRN
Tjp3	Tight junction protein 3 (zona occludens 3)		2.41	Adolescent	DRN

(Continued)

Table 3 Gene Expression, for Ancillary Proteins of the Excitatory Synapse, Differences Between P and NP Rats (i.e., Fold-Change (F-C) P vs. NP) or Changes Induced by Ethanol Consumption in P Rats, Relative to Ethanol-Naive P Rats (i.e., F-C Ethanol (E) vs. Control (C))—cont'd.

Gene ID	Gene Names	F-C P vs. NP	F-C E vs. C	Age of Rat	Brain Region
Mpp5	Membrane protein, palmitoylated 5 (MAGUK p55 subfamily member 5)		*−1.27*	Adolescent	DRN
Slc17a6/Vglut2	Solute carrier family 17, member 6 (vesicular glutamate transporter)		*−1.82*	Adolescent	DRN
Slc17a8/Vglut3	Solute carrier family 17, member 8 (vesicular glutamate transporter)		*−1.97*	Adolescent	DRN
Dlg1/Sap97	Discs, large homolog 1 (Sap97, AMPAR trafficking)		1.28	Adult	AcbSh
Dlg2/Psd93	Discs, large homolog 2 (Chapsyn-110/PSD-93 a MAGUK)		1.24	Adult	AcbSh
Gls2	Glutaminase 2	1.40		Adult	AcbSh
Glul	Glutamate-ammonia ligase (glutamine synthase)		1.14	Adult	AcbSh
Homer1	Homer homolog 1		3.49	Adult	AcbSh
Mpdz	Multiple PDZ domain protein	1.22		Adult	AcbSh
Mpp6	Membrane protein, palmitoylated 6 (MAGUK p55 subfamily member 6)	1.71		Adult	AcbSh
Shank1	SH3 and multiple ankyrin repeat domains 1	1.26		Adult	AcbSh
Shank1	SH3 and multiple ankyrin repeat domains 1		1.19	Adult	AcbSh
Shank2	SH3 and multiple ankyrin repeat domains 2		1.17	Adult	AcbSh
Slc6a9	Solute carrier family 6 (glycine transporter), member 9	1.40		Adult	AcbSh
Slc7a10	Solute carrier family 7 (glycine transporter), member 10 (glial)	1.42		Adult	AcbSh
Slc17a6/Vglut2	Solute carrier family 17, member 6 (vesicular glutamate transporter)	2.25		Adult	AcbSh
Slc25a13/Citrin	Solute carrier family 25 (mitochondrial glutamate carrier), member 13	1.23		Adult	AcbSh

Dlg1/Sap97	Discs, large homolog 1 (Sap97, AMPAR trafficking)		*−1.09*	Adult	AcbSh
Dlg2/Psd93	Discs, large homolog 2 (Chapsyn-110/PSD-93 a MAGUK)		*−1.17*	Adult	AcbSh
Dlg2/Psd93	Discs, large homolog 2 (Chapsyn-110/PSD-93 a MAGUK)	*−1.38*		Adult	AcbSh
Dlg4/Psd95	Discs, large homolog 4 (PSD95 or Sap90)	*−1.25*		Adult	AcbSh
Gls	Glutaminase	*−1.35*		Adult	AcbSh
Homer1	Homer homolog 1	*−1.68*		Adult	AcbSh
Mpp4	Membrane protein, palmitoylated 4 (MAGUK p55 subfamily member 4)		*−1.08*	Adult	AcbSh
Mpp7	Membrane protein, palmitoylated 7 (MAGUK p55 subfamily member 7)	*−1.58*		Adult	AcbSh
Pdlim7	PDZ and LIM domain 7	*−1.25*		Adult	AcbSh
Shank1	SH3 and multiple ankyrin repeat domains 1		*−1.14*	Adult	AcbSh
Shank2	SH3 and multiple ankyrin repeat domains 2	*−1.22*		Adult	AcbSh
Shank3	SH3 and multiple ankyrin repeat domains 3	*−1.30*		Adult	AcbSh
Slc1a3	Solute carrier family 1 (glial glutamate transporter), member 3	*−1.30*		Adult	AcbSh
Cask	Calcium/calmodulin-dependent serine protein kinase (MAGUK family)		1.14	Adult	CeA
Homer1	Homer homolog 1		2.19	Adult	CeA
Pdlim3	PDZ and LIM domain 3	1.28		Adult	CeA
Slc25a22	Solute carrier family 25 (mitochondrial glutamate carrier), member 22		1.12	Adult	CeA
Gls2	Glutaminase 2		*−1.17*	Adult	CeA
Homer2	Homer homolog 2		*−1.16*	Adult	CeA
Mpp3	Membrane protein, palmitoylated 3 (MAGUK p55 subfamily member 3)	*−1.45*		Adult	CeA
Tgm2	Transglutaminase 2, C polypeptide	*−1.30*		Adult	CeA

(*Continued*)

Table 3 Gene Expression, for Ancillary Proteins of the Excitatory Synapse, Differences Between P and NP Rats (i.e., Fold-Change (F-C) P vs. NP) or Changes Induced by Ethanol Consumption in P Rats, Relative to Ethanol-Naive P Rats (i.e., F-C Ethanol (E) vs. Control (C))—cont'd.

Gene ID	Gene Names	F-C P vs. NP	F-C E vs. C	Age of Rat	Brain Region
Glul	Glutamate-ammonia ligase (glutamine synthetase)	1.08		Adult	VTA
Dlg2/Psd93	Discs, large homolog 2 (Chapsyn-110/PSD-93 a MAGUK)	*−1.17*		Adult	VTA
Mpp6	Membrane protein, palmitoylated 6 (MAGUK p55 subfamily member 6)	*−1.10*		Adult	VTA
Pdlim7	PDZ and LIM domain 7	*−1.12*		Adult	VTA
Tgm2	Transglutaminase 2, C polypeptide	*−1.38*		Adult	VTA

Age of rat refers to whether the P rats had access to ethanol during (peri-)adolescence (PNDs, 30–50) or during adulthood>PND75. Brain regions published thus far: CeA, DRN, AcbSh, and VTA. To facilitate distinction between directions of expression (e.g., up- vs. downregulation), downregulation or lower level F-Cs are in italics and are below the findings of upregulation or higher-level F-Cs for that age and brain region.

increased expression levels of *Shank1* by ~20%[214] in the AcbSh. Some support for this dissociation can be deduced from the fact that operant ethanol self-administration also increased *Shank2* expression levels by ~20% in the AcbSh.[214]

In the adult CeA, operant ethanol self-administration upregulated *Homer1* expression levels by over two-fold; although operant ethanol self-administration also downregulated *Homer2* expression by ~20%, relative to controls.[214] This latter result is peculiar given that chronic free-choice access to ethanol upregulates CeA Homer2 protein expression in P rats[180] and C57BL/6J mice;[85] which, again, may be due to nonpharmacologic factors related to the operant-conditioning procedures employed in the mRNA study.

Glycine transporter expression levels in the adult AcbSh were higher and glutamate transporter expression levels were lower in P rats compared to NP rats.[211] Additionally, whenever ethanol intake modified membrane glutamate transporter expression levels, once each in the adult or adolescent CeA and multiple times in the adolescent DRN, it was always for an increase relative to control levels.[208,213] However, when examining vesicular glutamate transporter expression levels, *Vglut1* was increased by five-fold, whereas *Vglut2* and *Vglut3* were reduced by ~two-fold in the DRN following peri-adolescent binge-like drinking.[213] Also, in the adult AcbSh *Vglut2* expression levels were more than two-fold higher than that seen in NP rats.[211]

In general, adolescent binge-like drinking upregulated 14 of the 17 glutamate ancillary genes identified as significantly changed, suggesting that ethanol induces substantial increases in DRN glutamatergic activity during this stage of development. Of the genes identified as significantly different between P and NP rats, in the AcbSh of adult animals, half were higher and half were lower. In the CeA (9 of 17) and VTA (4 of 5), adult P rats generally had lower expression levels of glutamate ancillary genes than their NP counterparts. Regarding ethanol exposure, 6 of the 10 genes identified as significantly different relative to ethanol-naive controls were upregulated in the AcbSh of adult P rats. In ethanol-drinking/self-administering adult P rats, two of the five genes identified as significant in the CeA were downregulated. Overall, these findings of glutamate transporter and cytoskeleton/scaffolding protein gene expression level changes induced by ethanol parallel the existing literature indicating that ethanol exposure alters glutamate clearance from the synapse and induces neuroplastic changes in the PSD.

7. CONCLUSIONS

The findings discussed in this review include results from both preclinical as well as neuroimaging and postmortem clinical studies. Expression levels for a number of glutamate-associated genes and/or proteins are modulated by alcohol abuse and dependence. These changes in expression include metabotropic receptors and ionotropic receptor subunits as well as different glutamate transporters. Moreover, these changes in gene expression parallel pharmacologic manipulation of these same receptors and transporters. Some of these gene expression changes may have predated alcohol abuse and dependence, because a number of glutamate-associated polymorphisms are related to a genetic predisposition to develop alcohol dependence. Other glutamate-associated polymorphisms have been linked to age at the onset of alcohol-dependence and/or initial level of response/sensitivity to alcohol. Finally, findings of innate and/or ethanol-induced glutamate-associated gene expression differences/changes observed in a genetic animal model of alcoholism, the P rat, are highlighted. Overall, the existing literature indicates that changes in receptors, transporters, enzymes, and scaffolding proteins are crucial for the development of alcohol dependence and there is a substantial genetic component to these effects.

This review reveals that there are presently key limitations to our understanding of glutamate's role in the development of alcohol dependence and the impact that genetics has on this process. First, there are no studies examining glutamate-associated gene and/or protein expression changes across the juvenile, adolescent, emerging adult, and full adult stages of development. This information is crucial given the fact that the risky drinking age-of-onset is inversely associated with the probability of developing alcohol dependence (i.e., earlier onset leads to higher risk of developing alcoholism). And, while there is some evidence for an association between being a FHP individual and initiating risky drinking at a younger age, findings comparing the effects of ethanol on glutamate function between FHP individuals/models and FHN individuals/models are very limited. Second, studies thus far have been limited to gross examinations of regions and/or subregions of major structures in the central reward neurocircuitry. Also related to this point, a third limitation is the lack of publications examining multiple regions (i.e., putative circuits) within a single study. Despite these limitations, substantial progress has been made with new targets for medications development/screening identified, such as the GLT1/EAAT2 glutamate transporter or PKC-epsilon's modulation of mGlu5 activity. Nevertheless, the research community still has much to do in unraveling the role of glutamate-associated genes in the development of alcohol dependence, especially as it relates to pharmacogenomics and personalized pharmacologic interventions.

ACKNOWLEDGMENTS

This work was supported in part by AA13522 to RLB; AA020396 to RLB; AA020892 to HJE; AA016650 to KKS from the National Institutes of Health (NIH)/National Institute on Alcohol Abuse and Alcoholism (NIAAA). The views expressed in this manuscript are completely those of the authors and do not necessarily reflect the views of the funding agencies at the National Institutes of Health.

REFERENCES

1. Research Society on Alcoholism.*Impact of Alcoholism and Alcohol Induced Disease on America.* Austin, TX: Research Society on Alcoholism; 2009.
2. Research Society on Alcoholism.*Impact of Alcoholism and Alcohol Induced Disease on America.* Austin, TX: Research Society on Alcoholism; 2015.
3. Mokdad A, Marks J, Stroup D, Gerberding J. Actual causes of death in the United States, 2000. *J Am Med Assoc.* 2004;291:1238–1245.
4. Rehm J. The risks associated with alcohol use and alcoholism. *Alcohol Res Health.* 2011;34:135–143.
5. Rehm J, Room R, Graham K, Monteiro M, Gmel G, Sempos CT. The relationship of average volume of alcohol consumption and patterns of drinking to burden of disease—an overview. *Addict.* 2003;98:1209–1228.
6. Reed T, Page WF, Viken RJ, Christian JC. Genetic predisposition to organ-specific endpoints of alcoholism. *Alcohol Clin Exp Res.* 1996;20:1528–1533.
7. Cloninger CR. Neurogenetic adaptive mechanisms in alcoholism. *Science.* 1987;236: 410–416.
8. Cotton NS. The familial incidence of alcoholism. *J Stud Alcohol.* 1979;40:89–116.
9. Schuckit MA. Genetic aspects of alcoholism. *Ann Emerg Med.* 1986;15:991–996.
10. Reich T, Edenberg HJ, Goate A, Williams JT, Rice JP, VanEerdewegh P, Foroud T, Hesselbrock V, Schuckit MA, Bucholz K, Porjesz B, Li T-K, Conneally PM, Nurnberger Jr JI, Tischfield JA, Crowe R, Cloninger CR, Wu W, Shears S, Carr K, Crose C, Willig C, Begleiter H. Genome-wide search for genes affecting the risk for alcohol dependence. *Am J Med Genet.* 1998;81:207–215.
11. Agrawal A, Hinrichs AL, Dunn G, Bertelsen S, Dick DM, Saccone SF, Saccone NL, Grucza RA, Wang JC, Cloninger CR, Edenberg HJ, Foroud T, Hesselbrock V, Kramer J, Bucholz KK, Kuperman S, Nurnberger JI, Porjesz B, Schuckit MA, Goate AM, Bierut LJ. Linkage scan for quantitative traits identifies new regions of interest for substance dependence in the Collaborative Study on the Genetics of Alcoholism (COGA) sample. *Drug Alcohol Depend.* 2008;93:12–20.

12. Edenberg HJ, Foroud T. Genetics and alcoholism. *Nat Rev Gastroenterol Hepatol.* 2013;10:487–494.
13. Kapoor M, Wang J-C, Wetherill L, Le N, Bertelsen S, Hinrichs AL, Budde J, Agrawal A, Bucholz K, Dick D, Harari O, Hesselbrock V, Kramer J, Nurnberger JI, Rice J, Saccone N, Schuckit M, Tischfield J, Porjesz B, Edenberg HJ, Bierut L, Foroud T, Goate A. A meta-analysis of two genome-wide association studies to identify novel loci for maximum number of alcoholic drinks. *Hum Genet.* 2013;132: 1141–1151.
14. Levey DF, Le-Niculescu H, Frank J, Ayalew M, Jain N, Kirlin B, Learman R, Winiger E, Rodd Z, Shekhar A, Schork N, Kiefe F, Wodarz N, Muller-Myhsok B, Dahmen N, Consortium GESGASA, Nothen M, Sherva R, Farrer L, Smith AH, Kranzler HR, Rietschel M, Gelernter J, Niculescu AB. Genetic risk prediction and neurobiological understanding of alcoholism. *Transl Psychiatry.* 2014;4:e391.
15. Rietschel M, Treutlein J. The genetics of alcohol dependence. *Ann NY Acad Sci.* 2013;1282:39–70.
16. Wong CCY, Schumann G. Genetics of addictions: strategies for addressing heterogeneity and polygenicity of substance use disorders. *Philos Trans R Soc Lond B Biol Sci.* 2008;363:3213–3222.
17. Yan J, Aliev F, Webb BT, Kendler KS, Williamson VS, Edenberg HJ, Agrawal A, Kos MZ, Almasy L, Nunberger JI, Schuckit MA, Kramer JR, Rice JP, Kuperman S, Goate AM, Tischfield JA, Porjesz B, Dick DM. Using genetic information from candidate gene and genome-wide association studies in risk prediction for alcohol dependence. *Addict Biol.* 2014;19:708–721.
18. Richter CP, Campbell KH. Alcohol taste thresholds and concentrations of solutions preferred by rats. *Science.* 1940;9:507–508.
19. Williams RJ, Berry LJ, Beerstecher Jr E. Individual metabolic patterns, alcoholism, genotrophic diseases. *Proc Natl Acad Sci USA.* 1949;35:265–271.
20. Mardones J, Segovia-Riquelme N. Thirty-two years of selection of rats by ethanol preference: UChA and UChB strains. *Neurobehav Toxicol Teratol.* 1983;5:171–178.
21. Bell RL, Sable HJK, Colombo G, Hyytia P, Rodd ZA, Lumeng L. Animal models for medications development targeting alcohol abuse using selectively bred rat lines: neurobiological and pharmacological validity. *Pharmacol Biochem Behav.* 2012;103:119–155.
22. McBride WJ, Rodd ZA, Bell RL, Lumeng L, Li T-K. The alcohol-preferring (P) and high-alcohol-drinking (HAD) rats—animal models of alcoholism. *Alcohol.* 2014;48: 209–215.
23. Cicero TJ. A critique of animal analogues of alcoholism. In: Majchrowicz E, Noble EP, eds. *Biochemistry and Pharmacology of Ethanol.* Vol. 2. New York: Plenum Press; 1979: 533–560.
24. Lester D, Freed EX. Criteria for an animal model of alcoholism. *Pharmacol Biochem Behav.* 1973;1:103–107.
25. McBride WJ, Li T-K. Animal models of alcoholism: neurobiology of high alcohol-drinking behavior in rodents. *Crit Rev Neurobiol.* 1998;12:339–369.
26. Bell RL, Franklin KM, Hauser SR, Engleman EA. Next stop dependence. Binge drinking on the road to alcoholism: preclinical findings on its neurobiology from rat animal models. In: Harris SB, ed. *Binge Eating and Binge Drinking: Psychological, Social and Medical Implications.* New York: Nova Science Publishers; 2013:1–60.
27. Bell RL, Rodd ZA, Engleman EA, Toalston JE, McBride WJ. Scheduled access alcohol drinking by alcohol-preferring (P) and high alcohol-drinking (HAD) rats: modeling adolescent and adult binge-like drinking. *Alcohol.* 2014;48:225–234.
28. National Institute on Alcohol Abuse and Alcoholism (NIAAA) National Advisory Council. NIAAA Council approves definition of binge drinking. *NIAAA Newsl.* 2004;3:5.

29. Bell RL, Rodd ZA, Smith RJ, Toalston JE, Franklin KM, McBride WJ. Modeling binge-like ethanol drinking by peri-adolescent and adult P rats. *Pharmacol Biochem Behav.* 2011;100:90–97.
30. Crabbe JC, Bell RL, Ehlers CL. Human and laboratory rodent low response to alcohol: is better consilience possible? *Addict Biol.* 2010;15:125–144.
31. Henley JM, Wilkinson KA. AMPA receptor trafficking and the mechanisms underlying synaptic plasticity and cognitive aging. *Dialogues Clin Neurosci.* 2013;15:11–27.
32. Morris RGM. NMDA receptors and memory encoding. *Neuropharmacology.* 2013;74: 32–40.
33. Warburton EC, Barker GRI, Brown MW. Investigations into the involvement of NMDA mechanisms in recognition memory. *Neuropharmacology.* 2013;74:41–47.
34. Danbolt NC. Glutamate uptake. *Prog Neurobiol.* 2001;65:1–105.
35. Niciu MJ, Kelmendi B, Sanacora G. Overview of glutamatergic neurotransmission in the nervous system. *Pharmacol Biochem Behav.* 2012;100:656–664.
36. Traynelis SF, Wollmuth LP, McBain CJ, Menniti FS, Vance KM, Ogden KK, Hansen KB, Yuan H, Myers SJ, Dingledine R. Glutamate receptor ion channels: structure, regulation, and function. *Pharmacol Rev.* 2010;62:405–496.
37. Sari Y. Role of glutamate transporter 1 in the attenuation of alcohol intake. *Front Neurosci.* 2014;8:200.
38. Wang Y, Qin ZH. Molecular and cellular mechanisms of excitotoxic neuronal death. *Apoptosis.* 2010;15:1382–1402.
39. Abe T, Sugihara H, Nawa H, Shigemoto R, Mizuno N, Nakanishi S. Molecular characterization of a novel metabotropic glutamate receptor mGluR5 coupled to inositol phosphate/Ca2-signal transduction. *J Biol Chem.* 1992;267: 13361–13368.
40. Fotuhi M, Sharp AH, Glatt CE, Hwang PM, von Krosigk M, Snyder SH, Dawson TM. Differential localization of phosphoinositide-linked metabotropic glutamate receptor (mGluR1) and the inositol 1,4,5-trisphosphate receptor in rat brain. *J Neurosci.* 1993;13:2001–2012.
41. Shigemoto R, Nomura S, Ohishi H, Sugihara H, Nakanishi S, Mizuno N. Immunohistochemical localization of a metabotropic glutamate receptor, mGluR5, in the rat brain. *Neurosci Lett.* 1993;163:53–57.
42. Romano C, Sesma MA, McDonald CT, O'Malley K, Van den Pol AN, Olney JW. Distribution of metabotropic glutamate receptor mGluR5 immunoreactivity in rat brain. *J Comp Neurol.* 1995;355:455–469.
43. Tamaru Y, Nomura S, Mizuno N, Shigemoto R. Distribution of metabotropic glutamate receptor mGluR3 in the mouse CNS: differential location relative to pre- and postsynaptic sites. *Neuroscience.* 2001;106:481–503.
44. Bischoff S, Barhanin J, Bettler B, Mulle C, Heinemann S. Spatial distribution of kainate receptor subunit mRNA in the mouse basal ganglia and ventral mesencephalon. *J Comp Neurol.* 1997;379:541–562.
45. Isaac JTR, Nicoll RA, Malenka RC. Silent glutamatergic synapses in the mammalian brain. *Can J Physiol Pharmacol.* 1997;77:735–737.
46. Monaghan DT, Buller AL. Anatomical, pharmacological, and molecular diversity of native NMDA receptor subtypes. In: Collingridge GL, Watkins JC, eds. *The NMDA Receptor* 2nd ed New York: Oxford University Press; 1994:158–176.
47. Petralia RS, Wang Y-X, Wenthold RJ. Histological and ultrastructural localization of the kainite receptor subunits, KA2 and GluR6/7, in the rat nervous system using selective antipeptide antibodies. *J Comp Neurol.* 1994;349:85–110.
48. Tarazi FI, Baldessarini RJ. Regional localization of dopamine and ionotropic glutamate receptor subtypes in striatolimbic brain regions. *J Neurosci Res.* 1999;55:401–410.

49. Van Bockstaele EJ, Colago EEO. Ultrastructural localization of the kainite selective glutamate receptor in noradrenergic perikarya and dendrites of the nucleus locus coeruleus in the rat brain. *Brain Res.* 1996;732:223–231.
50. Ady V, Perroy J, Tricoire L, Piochon C, Dadak S, Chen X, Dusart I, Fagni L, Lambolez B, Levenes C. Type 1 metabotropic glutamate receptors (mGlu1) trigger the gating of GluD2 delta glutamate receptors. *EMBO Rep.* 2014;15:103–109.
51. Kohda K, Kakegawa W, Matsuda S, Yamamoto T, Hirano H, Yuzaki M. The delta2 glutamate receptor gates long-term depression by coordinating interactions between two AMPA receptor phosphorylation sites. *Proc Natl Acad Sci USA.* 2013;110: E948–E957.
52. Konno K, Matsuda K, Nakamoto C, Uchigashima M, Miyazaki T, Yamasaki M, Sakimura K, Yuzaki M, Watanabe M. Enriched expression of GluD1 in higher brain regions and its involvement in parallel fiber-interneuron synapse formation in the cerebellum. *J Neurosci.* 2014;34:7412–7424.
53. Orth A, Tapken D, Hollman M. The delta subfamily of glutamate receptors: characterization of receptor chimeras and mutants. *Eur J Neurosci.* 2013;37:1620–1630.
54. Swerdlow NR, Shilling PD, Breier M, Trim RS, Light GA, Marie RS. Fronto-temporal-mesolimbic gene expression and heritable differences in amphetamine-disrupted sensorimotor gating in rats. *Psychopharmacology.* 2012;224:349–362.
55. Wadiche JI, Amara SG, Kavanaugh MP. Ion fluxes associated with excitatory amino acid transport. *Neuron.* 1995;15:721–728.
56. Arriza JL, Eliasof S, Kavanaugh MP, Amara SG. Excitatory amino acid transporter 5, a retinal glutamate transporter coupled to a chloride conductance. *Proc Natl Acad Sci USA.* 1997;94:4155–4160.
57. Tanaka K. Functions of glutamate transporters in the brain. *Neurosci Res.* 2000;37: 15–19.
58. Patten AR, Brocardo PS, Sakiyama C, Wortman RC, Noonan A, Gil-Mohapel J, Christie BR. Impairments in hippocampal synaptic plasticity following prenatal ethanol exposure are dependent on glutathione levels. *Hippocampus.* 2013;23:1463–1475.
59. Sala C, Roussignol G, Meldolesi J, Fagni L. Key role of the postsynaptic density scaffold proteins Shank and Homer in the functional architecture of Ca^{2+} homeostasis at dendritic spines in hippocampal neurons. *J Neurosci.* 2005;25:4587–4592.
60. Sheng M, Kim E. The Shank family of scaffold proteins. *J Cell Sci.* 2000;113: 1851–1856.
61. O'Connor EC, Bariselli S, Bellone C. Synaptic basis of social dysfunction: a focus on postsynaptic proteins linking group-I mGluRs with AMPARs and NMDARs. *Eur J Neurosci.* 2014;39:1114–1129.
62. Funke L, Dakoji S, Bredt DS. Membrane-associated guanylate kinases regulate cell adhesion and plasticity at cell junctions. *Annu Rev Biochem.* 2005;74:219–245.
63. Garbett D, Bretscher A. The surprising dynamics of scaffolding proteins. *Mol Biol Cell.* 2014;25:2315–2319.
64. Shaw AS, Filbert EL. Scaffold proteins and immune-cell signaling. *Nat Rev Immunol.* 2009;9:47–56.
65. Bard L, Sainlos M, Bouchet D, Cousins S, Mikasova L, Breillat C, Stephenson FA, Imperiali B, Choquet D, Groc L. Dynamic and specific interaction between synaptic NR2-NMDA receptor and PDZ proteins. *Proc Natl Acad Sci USA.* 2010;107: 19561–19566.
66. Bockaert J, Perroy J, Bécamel C, Marin P, Fagni L. GPCR interacting proteins (GIPs) in the nervous system: roles in physiology and pathologies. *Annu Rev Pharmacol Toxicol.* 2010;50:89–109.
67. Tao-Cheng JH, Yang Y, Reese TS, Dosemeci A. Differential distribution of Shank and GKAP at the postsynaptic density. *PLoS One.* 2015;10:e0118750.

68. Béïque JC, Lin DT, Kang MG, Aizawa H, Takamiya K, Huganir RL. Synapse-specific regulation of AMPA receptor function by PSD-95. *Proc Natl Acad Sci USA*. 2006;103:19535–19540.
69. Schnell E, Sizemore M, Karimzadegan S, Chen L, Bredt DS, Nicoll RA. Direct interactions between PSD-95 and stargazin control synaptic AMPA receptor number. *Proc Natl Acad Sci USA*. 2002;99:13902–13907.
70. Chen J, Yu S, Fu Y, Li X. Synaptic proteins and receptors defects in autism spectrum disorders. *Front Cell Neurosci*. 2014;8:276.
71. Jaffrey SR, Snowman AM, Eliasson MJL, Cohen NA, Snyder SH. CAPON: a protein associated with neuronal nitric oxide synthase that regulates its interaction with PSD95. *Neuron*. 1998;20:115–124.
72. Garner CC, Nash J, Huganir RL. PDZ domains in synapse assembly and signaling. *Trends Cell Biol*. 2000;10:274–280.
73. Kornau HC, Seeburg PH, Kennedy MB. Interaction of ion channels and receptors with PDZ domain proteins. *Curr Opin Neurobiol*. 1997;7:368–373.
74. Ziff EB. Enlightening the postsynaptic density. *Neuron*. 1997;19:1163–1174.
75. El-Husseini AE, Schnell E, Chetkovich DM, Nicoll RA, Bredt DS. PSD-95 involvement in maturation of excitatory synapses. *Science*. 2000;290:1364–1368.
76. Malenka RC, Nicoll RA. Long-term potentiation—a decade of progress? *Science*. 1999;285:1870–1874.
77. Malinow R, Mainen ZF, Hayashi Y. LTP mechanisms: from silence to four-lane traffic. *Curr Opin Neurobiol*. 2000;10:352–357.
78. Beattie EC, Carroll RC, Yu X, Morishita W, Yasuda H, von Zastrow M, Malenka RC. Regulation of AMPA receptor endocytosis by a signaling mechanism shared with LTD. *Nat Neurosci*. 2000;3:1291–1300.
79. Ehlers MD. Reinsertion or degradation of AMPA receptors determined by activity-dependent endocytic sorting. *Neuron*. 2000;28:511–525.
80. Lin JW, Ju W, Foster K, Lee SH, Ahmadian G, Wyszynski M, Wang YT, Sheng M. Distinct molecular mechanisms and divergent endocytotic pathways of AMPA receptor internalization. *Nat Neurosci*. 2000;3:1282–1290.
81. Lüscher C, Xia H, Beattie EC, Carroll RC, von Zastrow M, Malenka RC, Nicoll RA. Role of AMPA receptor cycling in synaptic transmission and plasticity. *Neuron*. 1999;24:649–658.
82. Tomita S, Nicoll RA, Bredt DS. PDZ protein interactions regulating glutamate receptor function and plasticity. *J Cell Biol*. 2001;153:F19–F24.
83. Cozzoli DK, Goulding SP, Zhang P-W, Xiao B, Hu J-H, Ary AW, Obara I, Rahn A, Abou-Ziab H, Tyrrel B, Marini C, Yoneyama N, Metten P, Snelling C, Dehoff M, Crabbe JC, Finn DA, Klugmann M, Worley PF, Szumlinski KK. Binge drinking up-regulated accumbens mGluR5-Homer2-PI3K signaling: functional implications for alcoholism. *J Neurosci*. 2009;29:8655–8668.
84. Cozzoli DK, Courson J, Caruana AL, Miller BW, Thompson AB, Wroten M, Zhang PW, Xiao B, Hu J-H, Klugmann M, Metten P, Worley PW, Crabbe JC, Szumlinski KK. Accumbens shell metabotropic glutamate receptor 5-associated signaling regulates binge alcohol drinking: evidence from drinking-in-the-dark studies. *Alcohol Clin Exp Res*. 2012;36:1623–1633.
85. Cozzoli DK, Courson J, Wroten MG, Greentree DI, Lum EN, Campbell RR, Thompson AB, Worley PF, Jonquieres G, Klugmann M, Finn DA, Szumlinski KK. Binge alcohol drinking by mice requires intact Group1 metabotropic glutamate receptor signaling within the central nucleus of the amygdala. *Neuropsychopharmacology*. 2014;39:435–444.
86. Goulding SP, Obara I, Lominac KD, Gould AT, Miller BW, Klugmann M, Szumlinski KK. Accumbens Homer2-mediated signaling: a factor contributing to mouse strain differences in alcohol drinking? *Genes Brain Behav*. 2011;10:111–126.

87. Szumlinski KK, Dehoff MH, Kang SH, Frys KA, Lominac KD, Rohrer J, Griffin III W, Klugmann M, Toda S, Champtiaux NP, Berry T, Shealy S, During M, Middaugh LD, Worley PF, Kalivas PW. Homer proteins regulate vulnerability to cocaine. *Neuron.* 2004;43:401–413.
88. Szumlinski KK, Lominac KD, Oleson EB, Walker JK, Mason A, Dehoff MH, Klugmann M, Cagle S, Welt K, During MT, Worley PF, Middaugh LD, Kalivas PW. Homer2 is necessary for ethanol-induced neuroplasticity. *J Neurosci.* 2005;25: 7054–7061.
89. Szumlinski KK, Ary AW, Lominac KD, Klugmann M, Kippin TE. Accumbens Homer2 over-expression facilitates alcohol-induced neuroplasticity in C57BL/6J mice. *Neuropsychopharmacology.* 2008;33:1365–1378.
90. Harris BZ, Lim WA. Mechanism and role of PDZ domains in signaling complex assembly. *J Cell Sci.* 2001;114:3219–3231.
91. Jani K, Schöck F. Zasp is required for the assembly of functional integrin adhesion sites. *J Cell Biol.* 2007;179:1583–1597.
92. Maday S, Anderson E, Chang HC, Shorter J, Satoh A, Sfakianos J, Fölsch H, Anderson JM, Walther Z, Mellman I. A PDZ-binding motif controls basolateral targeting of syndecan-1 along the biosynthetic pathway in polarized epithelial cells. *Traffic.* 2008;9: 1915–1924.
93. Shiraishi-Yamaguchi Y, Furuichi T. The Homer family of proteins. *Genome Biol.* 2007;8:206–212.
94. Worley PF, Zeng W, Huang G, Kim JY, Shin DM, Kim MS, Yuan JP, Kiselyov K, Muallem S. Homer proteins in Ca^{2+} signaling by excitable and non-excitable cells. *Cell Calcium.* 2007;42:363–371.
95. Chandler LJ. Ethanol and brain plasticity: receptors and molecular networks of the postsynaptic density as targets of ethanol. *Pharmacol Ther.* 2003;99:311–326.
96. Grace AA, Floresco SB, Goto Y, Lodge DJ. Regulation of firing of dopaminergic neurons and control of goal-directed behaviors. *Trends Neurosci.* 2007;30:220–227.
97. Kalivas PW, Lalumiere RT, Knackstedt L, Shen H. Glutamate transmission in addiction. *Neuropharmacology.* 2009;56(suppl 1):169–173.
98. Koob GF. Alcoholism: allostasis and beyond. *Alcohol Clin Exp Res.* 2003;27:232–243.
99. McBride WJ. Central nucleus of the amygdala and the effects of alcohol and alcohol-drinking behavior in rodents. *Pharmacol Biochem Behav.* 2002;71:509–515.
100. McCool BA. Ethanol modulation of synaptic plasticity. *Neuropharmacology.* 2011;61: 1097–1108.
101. Steketee JD, Kalivas PW. Drug wanting: behavioral sensitization and relapse to drug-seeking behavior. *Pharmacol Rev.* 2011;63:348–365.
102. Spanagel R. Alcoholism: a systems approach from molecular physiology to addictive behavior. *Physiol Rev.* 2009;89:649–705.
103. Gass JT, Olive MF. Glutamatergic substrates of drug addiction and alcoholism. *Biochem Pharmacol.* 2008;75:218–265.
104. Quintero GC. Role of nucleus accumbens glutamatergic plasticity in drug addiction. *Neuropsychiatric Dis Treat.* 2013;9:1499–1512.
105. Rao PSS, Bell RL, Engleman EA, Sari Y. Targeting glutamate uptake to treat alcohol use disorders. *Front Neurosci.* 2015;9:144.
106. Reissner KJ, Kalivas PW. Using glutamate homeostasis as a target for treating addictive disorders. *Behav Pharmacol.* 2010;21:514–522.
107. Vengeliene V, Bilbao A, Molander A, Spanagel R. Neuropharmacology of alcohol addiction. *Br J Pharmacol.* 2008;154:299–315.
108. Lallemand F, Ward RJ, de Witte P, Verbanck P. Binge-drinking +/− chronic nicotine administration alters extracellular glutamate and arginine levels in the nucleus accumbens of adult male and female Wistar rats. *Alcohol Alcohol.* 2011;46:373–382.

109. Carlezon WA, Wise RA. Microinjections of phencyclidine (PCP) and related drugs into nucleus accumbens shell potentiate medial forebrain bundle brain stimulation reward. *Psychopharmacology.* 1996;128:413–420.
110. Ding ZM, Rodd ZA, Engleman EA, Bailey JA, Lahiri DK, McBride WJ. Alcohol drinking and deprivation alter basal extracellular glutamate concentrations and clearance in the mesolimbic system of alcohol-preferring (P) rats. *Addict Biol.* 2013;18: 297–306.
111. Szumlinski KK, Diab ME, Friedman R, Henze LM, Lominac KD, Bowers MS. Accumbens neurochemical adaptations produced by binge-like alcohol consumption. *Psychopharmacology.* 2007;190:415–431.
112. Floyd DW, Jung KY, McCool BA. Chronic ethanol ingestion facilitates *N*-methyl-D aspartate receptor function and expression in rat lateral/basolateral amygdala neurons. *J Pharmacol Exp Ther.* 2003;307:1020–1029.
113. Chandler LJ, Newsom H, Sumners C, Crews F. Chronic ethanol exposure potentiates NMDA excitotoxicity in cerebral cortical neurons. *J Neurochem.* 1993;60:1578–1581.
114. Chefer V, Meis J, Wang G, Kuzmin A, Bakalkin G, Shippenberg T. Repeated exposure to moderate doses of ethanol augments hippocampal glutamate neurotransmission by increasing release. *Addict Biol.* 2010;16:229–237.
115. Ward RJ, Colivicchi MA, Allen R, Schol F, Lallemand F, de Witte P, Ballini C, Corte LD, Dexter D. Neuro-inflammation induced in the hippocampus of 'binge drinking' rats may be mediated by elevated extracellular glutamate content. *J Neurochem.* 2009;111:1119–1128.
116. Xiao C, Shao XM, Olive MF, Griffin III WC, Li K-Y, Krnjevic K, Zhou C, Ye J-H. Ethanol facilitates glutamatergic transmission to dopamine neurons in the ventral tegmental area. *Neuropsychopharmacology.* 2009;34:307–318.
117. Ding ZM, Engleman EA, Rodd ZA, McBride WJ. Ethanol increases glutamate neurotransmission in the posterior ventral tegmental area of female Wistar rats. *Alcohol Clin Exp Res.* 2012;36:633–640.
118. McBride WJ, Murphy JM, Lumeng L, Li T-K. Effects of ethanol on monoamine and amino acid release from cerebral cortex slices of the alcohol-preferring P line of rats. *Alcohol Clin Exp Res.* 1986;10:205–208.
119. Selim M, Bradburry CW. Effect of ethanol on extracellular 5-HT and glutamate in the nucleus accumbens and prefrontal cortex: comparison between the Lewis and Fischer 344 rat strains. *Brain Res.* 1996;716:157–164.
120. Kapasova Z, Szumlinski KK. Strain differences in alcohol induced neurochemical plasticity: a role for accumbens glutamate in alcohol intake. *Alcohol Clin Exp Res.* 2008;32:617–631.
121. Gass JT, Sinclair CM, Cleva RM, Widholm JJ, Olive MF. Alcohol-seeking behavior is associated with increased glutamate transmission in basolateral amygdala and nucleus accumbens as measured by glutamate-oxidase-coated biosensors. *Addict Biol.* 2010;16: 215–228.
122. Nevo I, Hamon M. Neurotransmitter and neuromodulatory mechanisms involved in alcohol abuse and alcoholism. *Neurochem Int.* 1995;26:305–342.
123. Griffin III WC, Haun HL, Hazelbaker CL, Ramachandra VS, Becker HC. Increased extracellular glutamate in the nucleus accumbens promotes excessive ethanol drinking in ethanol dependent mice. *Neuropsychopharmacology.* 2014;39:707–717.
124. Bauer J, Pedersen A, Scherbaum N, Bening J, Patschke J, Kugel H, Heindel W, Arolt V, Ohrmann P. Craving in alcohol-dependent patients after detoxification is related to glutamatergic dysfunction in the nucleus accumbens and the anterior cingulate cortex. *Neuropsychopharmacology.* 2013;38:1401–1408.
125. Ende G, Hermann D, Demirakca T, Hoerst M, Tunc-Skarka N, Weber-Fahr W, Wichert S, Rabinstein J, Frischknecht U, Mann K, Vollstädt-Klein S. Loss of control

of alcohol use and severity of alcohol dependence in non-treatment-seeking heavy drinkers are related to lower glutamate in frontal white matter. *Alcohol Clin Exp Res.* 2013;37:1643–1649.

126. Hermann D, Weber-Fahr W, Sartorius A, Hoerst M, Frischknecht U, Tunc-Skarka N, Perreau-Lenz S, Hansson AC, Krumm B, Kiefer F, Spanagel R, Mann K, Ende G, Sommer WH. Translational magnetic resonance spectroscopy reveals excessive central glutamate levels during alcohol withdrawal in humans and rats. *Biol Psychiatry.* 2012;71:1015–1021.
127. Karkkainen O, Kupila J, Hakkinen M, Laukkanen V, Tupala E, Kautiainen H, Tiihonen J, Storvik M. AMPA receptors in post-mortem brains of Cloninger type 1 and 2 alcoholics: a whole-hemisphere autoradiography study. *Psychiatry Res Neuroimaging.* 2013;214:429–434.
128. Kryger R, Wilce PA. The effects of alcoholism on the human basolateral amygdala. *Neuroscience.* 2010;167:361–371.
129. Kupila J, Karkkainen O, Laukkanen V, Tupala E, Tiihonen J, Storvik M. mGluR1/5 receptor densities in the brains of alcoholic subjects: a whole hemisphere autoradiography study. *Psychiatry Res.* 2013;212:245–250.
130. Laukkainen V, Karkkainen O, Kupila J, Kautiainen H, Tiihonen J, Storvik M. Increased metabotropic glutamate 2/3 receptor binding in the perigenual anterior cingulate cortex of Cloninger type 2 alcoholics: a whole-hemisphere autoradiography study. *Alcohol Alcohol.* 2015;50:62–67.
131. Thoma R, Mullins P, Ruhl D, Monnig M, Yeo RA, Caprihan A, Bogenschutz M, Lysne P, Tonigan S, Kalyanam R, Gasparovic C. Perturbation of the glutamate–glutamine system in alcohol dependence and remission. *Neuropsychopharmacology.* 2011;36:1359–1365.
132. Hermens DF, Chitty KM, Lee RSC, Tickell A, Haber PS, Naismith SL, Hickie IB, Lagopoulas J. Hippocampal glutamate is increased and associated with risky drinking in young adults with major depression. *J Affect Disord.* 2015;186:95–98.
133. Cohen-Gilbert JE, Sneider JT, Crowley DJ, Rosso IM, Jensen JE, Silveri MM. Impact of family history of alcoholism on glutamine/glutamate ratio in anterior cingulate cortex in substance-naive adolescents. *Dev Cogn Neurosci.* 2015. 10.1016/j.dcn.2015.04.005.
134. Schroeder JP, Overstreet DH, Hodge CW. The mGluR5 antagonist MPEP decreases operant ethanol self-administration during maintenance and after repeated alcohol deprivations in alcohol-preferring (P) rats. *Psychopharmacology (Berl).* 2005;179: 262–270.
135. Besheer J, Faccidomo S, Grondin JJ, Hodge CW. Regulation of motivation to self-administer ethanol by mGluR5 in alcohol-preferring (P) rats. *Alcohol Clin Exp Res.* 2008;32:209–221.
136. Hodge CW, Miles MF, Sharko AC, Stevenson RA, Hillmann JR, Lepoutre V, Besheer J, Schroeder JP. The mGluR5 antagonist MPEP selectively inhibits the onset and maintenance of ethanol self-administration in C57BL/6J mice. *Psychopharmacology.* 2006;183:429–438.
137. Lominac KD, Kapasova Z, Hannun RA, Patterson C, Middaugh LD, Szumlinski KK. Behavioral and neurochemical interactions between Group 1 mGluR antagonists and ethanol: potential insight into their anti-addictive properties. *Drug Alcohol Depend.* 2006;85:142–156.
138. Olive MF, McGeehan AJ, Kinder JR, McMahon T, Hodge CW, Janak PH, Messing RO. The mGluR5 antagonist 6-methyl-2-(phenylethynyl)pyridine decreases ethanol consumption via a protein kinase C-epsilon dependent mechanism. *J Pharmacol Exp Ther.* 2005;67:349–355.
139. Schroeder JP, Spanos M, Stevenson JR, Besheer J, Salling M, Hodge CW. Cue-induced reinstatement of alcohol-seeking behavior is associated with increased ERK1/2

phosphorylation in specific limbic brain regions: blockade by the mGluR5 antagonist MPEP. *Neuropharmacology.* 2008;55:546–554.
140. Zamora-Martinez ER, Edwards S. Neuronal extracellular signal-regulated kinase (ERK) activity as marker and mediator of alcohol and opioid dependence. *Front Integr Neurosci.* 2014;8:24.
141. Besheer J, Grondin JJ, Cannady R, Sharko AC, Faccidomo S, Hodge CW. Metabotropic glutamate receptor 5 activity in the nucleus accumbens is required for the maintenance of ethanol self-administration in a rat genetic model of high alcohol intake. *Biol Psychiatry.* 2010;67:812–822.
142. Cowen MS, Djouma E, Lawrence AJ. The metabotropic glutamate 5 receptor antagonist 3-[(2-methyl-1,3-thiazol-4-yl)ethynyl]-pyridine reduces ethanol self-administration in multiple strains of alcohol-preferring rats and regulates olfactory glutamatergic systems. *J Pharmacol Exp Ther.* 2005;315:590–600.
143. Besheer J, Faccidomo S, Grondin JJ, Hodge CW. Effects of mGlu1-receptor blockade on ethanol self-administration in inbred alcohol-preferring rats. *Alcohol.* 2008;42:13–20.
144. Lum EN, Campbell RR, Rostock C, Szumlinski KK. mGluR1 within the nucleus accumbens regulates alcohol intake in mice under limited-access conditions. *Neuropharmacology.* 2014;79:679–687.
145. Rodd ZA, McKinzie DL, Bell RL, McQueen VK, Murphy JM, Schoepp DD, McBride WJ. The metabotropic glutamate 2/3 receptor agonist LY404039 reduces alcohol-seeking but not alcohol self-administration in alcohol-preferring (P) rats. *Behav Brain Res.* 2006;171:207–215.
146. Cannella N, Halbout B, Uhrig S, Evrard L, Corsi M, Corti C, Deroche-Gamonet V, Hansson AC, Spanagel R. The mGluR2/3 agonist LY379268 induced anti-reinstatement effects in rats exhibiting addition-like behavior. *Neuropsycholpharmacology.* 2013;38:2048–2056.
147. Enoch M-A, Rosser AA, Zhou Z, Mash DC, Yuan Q, Goldman D. Expression of glutamatergic genes in healthy humans across 16 brain regions; altered expression in the hippocampus after chronic exposure to alcohol or cocaine. *Genes Brain Behav.* 2014;13:758–768.
148. Jin Z, Bhandage AK, Bazov I, Kononenko O, Bakalkin G, Korpi ER, Birnir B. Selective increases of AMPA, NMDA, and kainite receptor subunit mRNAs in the hippocampus and orbitofrontal cortex but not in prefrontal cortex of human alcoholics. *Front Cell Neurosci.* 2014;8:11.
149. Cannady R, Fisher KR, Durant B, Besheer J, Hodge CW. Enhanced AMPA receptor activity increases operant alcohol self-administration and cue-induced reinstatement. *Addict Biol.* 2013;18:54–65.
150. Bäckström P, Hyytiä P. Ionotropic glutamate receptor antagonists modulate cue-induced reinstatement of ethanol-seeking behavior. *Alcohol Clin Exp Res.* 2004;28:558–565.
151. Watterson LR, Olive MF. Are AMPA receptor positive allosteric modulators potential pharmacotherapeutics for addiction? *Pharmaceuticals.* 2014;7:29–45.
152. Krystal JH, Petrakis IL, Mason G, Trevisan L, D'Souza DC. *N*-methyl-D-aspartate glutamate receptors and alcoholism: reward, dependence, treatment, and vulnerability. *Pharmacol Ther.* 2003;99:79–94.
153. Alhaddad H, Kim NT, Aal-Aaboda M, Althobaiti YS, Leighton J, Boddu SH, Wei Y, Sari. Effects of MS-153 on chronic ethanol consumption and GLT1 modulation of glutamate levels in male alcohol-preferring rats. *Front Behav Neurosci.* 2014;8:366.
154. Melendez RI, Hicks MP, Cagle SS, Kalivas PW. Ethanol exposure decreases glutamate uptake in the nucleus accumbens. *Alcohol Clin Exp Res.* 2005;29:326–333.
155. Sari Y, Sreemantula SN. Neuroimmunophilin GPI-1046 reduces ethanol consumption in part through activation of GLT1 in alcohol-preferring rats. *Neuroscience.* 2012;227:327–335.

156. Sari Y, Franklin KM, Alazizi A, Rao PS, Bell RL. Effects of ceftriaxone on the acquisition and maintenance of ethanol drinking in peri-adolescent and adult female alcohol-preferring (P) rats. *Neuroscience.* 2013;241:229–238.
157. Rothstein JD, Van Kammen M, Levey AI, Martin LJ, Kuncl RW. Selective loss of glial glutamate transporter GLT-1 in amyotrophic lateral sclerosis. *Ann Neurol.* 1995;38:73–84.
158. Abulseoud OA, Camsari UM, Ruby CL, Kasasbeh A, Choi S, Choi DS. Attenuation of ethanol withdrawal by ceftriaxone-induced upregulation of glutamate transporter EAAT2. *Neuropsychopharmacology.* 2014;39:1674–1684.
159. Das SC, Yamamoto BK, Hristov AM, Sari Y. Ceftriaxone attenuates ethanol drinking and restores extracellular glutamate concentration through normalization of GLT-1 in nucleus accumbens of male alcohol-preferring rats. *Neuropharmacology.* 2015;97:67–74.
160. Qrunfleh AM, Alazizi A, Sari Y. Ceftriaxone, a beta-lactam antibiotic, attenuates relapse-like ethanol-drinking behavior in alcohol-preferring rats. *J Psychopharmacol.* 2013;27:541–549.
161. Sari Y, Sakai M, Weedman JM, Rebec GV, Bell RL. Ceftriaxone, a beta-lactam antibiotic, reduces ethanol consumption in alcohol-preferring rats. *Alcohol Alcohol.* 2011;46:239–246.
162. Sari Y, Sreemantula SN, Lee MR, Choi DS. Ceftriaxone treatment affects the levels of GLT1 and ENT1 as well as ethanol intake in alcohol-preferring rats. *J Mol Neurosci.* 2013;51:779–787.
163. Aal-Aaboda M, Alhaddad H, Osowik F, Nauli SM, Sari Y. Effects of (R)-(−)-5-methyl-1 nicotinoyl-2-pyrazoline on glutamate transporter 1 and cysteine/glutamate exchanger as well as ethanol drinking behavior in male, alcohol-preferring rats. *J Neurosci Res.* 2015;93:930–937.
164. Alasmari F, Abuhamdah S, Sari Y. Effects of ampicillin on cystine/glutamate antiporter and glutamate transporter 1 isoforms as well as ethanol drinking in male P rats. *Neurosci Lett.* 2015;600:148–152.
165. Goodwani S, Rao PSS, Bell RL, Sari Y. Amoxicillin and amoxicillin/clavulanate reduce ethanol intake and increase GLT-1 expression as well as AKT phosphorylation in mesocorticolimbic regions. *Brain Res.* 2015;1622:397–408.
166. Rao PSS, Goodwani S, Bell RL, Wei Y, Boddu SH, Sari Y. Effects of ampicillin, cefazolin and cefoperazone treatments on GLT1 expressions in the mesocorticolimbic system and ethanol intake in alcohol-preferring rats. *Neuroscience.* 2015;295:164–174.
167. Lin X, Yang H, Zhang H, Zhou L, Guo Z. A novel transcription mechanism activated by ethanol: induction of Slc7a11 gene expression via inhibition of the DNA-binding activity of transcriptional repressor octamer-binding transcription factor 1 (OCT-1). *J Biol Chem.* 2013;288:14815–14823.
168. Zhou FC, Sahr RN, Sari Y, Behbahani K. Glutamate and dopamine synaptic terminals in extended amygdala after 14-week chronic alcohol drinking in inbred alcohol-preferring rats. *Alcohol.* 2006;39:39–49.
169. Rani VA, Nadiger HA, Marcus SR, Chandrakala MV, Sadasivudu B. Acute and short term effects of ethanol on the metabolism of glutamaic acid and GABA in rat brain. *Neurochem Res.* 1985;10:297–306.
170. Falco AM, Bergstrom HC, Bachus SE, Smith RF. Persisting changes in basolateral amygdala mRNAs after chronic ethanol consumption. *Physiol Behav.* 2009;96:169–173.
171. Bondy SC, Guo SX. Regional selectivity in ethanol-induced pro-oxidant events within the brain. *Biochem Pharmacol.* 1995;49:69–72.
172. Babu PP, Kumari R, Vemuri MC. Differential changes in cell morphology, macromolecular composition and membrane protein profiles of neurons and astrocytes in chronic ethanol treated rats. *Mol Cell Biochem.* 1994;130:29–40.
173. Matsumoto I. Proteomics approach in the study of the pathophysiology of alcohol-related brain damage. *Alcohol Alcohol.* 2009;44:171–176.

174. Hargreaves GA, Quinn H, Kashem MA, Matsumoto I, McGregor IS. Proteomic analysis demonstrates adolescent vulnerability to lasting hippocampal changes following chronic alcohol consumption. *Alcohol Clin Exp Res.* 2009;33:86–94.
175. Swartzwelder HS, Wilson WA, Tayyeb MI. Age-dependent inhibition of long-term potentiation by ethanol in immature versus mature hippocampus. *Alcohol Clin Exp Res.* 1995;19:1480–1485.
176. Swartzwelder HS, Wilson WA, Tayyeb MI. Differential sensitivity of NMDA receptor-mediated synaptic potentials to ethanol in immature versus mature hippocampus. *Alcohol Clin Exp Res.* 1995;19:320–323.
177. Risher ML, Fleming RL, Risher WC, Miller KM, Klein RC, Wills T, Acheson SK, Moore SD, Wilson WA, Eroglu C, Swartzwelder HS. Adolescent intermittent alcohol exposure: persistence of structural and functional hippocampal abnormalities into adulthood. *Alcohol Clin Exp Res.* 2015;39:989–997.
178. Staples MC, Kim A, Mandyam CD. Dendritic remodeling of hippocampal neurons is associated with altered NMDA receptor expression in alcohol dependent rats. *Mol Cell Neurosci.* 2015;65:153–162.
179. Klugmann M, Szumlinski KK. Targeting Homer genes using AAV: lessons learned from behavioural and neurochemical studies. *Behav Pharmacol.* 2008;19:485–500.
180. Obara I, Bell RL, Goulding SP, Reyes CM, Larson LA, Ary AW, Truitt WA, Szumlinski KK. Differential effects of chronic ethanol consumption and withdrawal on homer/glutamate receptor expression in subregions of the accumbens and amygdala of P rats. *Alcohol Clin Exp Res.* 2009;33:1924–1934.
181. Joslyn G, Ravindranathan A, Brush G, Schuckit M, White RL. Human variation in alcohol response is influenced by variation in neuronal signaling genes. *Alcohol Clin Exp Res.* 2010;34:800–812.
182. Petrakis IL, Limoncelli D, Gueorguieva R, Jatlow P, Boutros NN, Trevisan L, Gelernter J, Krystal JH. Altered NMDA glutamate receptor antagonist response in individuals with a family vulnerability to alcoholism. *Am J Psychiatry.* 2004;161:1776–1782.
183. Reimers MA, Riley BP, Kalsi G, Kertes DA, Kendler KS. Pathway based analysis of genotypes in relation to alcohol dependence. *Pharmacogenomics J.* 2012;12:342–348.
184. Karpyak VM, Geske JR, Colby CL, Mrazek DA, Biernaka JM. Genetic variability in the NMDA-dependent AMPA trafficking cascade is associated with alcohol dependence. *Addict Biol.* 2011;17:798–806.
185. Schumann G, Johann M, Frank J, Preuss U, Dahmen N, Laucht M, Rietschel M, Rujescu D, Lourdusamy A, Clarke T-K, Krause K, Dyer A, Depner M, Wellek S, Treutlein J, Szegedi A, Giegling I, Cichon S, Blomeyer D, Heinz A, Heath S, Lathrop M, Wodarz N, Soyka M, Spanagel R, Mann K. Systematic analysis of glutamatergic neurotransmission genes in alcohol dependence and adolescent risky drinking behavior. *Arch Gen Psychiatry.* 2008;65:826–838.
186. Domart M-C, Benyamina A, Lemoine A, Bourgain C, Blecha L, Debuire B, Reynaud M, Saffroy R. Association between a polymorphism in the promotor of a glutamate receptor subunit gene (GRIN2A) and alcoholism. *Addict Biol.* 2011;17:783–785.
187. Preuss UW, Zill P, Koller G, Bondy B, Hesselbrock V, Soyka M. Ionotropic glutamate receptor gene GRIK3 ser310ala functional polymorphism is related to delirium tremens in alcoholics. *Pharmacogenomics J.* 2006;6:34–41.
188. Kranzler HR, Gelernter J, Anton RF, Arias AJ, Herman A, Zhao H, Burian L, Covault J. Association of markers in the 3' region of the GluR5 kainate receptor subunit gene to alcohol dependence. *Alcohol Clin Exp Res.* 2009;33:925–930.
189. Begleiter H, Porjesz B, Reich T, Edenberg HJ, Goate A, Blangero J, Almasy L, Foroud T, Van Eerdewegh P, Polich J, et al. Quantitative trait loci analysis of human event-related brain potentials: P3 voltage. *Electroenceph Clin Neurophysiol.* 1998;108: 244–250.

190. O'Connor S, Morzorati S, Christian JC, Li T-K. Heritable features of the auditory oddball event-related potential: peaks, latencies, morphology and topography. *Electroenceph Clin Neurophysiol.* 1994;92:115–125.
191. Chen ACH, Tang Y, Rangaswamy M, Wang JC, Almasy L, Foroud T, Edenberg HJ, Hesselbrock V, Nurnberger Jr JI, Kuperman S, O'Connor SJ, Schuckit MA, Bauer LO, Tischfield J, Rice JP, Bierut L, Goate A, Porjesz B. Association of single nucleotide polymorphisms in a glutamate receptor gene [GRM8] with theta power of event-related oscillations and alcohol dependence. *Am J Med Genet Part B.* 2008;150B:359–368.
192. Long EC, Aliev F, Wang J-C, Edenberg HJ, Nurnberger Jr JI, Hesselbrock V, Porjesz B, Dick DM. Further analyses of genetic association between GRM8 and alcohol dependence symptoms among young adults. *J Stud Alcohol Drugs.* 2015;76:414–418.
193. Xia X, Wu Z, Ma D, Tang C, Liu L, Xin F, Zhu D, Hu J. Association of single-nucleotide polymorphisms in a metabotropic glutamate receptor GRM3 gene subunit to alcohol-dependent male subjects. *Alcohol Alcohol.* 2014;49:256–260.
194. Comasco E, Hallman J, Wallen-Mackenzie A. Haplotype tag single nucleotide polymorphism analysis of the vesicular glutamate transporter (VGLUT) genes in severely alcoholic women. *Psychiatry Res.* 2014;219:403–405.
195. Kuo P-H, Kalsi G, Prescott CA, Hodgkinson CA, Goldman D, Alexander J, van den Oord EJ, Chen X, Sullivan PF, Patterson DG, Walsh D, Kendler KS, Riley BP. Associations of glutamate decarboxylase genes with initial sensitivity and age-at-onset of alcohol dependence in the Irish Affected Sib Pair Study of Alcohol Dependence. *Drug Alcohol Depend.* 2009;101:80–87.
196. Terranova C, Tucci M, Forza G, Barzon L, Palu G, Ferrara SD. Alcohol dependence and glutamate decarboxylase gene polymorphisms in an Italian male population. *Alcohol.* 2010;44:407–413.
197. Lappalainen J, Krupitsky E, Kranzler HR, Luo X, Remizov M, Pchelina S, Taraskina A, Zvartau E, Rasanen P, Makikyro T, Somberg LK, Krystal JH, Stein MB, Gelernter J. Mutation screen of the GAD2 gene and association study of alcoholism in three populations. *Am J Med Genet Part B.* 2007;144B:183–192.
198. Loh E-W, Lane H-Y, Chen C-H, Chang P-S, Ku L-W, Wang KHT, Cheng ATA. Glutamate decarboxylase genes and alcoholism in Han Taiwanese men. *Alcohol Clin Exp Res.* 2006;30:1817–1823.
199. Zhao R, Zhang R, Li W, Liao Y, Tang J, Miao Q, Hao W. Genome-wide DNA methylation patterns in discordant sib pairs with alcohol dependence. *Asia Pacific Psychiatry.* 2013;5:39–50.
200. Meinhardt MW, Hansson AC, Perreau-Lenz S, Bauder-Wenz C, Stahlin O, Heilig M, Harper C, Drescher KU, Spanagel R, Sommer WH. Rescue of infralimbic mGluR2 deficit restores control over drug-seeking behavior in alcohol dependence. *J Neurosci.* 2013;33:2794–2806.
201. Flatscher-Bader T, Zuvela N, Landis N, Wilce PA. Smoking and alcoholism target genes associated with plasticity and glutamate transmission in the human ventral tegmental area. *Hum Mol Genet.* 2008;17:38–51.
202. Zhou Z, Karlsson C, Liang T, Xiong W, Kimura M, Tapocik JD, Yuan Q, Barbier E, Feng A, Flanigan M, Augier E, Enoch MA, Hodgkinson CA, Shen PH, Lovinger DM, Edenberg HJ, Heilig M, Goldman D. Loss of metabotropic glutamate receptor 2 escalates alcohol consumption. *Proc Natl Acad Sci USA.* 2013;110:16963–16968.
203. Bell RL, Kimpel MW, Rodd ZA, Strother WN, Bai F, Peper CL, Mayfield RD, Lumeng L, Crabb DW, McBride WJ, Witzmann FA. Protein expression changes in the nucleus accumbens and amygdala of inbred alcohol-preferring rats given either continuous or scheduled access to ethanol. *Alcohol.* 2006;40:3–17.
204. Bell RL, Kimpel MW, McClintick JN, Strother WN, Carr LG, Liang T, Rodd ZA, Mayfield RD, Edenberg HJ, McBride WJ. Gene expression changes in the nucleus

accumbens of alcohol-preferring rats following chronic ethanol consumption. *Pharmacol Biochem Behav.* 2009;94:131–147.
205. Edenberg HJ, Strother WN, McClintick JN, Tian H, Stephens M, Jerome RE, Lumeng L, Li T-K, McBride WJ. Gene expression in the hippocampus of inbred alcohol-preferring (iP) and -nonpreferring (iNP) rats. *Genes Brain Behav.* 2005;4:20–30.
206. Kimpel MW, Strother WN, McClintick JN, Carr LG, Liang T, Edenberg HJ, McBride WJ. Functional gene expression differences between inbred alcohol-preferring and -nonpreferring rats in five brain regions. *Alcohol.* 2007;41:95–132.
207. McBride WJ, Schultz JA, Kimpel MW, McClintick JN, Wang M, You J, Rodd ZA. Differential effects of ethanol in the nucleus accumbens shell of alcohol-preferring (P), alcohol non-preferring (NP) and Wistar rats: a proteomic study. *Pharmacol Biochem Behav.* 2009;92:304–313.
208. McBride WJ, Kimpel MW, Schultz JA, McClintick JN, Edenberg HJ, Bell RL. Changes in gene expression in regions of the extended amygdala of alcohol-preferring rats after binge-like alcohol drinking. *Alcohol.* 2010;44:171–183.
209. McBride WJ, Kimpel MW, McClintick JN, Ding ZM, Hyytia P, Colombo G, Edenberg HJ, Lumeng L, Bell RL. Gene expression in the ventral tegmental area of 5 pairs of rat lines selectively bred for high or low ethanol consumption. *Pharmacol Biochem Behav.* 2012;102:275–285.
210. McBride WJ, Kimpel MW, McClintick JN, Ding ZM, Hauser SR, Edenberg HJ, Bell RL, Rodd ZA. Changes in gene expression within the ventral tegmental area following repeated excessive binge-like alcohol drinking by alcohol-preferring (P) rats. *Alcohol.* 2013;47:367–380.
211. McBride WJ, Kimpel MW, McClintick JN, Ding Z-M, Hyytia P, Colombo G, Liang T, Edenberg HJ, Lumeng L, Bell RL. Gene expression within the extended amygdala of 5 pairs of rat lines selectively bred for high or low ethanol consumption. *Alcohol.* 2013;47:517–529.
212. McBride WJ, Kimpel MW, McClintick JN, Ding Z-M, Edenberg HJ, Liang T, Rodd ZA, Bell RL. Changes in gene expression within the extended amygdala following binge-like alcohol drinking by adolescent alcohol-preferring (P) rats. *Pharmacol Biochem Behav.* 2014;117:52–60.
213. McClintick JN, McBride WJ, Bell RL, Ding Z-M, Liu Y, Xuei X, Edenberg HJ. Gene expression changes in serotonin, GABA-A receptors, neuropeptides and ion channels in the dorsal raphe nucleus of adolescent alcohol-preferring (P) rats following binge-like alcohol drinking. *Pharmacol Biochem Behav.* 2015;129:87–96.
214. Rodd ZA, Kimpel MW, Edenberg HJ, Bell RL, Strother WN, McClintick JN, Carr LG, Liang T, McBride WJ. Differential gene expression in the nucleus accumbens with ethanol self-administration in inbred alcohol-preferring rats. *Pharmacol Biochem Behav.* 2008;89:481–498.
215. Witzmann FA, Li J, Strother WN, McBride WJ, Hunter L, Crabb DW, Lumeng L, Li T-K. Innate differences in protein expression in the nucleus accumbens and hippocampus of inbred alcohol-preferring and -nonpreferring rats. *Proteomics.* 2003;3:1335–1344.
216. Gould AT, Sacramento AD, Wroten MG, Miller BW, Klugmann M, Ben-Shahar O, Szumlinski KK. Extended access to intravenous cocaine imbalances ventromedial prefrontal cortex Homer1 versus Homer2 expression: implications for relapse. *Addict Biol.* 2015;20:148–157.

CHAPTER THREE

S-Glutathionylation and Redox Protein Signaling in Drug Addiction

Jacqueline S. Womersley, Joachim D. Uys[1]
Department of Cellular and Molecular Pharmacology and Experimental Therapeutics, Medical University of South Carolina, Charleston, South Carolina, USA
[1] Corresponding author: e-mail address: uys@musc.edu.

Contents

Progress in Molecular Biology and Translational Science, Volume 137
ISSN 1877-1173
http://dx.doi.org/10.1016/bs.pmbts.2015.10.001

Abstract

Drug addiction is a chronic relapsing disorder that comes at a high cost to individuals and society. Therefore understanding the mechanisms by which drugs exert their effects is of prime importance. Drugs of abuse increase the production of reactive oxygen and nitrogen species resulting in oxidative stress. This change in redox homeostasis increases the conjugation of glutathione to protein cysteine residues; a process called S-glutathionylation. Although traditionally regarded as a protective mechanism against irreversible protein oxidation, accumulated evidence suggests a more nuanced role for S-glutathionylation, namely as a mediator in redox-sensitive protein signaling. The reversible modification of protein thiols leading to alteration in function under different physiologic/pathologic conditions provides a mechanism whereby change in redox status can be translated into a functional response. As such, S-glutathionylation represents an understudied means of post-translational protein modification that may be important in the mechanisms underlying drug addiction. This review will discuss the evidence for S-glutathionylation as a redox-sensing mechanism and how this may be involved in the response to drug-induced oxidative stress. The function of S-glutathionylated proteins involved in neurotransmission, dendritic spine structure, and drug-induced behavioral outputs will be reviewed with specific reference to alcohol, cocaine, and heroin.

1. INTRODUCTION

Substance use disorders are chronic, relapsing conditions, which exert deleterious consequences on individuals and society. Regrettably, despite extensive research and a continually evolving understanding of these conditions, current treatment options are limited and ineffective. Therefore, investigating common mechanisms that underlie addictive behavior in the search for novel therapies is of prime importance. In this review, we will present evidence that despite distinct neurochemical mechanisms, abuse of ethanol, cocaine, and heroin all produce oxidative stress, which in turn may induce S-glutathionylation of proteins. We contend that this redox-sensitive epigenetic modification alters protein signaling and may contribute to an addictive phenotype.

2. OXIDATIVE STRESS

Reactive oxygen species (ROS) and reactive nitrogen species (RNS) are products of aerobic metabolism generated during normal physiology and are important in modulating cellular physiologic processes including cell survival, proliferation, differentiation, and apoptosis.[1–4] However, if local or systemic stress increases ROS or RNS production such that endogenous antioxidant defense mechanisms are overwhelmed, oxidative stress occurs.[1] This cytotoxic process has the potential to damage proteins, DNA, and lipids as well as to activate signaling pathways leading to apoptosis.[1,2,5,6] Therefore the balance between oxidants and antioxidants, or redox status, of a cell is critical to healthy cell survival and function. Redox status is highly compartmentalized within cells and it reflects different cellular activities, for example, the high metabolic activity in mitochondria requires a relatively more reducing environment than the cytosol.[7] Maintaining redox homeostasis is especially critical in the brain, an organ which consumes approximately 20% of the oxygen requirement of the body; has a high concentration of oxidation-prone polyunsaturated lipids and has lower levels of antioxidant enzymes.[8–11] Indeed, imbalance between ROS production and antioxidant capacity resulting in oxidative stress has been implicated in the etiology of neurologic disorders including those due to neuroinflammatory and neurodegenerative processes.[4,10,12]

3. S-GLUTATHIONYLATION OF PROTEINS OCCURS IN RESPONSE TO OXIDATIVE STRESS

Protection against oxidative and nitrosative stress is partially mediated by glutathione (GSH), a small, highly abundant hydrophilic γ-glutamyl-cysteine-glycine tripeptide. Oxidative-stress-induced reduction in GSH content in specific brain areas has previously been implicated in neurologic disorders including Parkinson's disease, schizophrenia, Alzheimer's, and epilepsy.[5,11,13] Due to the presence of the central cysteine, GSH exists in either a reduced form or, under oxidative conditions, as glutathione disulfide (GSSG).[14] Therefore the GSH:GSSG ratio provides an indication of redox metabolism within a particular cellular compartment.[14–16] Reduced GSH exists at high (2–3 mM) concentrations in brain tissues and contributes to the reducing status of the cell during normal homeostasis.[1,11,17] As a result of this reducing intracellular environment, many cytoplasmic proteins are rich in free cysteine thiols, which are available to undergo oxidative modification.[1,18] The conjugation of the cysteine residue in GSH with partially oxidized reactive protein thiols, including thiyls and sulfenic acid to form a mixed disulfide is known as S-glutathionylation.[3,11,19,20] ROS-mediated changes to protein thiols move sequentially via oxidation from cysteine sulfenic, sulfinic, and sulfonic acids, of which the latter two are irreversible reactions.[1] As the S-glutathionylation of a protein sulfhydryl group prevents its irreversible oxidation, S-glutathionylation is considered a protective mechanism in cells.[21] Oxidative stress results in an order of magnitude increase in protein S-glutathionylation clearly implicating S-glutathionylation as a redox-sensitive process.[22] In addition, S-glutathionylated proteins can be formed under conditions of nitrosative stress via the intermediate production of S-nitrosglutathione.[5,23] The capacity of a protein to be S-glutathionylated is determined by the accessibility of the cysteine residue in the three-dimensional protein structure and the presence of surrounding basic amino acids, that is, a low pK_a value.[5,24] S-glutathionylation may occur either spontaneously or may be enzymatically catalyzed by glutaredoxin or members of the glutathione S-transferase (GST) family, which are widely expressed in the brain.[10,22,25–28] Deglutathionylation may occur spontaneously via reversal of the thiol-disulfide exchange reaction once excess oxidative stress has terminated and the GSH:GSSH ratio has increased, or via enzymatic action of thiol-disulfide oxireductases: glutaredoxin, sulfiredoxin, thioredoxin, and protein-disulfide isomerase.[1,18,23,25,29,30] When combined, the S-glutathionylation and deglutathionylation of proteins compose the S-glutathionylation cycle (Fig. 1).[16] The low molecular weight of GSH, which increases its capacity to interact with cysteine residues, combined with its abundance makes it a critical regulator of redox homeostasis.[1]

Although S-glutathionylation of proteins may be constitutive, that is, occurring under basal conditions, they are increased under conditions of increased ROS or RNS, which suggests that they are important in mediating the cellular response to oxidative and nitrosative stress.[1,21]

4. S-GLUTATHIONYLATION OF PROTEINS IS A REDOX-SENSITIVE SIGNALING MECHANISM IN CELLS

In addition to protecting proteins from irreversible oxidation, S-glutathionylation is increasingly being recognized as a means of redox-sensitive protein signaling.[31] The thiol groups of cysteines form intra- and intermolecular disulfide bridges and therefore play an important role in maintaining proper protein folding and structural stability.[14,18] Consequently, the binding of GSH to cysteine residues in proteins may alter the catalytic site or, even if not near the active site, may produce a conformational change such that it

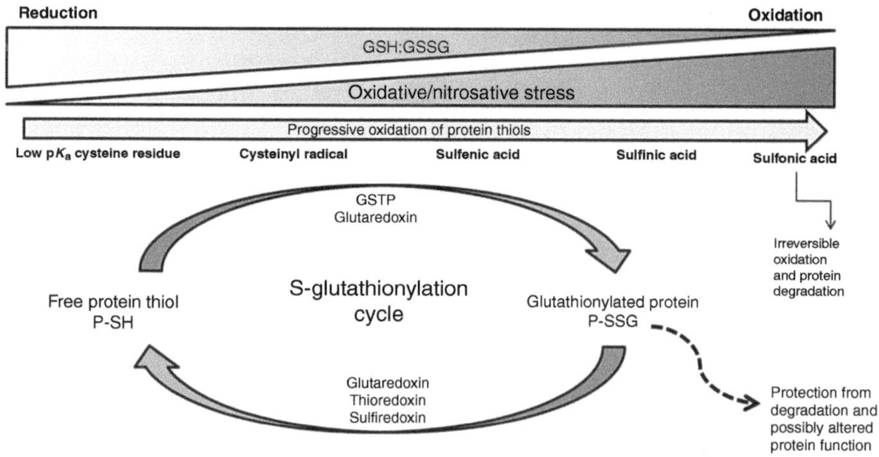

Figure 1 S-glutathionylation is a reversible oxidative modification of cysteine residues. In response to increased oxidative stress, GSH is conjugated to low pK_a cysteine residues (P-SH) by GSTP or glutaredoxin to form a glutathionylated protein (P-SSG). This process is referred to as S-glutathionylation. Upon return to a more reducing environment, the reverse process is catalyzed by glutaredoxin, thioredoxin, or sulfiredoxin. S-glutathionylation protects against irreversible oxidation of protein thiols and may alter protein function.

results in loss or gain of protein function.[21,32,33] In addition, S-glutathionylation imparts a negative charge to the protein and as a result may produce structural and functional changes.[16] Importantly, modification of cysteine residues via S-glutathionylation is a reversible process and any loss- or gain-of-function will be recovered following normalization of cellular redox status.[1] Therefore S-glutathionylation represents a mechanism by which changes in cellular redox status can be transduced into a functional response via post-translational protein modification, that is, S-glutathionylation is a means of cellular redox siganling.[1,20,34,35]

The mammalian genome encodes 214,000 cysteine residues of which at least 10–20% are redox sensitive under biologic conditions.[36] Proteins susceptible to S-glutathionylation have multiple functions including signal transduction, calcium homeostasis, nitric oxide regulation, energy metabolism, cytoskeletal rearrangement, apoptosis, inflammation, and the expression, folding, trafficking, and degradation of proteins.[3,16,22,28,37–40] The diverse functions included in this list as well as the sheer number of S-glutathionylated proteins suggest both modification of individual protein function as well as modification of coordinated protein pathways in the cellular response to oxidative stress.[41,42] In addition, evidence suggests that

S-glutathionylation impacts other important post-translational modifications including phosphorylation and S-nitrosylation.[2,16] Although the physiologic significance of these modifications are not yet fully known, as S-glutathionylation has the capacity to increase, decrease, or leave unaltered protein function, it is clear that protein thiol redox state may act as a molecular switch by altering protein structure and/or function.[43]

S-glutathionylation fulfills the criteria of specificity, reversibility, and sensitivity under physiologic conditions necessary to be recognized as a regulatory mechanism.[1] S-glutathionylation occurs only on locally available and reactive thiols produce changes in protein function ensuring specificity in signaling.[36] Return to a reducing cellular environment results in a rapid and efficient deglutathionylation characteristic of a reversible system.[44] The degree of GSH conjugation to cysteine thiols is proportional to the aberration in redox status suggesting sensitivity of the system.[44] In addition, these criteria are fulfilled under a high GSH:GSSG ratio in intact cells suggesting that it is a physiologically relevant mechanism.[44] This suggests that cysteine residues in proteins serve as redox sensors capable of inducing changes in redox balance into integrated biologic outcomes.[36]

5. S-GLUTATHIONYLATION IN ADDICTION-RELATED PROTEIN SIGNALING

5.1 Drug Addiction, Dopamine, and Oxidative Stress

Drug addiction is a chronic relapsing disorder characterized by compulsive drug use, loss of control in drug intake despite potential negative consequences, and emergence of negative affect following drug withdrawal.[45] Acute impulsive drug use, defined as a tendency toward rapid unplanned actions and preference for immediate over delayed gratification, progresses to compulsive drug-seeking behavior, defined as the perseveration of maladaptive behavior.[45] This process produces cycles of preoccupation and anticipation of drug use, drug binge, and intoxication following drug consumption, and subsequent withdrawal and negative affect after drug-induced effects have diminished.[46]

Although drugs of abuse produce divergent effects in the brain, increased dopaminergic signaling primarily in the mesocortical and mesolimbic pathways, is a common underlying mechanism.[47–49] At high concentrations, autooxidation of the dopamine catechol ring produces quinone and semiquinone species, which reduces the GSH:GSSG ratio.[9,50–52] Alternatively, enzymatic oxidation of dopamine via monoamine oxidases results in the formation of hydrogen peroxide.[9] Due to the high oxidative potential of dopamine, it has been hypothesized that dopaminergic neuron terminals are more likely working closer to the limits of their antioxidant potential due to the need to store dopamine in a reduced state and to counteract ROS generated by dopamine metabolism.[53] Increased ROS production due to dopamine metabolism results in potentially neurotoxic effects especially at high dopamine concentrations such as those elicited by the use of drugs of abuse.[47] Evidence for the potentially neurotoxic role of endogenous dopamine in the striatum has been shown by measuring free and protein-bound cysteinyl-catechols, markers of oxidative damage, in response to intrastriatal injection of dopamine.[53] The resulting lesion and amount of cysteinyl-catechols produced were proportional to the concentration of dopamine injected, suggesting a causal and proportional relationship between striatal dopamine, modification of protein thiols, and neurotoxicity.[53] This relationship between dopamine toxicity and cellular redox status is at least partially attributable to GSH, as SK-N-SH neuroblastoma cells pretreated with a GSH-depleting compound showed elevated markers of apoptosis.[54] In addition, incubation of astroglia-rich primary cultures in a dopamine-rich solution reduced extracellular GSH, indicating that GSH has a key role in combating the oxidative stress generated by elevated dopamine.[50] A similar effect was found in an experiment where incubation of astroglial cultures in dopamine showed an inverse dose-dependent relationship between dopamine concentration and GSH content.[50] Evidence for dopamine-induced oxidative stress specifically due to drugs of abuse has been shown in a binge methamphetamine paradigm in rats where methamphetamine-induced reductions in striatal GSH were recorded.[52] In addition to having potential neurotoxic effects, generation of ROS in response to drugs of abuse has been implicated in memory and learning processes, which underlie drug addiction behavior.[55]

Changes in protein S-glutathionylation can be expected under any conditions where GSH concentrations are decreased and therefore the oxidation of dopamine has also been implicated in protein S-glutathionylation.[1] In addition, GSTs have been implicated in the metabolism of dopamine.[9] Humans may possess one of four glutathione-S-transferase pi (GSTP) polymorphisms that arise from nucleotide transitions that change codon 105 from Ile to Val and codon 114 from Ala to Val.[22] These polymorphisms are associated with varying levels of functionality due to steric changes in the active site of the enzyme.[22] This implies that individual reactions to oxidative stress will be different and therefore by extension, this may modulate the varied effects of drugs of abuse on humans.[39] In support of this theory, evidence for GSTP polymorphisms and elevated risk of drug use has been found in both cocaine- and methamphetamine-using populations.[56,57] This suggests that dopamine-induced oxidative stress disturbs thiol homeostasis with potential downstream effects on protein function.[58]

The number of proteins that are recognized to undergo S-glutathionylation are increasing rapidly and many of these are implicated in drug addiction (Table 1). However, the physiologic relevance of whether these proteins exist in an oxidized or reduced state and the effect that this may have on addictive processes is not yet clearly understood. For this reason, the effect of S-glutathionylation on a limited number of proteins central to the effects relevant to psychostimulant, ethanol, and opioid addiction will be reviewed.

5.2 Altered Neurotransmission in Response to Ethanol, Cocaine, and Opioids Share Common Signaling Mechanisms

Drug addiction involves a sensitization in the neural processes that control drug reward such that the motivation to consume drug is increased as addiction progresses.[90] Although the mechanisms underlying drug of abuse signaling are complex and not yet fully elucidated, several common substrates exist (Fig. 2). The rewarding effects of drugs are mediated by the mesolimbic dopamine system where increased dopamine release results in binding to postsynaptic dopamine receptors, which are broadly classified into type 1 (D1R) and type 2 (D2R) receptors.[80,90] D1R stimulation activates the cyclic AMP (cAMP)-dependent protein kinase pathway leading to transcription factor induction and phosphorylation of downstream proteins.[91] Briefly, binding to D1Rs increases cAMP with subsequent activation of cAMP-dependent protein kinase or protein kinase A (PKA).[80] Activation of PKA induces phosphorylation of dopamine and cAMP-regulated phosphoprotein 32 kDa (DARPP-32) at threonine 34, which subsequently inhibits protein phosphatase-1, a regulator of gene transcription.[79] D1R-mediated signaling also increases the translocation of PKA to the nucleus to evoke a nuclear cascade including activation of the transcription factor ΔFosB, activation of which increases cyclin-dependent protein kinase 5 (Cdk5).[79] Cdk5 then phosphorylates DARPP-32 at threonine 75 (at the expense of threonine 34 phosphorylation) to dampen PKA signaling and disinhibit protein phosphatase-1 in a negative feedback mechanism.[66,79,92] Conversely, D2Rs are negatively coupled to adenylyl cyclase with a resultant decrease in cAMP and instead signal via a cAMP-independent process of Akt/GSK3 activation.[79,90] Binding to D2Rs

Table 1 Proteins Susceptible to S-Glutathionylation are Involved in Drug Addiction

Protein	Function	Relevance to Addiction	Effect of Glutathionylation
Actin	Structural cytoskeletal protein[59]	Actin polymerization is required for dendritic spine remodeling in response to drug use[59,60]	Polymerization is inhibited[61]
Alcohol dehydrogenase	Catalyzes alcohol metabolism	Reduced alcohol dehydrogenase activity results in increased alcohol levels in brain tissue[62]	Glutathionylation has been confirmed but effect is unknown[63]
Caspase 3	Proapoptotic factor[47]	May induce apoptosis in response to drug-induced oxidative stress[47]	Reduced[64]
Cdk5	Neuronal development, survival, trafficking and transport[65]	Phosphorylates DARPP-32 to reduce dopaminergic signaling via D1Rs[66]	Reduced[67]
Cofilin	Depolymerization of filamentous actin[68]	Important for actin cycling, which is required for dendritic spine remodeling in response to drug use[68,69]	Glutathionylation has been confirmed but effect is unknown[70]
COMT	Catecholamine catabolism	COMT is responsible for the degradation of dopamine, the neurotransmitter responsible for drug reward. Increased COMT activity is positively associated with drug abuse[71,72]	Reduced[73]
GSTP	Phase II detoxification enzyme[74]	Conjugates GSH to target proteins in response to oxidative stress to modulate target protein function[28]	Reduced[28]
Myosin	Cytoskeletal structural protein	Involved in dendritic spine morphogenesis and controls the dendritic spine actin cytoskeleton.[75] Dendritic spine remodeling occurs in response to drug use[69]	Reduced[76]
NF-κB	Redox-sensitive transcription factor[4]	Required for dendritic spine remodeling in response to drug use and plays a role in the induction of long-term potentiation[77]	Reduced[78]

(*Continued*)

Table 1 Proteins Susceptible to S-Glutathionylation are Involved in Drug Addiction—cont'd.

Protein	Function	Relevance to Addiction	Effect of Glutathionylation
PKA	Kinase responsible for phosphorylation of multiple downstream targets including DARPP-32[79]	Involved in AMPAR:NMDAR trafficking and localization.[80] Increased PKA activity correlates with increased motivation to consume drugs[81]	Reduced[82]
PKC	Kinase responsible for phosphorylation of multiple targets including G-protein-coupled receptors[83]	Increased PKC activity correlates with increased drug-related memory, drug seeking, and reinstatement behavior[83,84]	Reduced[85]
Profilin	Polymerization of globular actin[86]	Important for actin cycling, which is required for dendritic spine remodeling in response to drug use[69]	S-glutathionylation has been confirmed but effect is unknown[70]
Tyrosine hydroxylase	Rate-limiting enzyme in dopamine synthesis[87]	Dopamine is responsible for the acute rewarding effects of drugs[88]	Reduced[89]

A small subset of the proteins that have been identified as susceptible to S-glutathionylation are included in the table and their relevance to drug addiction is also included.

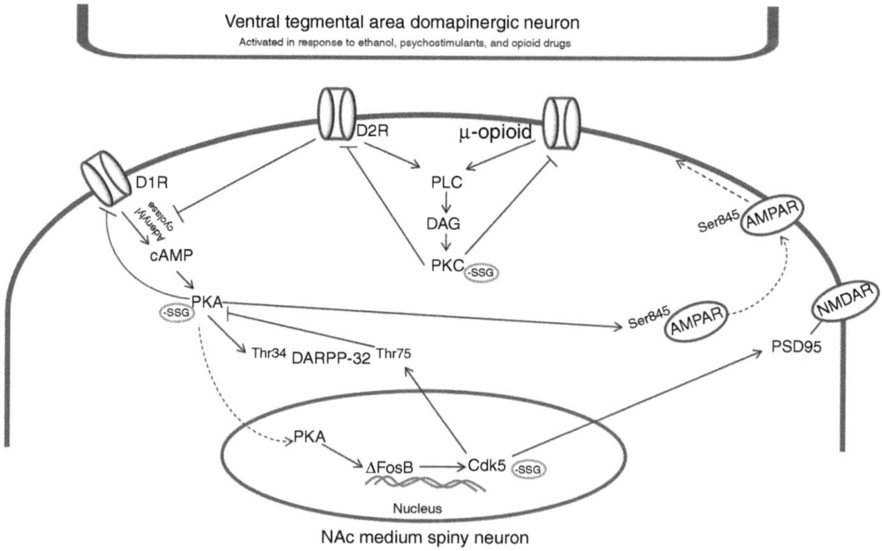

Figure 2 Drugs of abuse share common signaling pathways. Ethanol, cocaine, and heroin increase dopamine release from ventral tegmental area dopaminergic neurons. Stimulation of D1Rs increases cAMP levels and activates PKA, which phosphorylates threonine34 of DARPP-32. Translocation of PKA to the nucleus increases ΔFosB-mediated transcription, which in turn elevates Cdk5. Cdk5 phosphorylates threonine75 of DARPP-32, which subsequently inhibits PKA activity. Cdk5 also phosphorylates PSD95, a cytoskeletal protein responsible for stabilizing NMDARs in the membrane. Phosphorylation decreases the size of PSD95 aggregates and thereby reduces synaptic NMDAR levels. PKA increases phosphorylation of serine845 of AMPARs, which increases their insertion into the neuronal membrane and their lateral diffusion into the synaptic field. Binding to D2 and μ-opioid receptors increases PLC activity and thereby DAG levels, which activates PKC. PKC-mediated phosphorylation of D2 and μ-opioid receptors results in desensitization. PKA, PKC, and Cdk5 are targets for S-glutathionylation, which decreases their activity.

activates phospholipase C, which hydrolyzes phosphatidylinositol-4,5-bisphosphate to form diacylglycerol and inositol triphosphate after which diacylglycerol binds to most of the isoenzymes of PKC to activate them.[83] PKC then acts to phosphorylate serine/threonine residues of the target protein.[83]

D1Rs and D2Rs are colocalized with glutamate receptors in medium spiny neurons of the nucleus accumbens (NAc) and as such the two interact to produce the long-term changes in synaptic plasticity and structure that are associated with drug addiction.[80] Dopaminergic neurotransmission in response to drugs produces reward while glutamatergic signaling codes the context of this reward to result in drug-related learning and memory.[79,80] Glutamatergic neurons from the prefrontal cortex and other limbic areas innervate the ventral tegmental area and NAc to drive dopaminergic neuronal activity or to modulate the activity of dopaminoceptive neurons.[79] Increased dopamine-dependent activity of PKA affects glutamatergic receptors and neurotransmission. PKA-mediated phosphorylation of serine 897 of the GluN1 subunit of glutamatergic *N*-methyl-D-aspartate receptors (NMDAR) controls their trafficking to the membrane surface.[80] In addition, phosphorylation of α-amino-3-hydroxy-5-methyl-4-isoxazolepropionic acid receptors (AMPAR) at Ser845 by PKA increases binding of AMPARs to anchor proteins required for insertion into the cell membrane.[93,94] Phosphorylation at Ser845 of AMPARs also primes extrasynaptic receptors for insertion into the synaptic space.[95] PKA-induced increase in Cdk5 affects glutamatergic signaling via phosphorylation of postsynaptic density 95 (PSD95), a postsynaptic scaffolding protein that links NMDARs to the cytoskeleton.[96] Cdk5 activation results in smaller PSD95 clusters and therefore reduced NMDAR levels in the synapse.[96] The net result is an increase in the AMPAR:NMDAR ratio, which underlies many forms of plasticity and is required for the induction of long-term potentiation learning.[49,97] Recordings from midbrain slices show that ethanol, morphine, and cocaine produce a significant increase in AMPAR:NMDAR ratio reflecting an increase in basal excitatory strength.[49,97]

The rewarding properties of opiates are mediated primarily by μ-opioid receptors found in the ventral tegmental area and NAc.[49] Binding to μ-opioid receptors of GABAergic interneurons in the VTA disinhibits dopaminergic projection neurons to increase dopamine release in the NAc.[49] Heroin, an opioid drug, exerts effects on the brain by binding to postsynaptic $G_{i/o}$ opioid receptors located postsynaptically on NAc medium spiny neurons with the result that $G_{i/o}$ dissociates into α and βγ subunits.[98] $G\alpha_i$ inhibits adenylyl cyclase activity and thereby cAMP production while $G\alpha_5$ and Gβγ dimers activate adenylyl cyclase, to increase cAMP and PKA.[98] Increased Gβγ activates phospholipase C leading to the activation of PKC and increasing downstream target signaling in an extracellular signal-related kinase (ERK) dependent process.[99] PKC is capable of acutely desensitizing dopamine receptors.[100] PKA and PKC are implicated in the long-term changes to opioid abuse with PKC mediating the phosphorylation and subsequent internalization of opioid receptors.[101] In addition, the chronic opioid administration upregulation of PKC correlates with the development of tolerance to opioids.[101] This effect is possibly due to binding to opioid receptors, increasing diacylglycerol, and thereby PKC activity.[101] Moreover activation of PKA following chronic opioid use phosphorylates μ-opioid receptors disrupting coupling of these receptors to G_i proteins.[101]

6. S-GLUTATHIONYLATION OF PKA, PKC, Cdk5, AND ACTIN MAY INFLUENCE DRUG BEHAVIOR

In keeping with their role in drug-induced neurotransmission, PKA, PKC, and Cdk5 mediate behavioral responses to drugs of abuse, while alterations in the cytoskeletal protein actin may underlie the effects of drugs of abuse on synaptic strength.

PKA signaling in the NAc is upregulated in response to repeated cocaine use.[102] A study examining the effect of activation or inhibition of the PKA system in the rat NAc found that inhibition of PKA signaling reduced self-administration and induced relapse of cocaine-seeking behavior; a finding explained by a heightened sensitivity to cocaine.[102] In contrast, a PKA activator resulted in increased cocaine self-administration and produced nonspecific motor activity on both drug paired and inactive levers, that is, increased nonreinforced locomotor responses.[102] The effect of PKA inhibition on drug self-administration was repeated in a study where bilateral infusion of PKA inhibitors into the NAc in rat reduced cocaine self-administration and induced relapse after withdrawal suggesting that the effects to cocaine were sensitized.[103] In a further study, administration of a PKA activator to rat NAc increased the motivation for drug consumption measured by an increase in the progressive ratio responding (the effort that the rat was willing to expend in order to obtain a cocaine reward) while the opposite was true following administration of a PKA inhibitor.[81] When combined these results suggest that cocaine-induced alterations in PKA activity have direct relevance to drug-seeking behavior.[81] Of relevance to this review, the C subunit of PKA can be inactivated by the S-glutathionylation of cysteine 199 located in the activation loop, which inactivates the kinase in a redox-sensitive manner and which may allow dephosphorylation of threonine197, which is responsible for coordinating the active conformation and optimal enzymatic activity.[82] This suggests that PKA activity is a potential target of S-glutathionylation-mediated redox signaling control.[104]

Protein kinase C (PKC) isozymes phosphorylate and regulate the activity of some G-protein-coupled receptors and altered levels of PKC have been associated with opiate drugs, ethanol, and cocaine.[83,105,106] While many functions of PKC are protective and required for normal cellular functioning, excessive or prolonged increase in PKC activity may be deleterious.[88,107] Cocaine reinstatement (a model of relapse) following withdrawal is at least partially due to D2R-dependent increases in PKC signaling in the NAc shell with increased D1R or direct pathway signaling to promote relapse.[108] Indeed, a study of cocaine-primed reinstatement of cocaine-seeking behavior showed an increase in phosphorylated PKCγ.[109] Injection of PKC inhibitors into the NAc core inhibits cocaine-seeking in rats while injection of inhibitors into the ventral tegmental area reduces dopamine release in the NAc as well as cocaine-induced hyperlocomotion. ZIP-mediated inhibition of PKC reduced cocaine sensitization, suggesting that the λ and ξ isozymes specifically play an important role in nonassociative cocaine memory and cocaine addiction.[84] A role for PKC in drug-associated memory was also found in a study where administration of a PKC inhibitor after cocaine administration prevented the development of cocaine preference measured in a conditioned place-preference paradigm.[83] A role for PKC in the rewarding effects of ethanol has also been found with PKCε knockout mice consuming less ethanol than their wild-type counterparts.[107] S-glutathionylation of PKC has been shown to negatively regulate its function.[85]

Chronic cocaine upregulates Cdk5, a serine/threonine kinase, as well as its activator p35.[65] This Cdk5/p35 complex has been implicated in neuronal development and survival, trafficking, transport, and dopaminergic neurotransmission.[65] Cdk5 activation following drug use follows the action of ΔFosB transcription factor, which is upregulated in response to drugs of abuse such as psychostimulants.[65] The relationship between ΔFosB and Cdk5 in the context of cocaine addiction was investigated in mice with inducible and targeted striatal expression of ΔFosB.[66] Adult rats injected with cocaine increased Cdk5 expression in the caudate putamen and NAc[66] while intra-accumbal injection of a Cdk5 inhibitor increased cocaine-induced locomotor sensitization.[66] In a separate study, infusion of a Cdk5 inhibitor into the NAc increased behavioral sensitization to cocaine and cocaine self-administration with increased lever presses for cocaine administration.[110] Increased GSTP decreases Cdk5 function suggesting that S-glutathionylation of Cdk5 may be inhibitory.[67]

Excitatory synapses modify the strength of synapses due to previous experience in a neuroadaptive process known as synaptic plasticity, which allows for the integration of experience-dependent learning.[94,111,112] However, this response is hijacked by drugs of abuse resulting in aberrant plasticity where drugs can produce long-lasting changes in motivation and reward behavior via changes in synaptic plasticity, structure, and function.[45,79,113] The persistence of addiction even following prolonged withdrawal suggests that drugs induce long-term changes in neuronal structure and function that extend beyond the period of intoxication producing a long-lasting drug-induced plasticity that might underlie addictive behavior.[91,114] As the majority of excitatory synapses are formed with dendritic spines, change in the shape or numbers of dendritic spines reflects changes in experiential learning.[111] Rodents chronically treated with cocaine display increased density of immature spines on medium spiny neurons and these structural changes in dendritic spine morphology may persist for several months following the onset of withdrawal.[115] A chronic intermittent ethanol exposure model of alcohol use disorder found that alcohol increased dendritic spine density.[116] Changes in the actin cytoskeleton mediate changes in spine morphogenesis. S-glutathionylation of G-actin on cysteine 374 inhibits its polymerization into F-actin, the form of the protein responsible for increased length and spine head diameter required for the induction of LTP.[21,59,61] In addition, Cdk5 is critically involved in neuronal activity–-mediated spine morphogenesis due to phosphorylation of S6K, a signaling protein which increases protein synthesis.[117]

7. S-GLUTATHIONYLATION IN ALCOHOL ADDICTION

Alcohol is a widely abused drug with 17.7 million people in the United States meeting the criteria for alcohol dependence or abuse in 2012.[118] The effects of alcohol on the brain are complex and varied with ethanol consumption affecting norepinephrine, serotonin, dopamine, opioid, and GABA signaling.[119,120] The primary effects of ethanol are exerted via $GABA_A$ and NMDARs to enhance GABA and inhibit glutamate signaling, which when combined produces an overall depressive effect.[45,121] However, with increasing use, NMDAR function is increased and this is further exacerbated in withdrawal to produce cognitive effects such as delirium.[121] In addition, chronic alcohol use and repeated withdrawal downregulates the sensitivity of $GABA_A$ receptor signaling to reduce the locomotor and cognitive effects of ethanol while enhancing the action of $GABA_B$ receptors, which are responsible for decreasing neuronal excitability.[121] Ethanol results in excitation of ventral tegmental area dopamine neurons, increasing dopamine release in NAc projections, which when combined with activation of the endogenous opioid reward system to increase endorphin content in the

NAc, is responsible for the rewarding properties of alcohol.[121] Furthermore, ethanol also increases serotonin levels in the NAc.[121]

7.1 Ethanol Metabolism Increases ROS Production

Strong evidence exists for ethanol-induced increases in ROS production with both chronic exposure to ethanol and subsequent withdrawal resulting in oxidatively modified lipids and proteins that may damage glia and neurons.[122] Animal studies have found increased brain lipid peroxidation products in ethanol-exposed rats while stimulation of hippocampal astrocytes with ethanol has been shown to increase ROS generation in the mitochondria.[123,124] Studies of alcoholic patients have found a strong correlation between increased levels of plasma malondialdehyde, a marker of lipid peroxidation, and blood alcohol content at the time of admission.[125] In addition, alcoholic patients have increased blood hydroxyl ion levels and increased levels of the antioxidant enzyme Cu, Zn-superoxide dismutase, and Mn-superoxide dismutase, indicative of oxidative stress.[126,127] Evidence for oxidative-stress-induced damage, as measured by TUNEL-staining, has been found in postmortem superior frontal cortex and hippocampus samples of alcoholic patients.[128] Evidence for the importance of oxidative stress in the metabolism of alcohol has also been found in models of fetal alcohol spectrum disorder where reduced antioxidant defenses during embryonic and fetal stages may explain the sensitivity of the developing brain to the toxic effects of ethanol.[129] When combined this suggests that ethanol increases free radical generation and that the deleterious effects of chronic alcoholism may be associated with oxidative-stress-induced damage. Given the relationship between ethanol and ROS production, it is perhaps unsurprising that acute and chronic ethanol consumption has been associated with decreased GSH levels.[130–132] Postmortem samples from patients with a history of chronic alcohol use found an overall reduction in brain GSH content compared to controls, which may be explained by increased conjugation of GSH to the oxidative metabolites of ethanol.[133] This ethanol-induced reduction in GSH may produce a compensatory increase in serum gamma glutamyltransferase, which occurs following a history of prolonged ethanol consumption and has been used as a biomarker of alcoholic behavior.[134,135]

7.2 Increased ROS Production May Contribute to the Behavioral Effects of Ethanol

Ethanol is metabolized primarily by astrocytic catalase and cytochrome P450 IIE1 in the brain to produce acetaldehyde in a process that leads to increased generation of ROS including hydroxyl and hydroxyethyl radicals.[124,125,136–138] In addition to being potentially cytotoxic, oxidative stress might underlie some of the behavioral features associated with ethanol consumption.[139] Acetaldehyde reacts with 5-hydroxytryptamine in serotonergic neurons to produce 1-methyl-6-hydroxy-1,2,3,4-tetrahydro-β-carboline.[120] The oxidation of this product by hydroxyl radicals produces greater flux in ROS with the potential to cause oxidative damage.[120] The resultant damage to serotonergic neurons and concomitant decrease in 5-hydroxytryptamine is believed to contribute to increased alcohol preference and ultimately addiction.[120] In addition, the GSH conjugates of 1-methyl-6-hydroxy-1,2,3,4-tetrahydro-β-carboline are oxidized to form 8-S-glutathionyl-1-methyl-1,2,3,4-tetrahydro-β-carboline-5,6-dione, which binds to $GABA_B$ receptors, stimulation of which results in increased ethanol-addictive-type behaviors such as locomotor stimulation and increased risk of relapse.[120] Mice administered alcohol and ebselen, a hydrogen peroxide scavenger, showed reduced locomotor activity when allowed to explore an open-field apparatus suggesting that there is a proportional relationship between brain catalase-H_2O_2 activity and ethanol-induced locomotor stimulatory effects.[139] A further study found that administration of nitropropionic acid, an inhibitor of mitochondrial complex II which increases free radical production, increased locomotor activity in ethanol-treated mice.[140] In addition, mice exposed to hyperoxic conditions to increase brain hydrogen peroxide content, showed increased locomotor activity.[141] These results suggest that ethanol-induced redox signaling may contribute to the behavioral and molecular effects of alcohol in alcohol addiction.[120]

7.3 Protein S-Glutathionylation in Response to Ethanol

While direct evidence for the role of protein S-glutathionylation in the response to ethanol-induced oxidative stress is scarce, a number of studies point to the possible involvement of S-glutathionylating enzymes in alcohol addiction. Ethanol administered in drinking water to rats over 4 weeks increased GST levels in the cerebellum suggesting that increased oxidative stress in response to ethanol consumption increased protein S-glutathionylation.[142] Alpha-lipoic acid, a redox-modulating agent and thiol replenisher, scavenges free radicals and was used to mitigate the effects of ethanol-induced oxidative damage produced in a mouse hippocampal cell line.[143] Rodent studies have also indicated that the transcription factor, Nrf2, is activated by ethanol administration and in turn activates several antioxidative cytoprotective gene clusters including the enzymes involved in GSH homeostasis: γ-glutamyl transpeptidase, GSH synthase, and γ-glutamyl cysteine ligase.[144] In human studies, proteomic analysis of the corpus callosum of healthy and alcoholic postmortem corpus callosum found an alcohol-induced increase in GSTP in the corpus callosum genu.[145] In addition, elevated levels of GST mu were found in the corpus callosum body of alcoholic patients.[146] Reduced levels of GST have been indicated as a potential contributor to the sensitivity of the developing fetal brain to the toxic effects of ethanol and may be a contributor to the enhanced sensitivity of the fetus to oxidative stress.[147] When combined this provides preliminary evidence for the involvement of S-glutathionylation in the response to ethanol-induced oxidative stress.

8. S-GLUTATHIONYLATION IN COCAINE ADDICTION

In 2012, 1.1 million Americans met the criteria for dependence/abuse of the psychostimulant cocaine.[118] Cocaine activates dopaminergic systems in the NAc and amygdala via inhibition of the dopamine transporter.[45] In addition, cocaine blocks serotonin and noradrenaline transporters to glutamate and GABA in the NAc.[45,148]

8.1 Cocaine Metabolism Increases ROS Production

Cocaine is metabolized by either spontaneous or enzymatic hydrolysis or alternatively via oxidation by P450 with the sequential production of N-oxidative metabolites: norcocaine, *N*-hydroxynorcocaine, and norcocaine nitroxide.[149] ROS including hydrogen peroxide and superoxide anion radicals are produced during the oxidative metabolism of cocaine and are believed to induce cellular redox cycling with the end product of increased oxidative stress.[150,151] The positive charge of cocaine at physiologic pH may result in accumulation of cocaine in the mitochondria and its capacity to dissipate mitochondrial membrane potential.[152,153] Cocaine appears to directly inhibit mitochondrial complex I activity to increase ROS accumulation and potentially generate oxidative stress.[153,154] In addition, cocaine has been found to downregulate complex I subunit expression as determined by differential display of genes induced by cocaine when measured in the cingulate cortex.[155] Cocaine administration has also been suggested to increase metabolic activity in the brain, evidenced by a significant increase in temperature, that might contribute to the generation of ROS, as they are a product of physiologic respiratory processes.[156] Evidence for cocaine-induced oxidative stress is clear. Increased ROS levels are found in the striatum and prefrontal cortex of cocaine-treated rats while administration of cocaine to mice has been studied primarily as a means of determining cocaine-induced toxicity.[156] Cocaine administration to mice also produces elevated levels of lipid peroxidation products in the striatum and cortex.[157] Single or repeated cocaine administration in rats increases the formation of hydrogen peroxide in prefrontal cortex mitochondria and increased lipid peroxide production in the striatum and prefrontal cortex.[154] A concomitant increase in the antioxidant superoxide dismutase and GSH peroxidase that co-occurred with the increase in ROS was also found.[154] Therefore acute or chronic cocaine administration was shown to increase ROS production in the dopaminergic brain areas of the striatum and prefrontal cortex and produce a compensatory increase in antioxidant activity.[154] The resulting increase in ROS generation in the brain after cocaine administration, affects the antioxidant status as evidence by decreased GSH concentration and reduced catalase activity.[158,159] Human studies have also found that increased ROS production may produce harmful effects with a study examining cocaine-induced changes in the prefrontal cortex of human cocaine users and in rats administered cocaine finding increased degradation of nuclear PARP-1, a sensor of DNA damage and marker of apoptosis, which was considered potentially indicative of oxidative-stress-induced apoptotic signaling.[113]

8.2 Increased ROS Production May Contribute to the Behavioral Effects of Cocaine

Several studies have found evidence for an association between oxidative stress and behavioral outcomes following cocaine administration. To determine the effects of ROS on cocaine-induced signaling in reinforcement processes, a nonspecific ROS scavenger, PBN, was administered to rats trained to self-administer cocaine and produced a significant reduction in the number of active lever presses compared to saline-treated controls.[156] This effect was found to be mediated by superoxide anions, as administration of the superoxide dismutase mimetic TEMPOL, a compound shown to reduce superoxide radical generation, produced a similar dose-dependent reduction in cocaine self-administration.[156,160] Both TEMPOL and PBN were also shown to alleviate ROS-induced damage by reducing 8-hydroxyguanosine staining in NAc neurons.[156] Peripheral administration of TEMPOL also reduced cocaine-enhanced dopamine release measured by fast scan cyclic voltammetry.[156] When combined this suggests that the rewarding and reinforcing properties of drugs are at least in part mediated via the effects of ROS.[156]

TEMPOL was also investigated in a study examining the link between the attenuation of oxidative stress and change in behavioral response to cocaine.[161] Cocaine increased lipid peroxidation and nitric oxide radical content in the prefrontal cortex and NAc in both *in vitro* and *in vivo* experiments that were accompanied by a decrease in the total antioxidant capacity *in vitro*. These oxidative-stress-induced changes were attenuated by administration of TEMPOL.[161] The behavioral response to cocaine was measured by locomotor sensitization where cocaine produced a persistent increase in locomotor activity.[161] This was attenuated by the administration of TEMPOL prior to the administration of cocaine while TEMPOL alone had no effect on locomotor activity.[161] These results suggest that the attenuation of cocaine-induced oxidative stress reduces the induction and expression of behavioral sensitization to cocaine providing a link between oxidative stress and behavioral measures.[161]

The association between cocaine-induced oxidative stress in the brain and the oxidative sensor nuclear factor kappa-light-chain-enhancer of activated B cells (NF-κB), which is involved in learning and memory, was investigated in an experimental model of cocaine administration in rats.[162] Changes in synaptic plasticity are required for the memory and learning processes that underlie drug addiction and NF-κB has been implicated in experience-dependent synaptic plasticity as well as the response to oxidative stress.[162] Administration of cocaine to rats with subsequent spatial memory training in the Morris water maze found reduced GSH and GSH-peroxidase levels in the hippocampus although there were no changes in caspase-3 or NF-κB activity.[162] NF-κB activity was decreased by cocaine in the frontal cortex, and was negatively correlated with memory retrieval experiments whereas training in the Morris water maze occurred prior to the administration of cocaine.[162] When combined these results suggest an additional role for NF-κB in cocaine-related learning and memory, separate to that involved in the response to toxic levels of cocaine as suggested by the lack of increase in caspase-3 and the involvement of the prefrontal cortex, an area implicated in addiction-related learning and memory.[162]

A study investigating the effects of cocaine self-administration on markers of oxidative stress differentiated between the motivational aspects of cocaine and the purely pharmacologic responses to cocaine by using a yoked triad scheme whereby contingent responding to cocaine by one rat resulted in a similar administration of cocaine to another rat and saline to a third rat.[163] In this paradigm, active cocaine administration produced an increase in superoxide dismutase activity in the hippocampus, frontal cortex, and dorsal striatum and a corresponding decrease in malondialdehyde formation with no change in superoxide dismutase activity in those rats that passively received cocaine at exactly the same dose.[163] Withdrawal from cocaine produced elevations in malondialdehyde levels in the hippocampus and prefrontal cortex in rats both actively and passively receiving cocaine suggesting that oxidative stress may be involved in the process of craving.[163] Therefore, oxidative stress occurs during both drug taking and withdrawal with oxidative-stress-induced changes occurring in brain regions associated with motivational use of cocaine and cocaine-induced learning and memory.[163] These results may be partially explained by the motivation aspect of active cocaine administration, which results in greater dopamine release than those rats that receive cocaine passively.[163] This study concludes that oxidative stress is involved in the motivational aspects of voluntary cocaine use possibly via an effect on learning and memory processes.[163]

8.3 Protein S-Glutathionylation in Response to Cocaine

The potential role of GSTP in cocaine-mediated neuroplasticity and related addictive behavior has been previously suggested and was investigated in a study examining protein S-glutathionylation in response to chronic and acute cocaine.[39,164] Chronic cocaine treatment as well as an additional acute challenge increased S-glutathionylation of proteins as measured by a decrease in free sulfhydryl groups, an indirect measure of the amount of GSH conjugated to cysteine residues.[164] In contrast, GSTP levels were reduced by daily cocaine administration but were increased after acute cocaine administration in previously chronic cocaine-treated animals.[164] GSTP knockout mice showed enhanced behavioral sensitization to cocaine and greater preference for cocaine measured in the conditioned place-preference paradigm, an effect replicated by the pharmacologic inhibition of GSTP by ketoprofen injection. Therefore it was concluded that reductions in GSTP may lead to enhanced synaptic plasticity in the response to cocaine.[164]

9. S-GLUTATHIONYLATION IN HEROIN ADDICTION

Heroin is one of the most dangerous of the abused drugs producing a wide range of deleterious effects on the brain including gray matter loss, neuronal apoptosis, mitochondrial dysfunction, synaptic defects, and reduced adult neurogenesis.[165] Although heroin is not widely abused (approximately 467,000 Americans were classified as heroin dependent in 2012), it has a high abuse potential.[118] Heroin, a 3,6-diacetyl ester of morphine, is an opioid analgesic, which is metabolized to morphine in the body.[165] Heroin activates opioid receptors in the ventral tegmental nucleus, nucleus, and amygdala through direct or indirect actions via interneurons.[45] Although the μ-opiate receptor is the primary target of heroin in the brain, heroin also affects dopaminergic, GABAergic, glutamatergic, and serotonergic signaling.[165] Heroin has been shown to upregulate neuronal nitric oxide synthase, which increases brain nitric oxide levels, in the right temporal cortex and left hypothalamic paraventricular nucleus, an area of the classical reward system, of individuals who died from heroin abuse.[165]

9.1 Heroin Metabolism Increases ROS Production

Heroin metabolism occurs via deactylation in the active metabolites morphine and 6-monoacetylmorphine and it is believed that the effects of heroin on opioid receptors occurs primarily through its metabolites via 6-monoacetylmorphine.[166] Long-term heroin abuse induces oxidative stress and reduces total antioxidant capacity, possibly due to elevated dopamine and autoxidation of catecholamines.[167] In addition, incubation of primary cell culture rat cortical neurons in a heroin-containing medium indicated that heroin decreases the mitochondrial membrane potential, which may increase ROS generation.[166] Evidence for increased heroin-induced ROS production has also been found in human studies with heroin users having a greater pro-oxidative than antioxidative balance measured in blood.[168,169] In addition, malondialdehyde levels are increased and total antioxidant capacity is decreased in the cortex of heroin users in comparison to controls with a proportional relationship between malondialdehyde and blood opioid concentration.[170]

9.2 Increased ROS Production May Contribute to the Behavioral Effects of Heroin

Several studies have found evidence that heroin-induced ROS production may be involved in the behavioral effects of the drug. A study using a mouse model of heroin examined the effect of antioxidant treatment combined with heroin on markers of oxidative stress and withdrawal behavior.[171] Mice were administered heroin daily with dosages increasing over the course of 40 days, and were divided into three groups: those that received only heroin, those that received heroin coadministered with antioxidants, or those that received antioxidant supplementation for 10 days following the end of heroin administration.[171] Heroin administration increased oxidative stress as measured by increased percentage of cells with DNA damage and increased carbonyl and malondialdehyde content in brain cells.[171] A corresponding decrease in superoxide dismutase, catalase, and GSH peroxidase activity, and total antioxidant capacity with increasing heroin dose was also found.[171] Administration of antioxidants either concurrently or after heroin administration, reduced markers of oxidative damage.[171] Naloxone-precipitated withdrawal behavior in these mice were measured and suggested that the administration of heroin had produced a dependent phenotype.[171] Interestingly, antioxidant administration resulted in a significant reduction in withdrawal behaviors suggesting that oxidative stress mediated the withdrawal behavioral phenotype.[171] In a separate study, naloxone-precipitated withdrawal increased malondialdehyde and decreased GSH levels measured in whole blood suggesting that oxidative stress is also involved in withdrawal.[172] These changes in malondialdehyde and GSH were reduced by administration of the antioxidants vitamin E or melatonin.[172]

Evidence of a link between oxidative stress and the reinforcing properties of morphine, the primary metabolite of heroin, has also been found. Administration of TEMPOL and the subsequent reduction in oxidative stress in the NAc shell prevented the expression of morphine-conditioned place preference.[173] A study examining the relationship between morphine-induced oxidative stress and withdrawal behavior in rats found that morphine increased malondialdehyde levels, which was further increased by naloxone-induced withdrawal.[174] Treatment with alpha-lipoic acid lowered malondialdehyde levels and symptoms of withdrawal further suggesting that oxidative stress is involved in the morphine addiction and withdrawal phenotype.[174] Administration of the antioxidant hydrogen sulfide combined with naloxone-precipitated heroin withdrawal has also been shown to significantly decrease behavioral withdrawal symptoms as well as malondialdehyde levels while increasing GSH content and superoxide dismutase, GSH peroxidase, and catalase activity.[175]

9.3 Protein S-Glutathionylation in Response to Heroin

A study examining the effect of antioxidants on heroin-induced oxidative stress and withdrawal behavior may provide evidence for the role of protein S-glutathionylation in mediating the response to heroin. In the study, mice administered heroin at increasing dosages twice daily over the course of 15 days showed reduced total antioxidant capacity and increased ROS production measured in blood and increased thiobarbituric acid and carbonyl content measured in brain.[176] In addition, heroin-treated mice showed significantly more naloxone-precipitated heroin withdrawal behavioral signs than mice administered saline.[176] Treatment with the antioxidant, rosmarinic acid, reduced the levels of ROS in blood and thiobarbituric and carbonyl levels in brain while increasing total antioxidant capacity.[176] Moreover, antioxidant treatment, including treatment with rosmarinic acid, reduced exploring, shaking, and jumping behavior, markers of heroin withdrawal, suggesting that oxidative stress mediates heroin withdrawal-induced behavior.[176] Although this study did not directly measure protein S-glutathionylation, rosmarinic acid has been shown to increase GSTP.[177,178] Therefore, it is possible that S-glutathionylation may be involved in mediating the effects of rosmarinic acid on heroin withdrawal.

10. FUTURE RESEARCH

Further research into drug-induced S-glutathionylation may reveal targets for future treatment. Indeed, the antioxidant and *N*-acetyl prodrug of cysteine, has shown great promise as a treatment in animal models of alcohol and cocaine abuse, reducing both drug-seeking behavior and symptoms of withdrawal.[179–181] NAC is taken up into glia by the cystine/glutamate exchanger (X_c^-), which is responsible for providing the cysteine necessary for the synthesis of GSH and therefore the supply of GSH to neurons by astrocytes.[182–184] Interestingly, X_c^- is itself an unconfirmed target of S-glutathionylation, due to its low pK_a cysteine residues (cysteines 197 and 414). Oxidative-stress-induced reduction in intracellular GSH has been found to increase X_c^- activity, possibly as a mechanism of normalizing redox homeostasis.[185] In keeping with this, NAC was found to reverse GSH depletion in rat astroglial cells following cocaine-induced oxidative stress.[186,187] Regrettably, clinical trials of NAC efficacy as a treatment for substance use disorders have received mixed results.[188] However, these results do suggest that drugs affecting glutathionylation are valid areas of research in addiction pharmacotherapy. In line with this, development of drugs that increase GSTP activity or reversibly inhibit glutaredoxin might represent a novel and highly effective means of modulating drug-induced S-glutathionylation and addictive behavior. However, as the effects of

S-glutathionylation on protein function are varied, modification of S-glutathionylation status of specific proteins will be a further challenge.

11. CONCLUSIONS

S-glutathionylation is an understudied post-translational protein modification that allows for sensitive and reversible signaling in response to altered redox status. Increased generation of ROS with subsequent oxidative stress is a mechanism shared by all drugs of abuse despite their acting on diverse molecular targets. As a result, drug-induced protein S-glutathionylation may translate transient increases in ROS into enduring modifications in biochemistry, neural responses, and ultimately behavior. Indeed, evidence exists for S-glutathionylation of key proteins involved in drug-induced neurotransmission such as PKA, PKC, and Cdk5, which regulate AMPAR and NMDAR insertion into the postsynaptic membrane to influence drug-related learning and memory. Furthermore, S-glutathionylation also affects actin cycling to influence dendritic spine morphology and synaptic strength. As this field of study progresses, the number of proteins confirmed as targets of S-glutathionylation will likely grow and it is possible that many of these may be involved in addictive processes. For this reason, further research in targeting S-glutathionylation as a potential therapy for drug addiction is warranted.

Although a wide body of evidence suggests an important role for S-glutathionylation in drug addiction, this area of research is still relatively unexplored. The role of S-glutathionylation in drug addiction is complicated by its modification of multiple proteins, the function of each of which may be increased, decreased, or unaltered by this post-translational modification. Furthermore, research to date has not examined differences in drug-induced oxidative stress between different cellular compartments, an important line of research given that redox status compartmentalization allows for subcellular specificity in redox signaling.[7] However, previous research has linked neurodegeneration in Parkinson's disease with upregulated GSTP in neuronal synaptosomal fractions.[189] This suggests that GSTP-mediated S-glutathionylation can be modified in the postsynapse, an area where drugs of abuse exert their initial effects. Similarly, glutaredoxin has been found in neuronal cytoplasmic fractions.[190] However, the effect of drug-induced oxidative stress on GSTP and glutaredoxin activity within specific cellular compartments is unknown.

ACKNOWLEDGMENT

This work was supported by NIH grant GM103542.

REFERENCES

1. Giustarini D, Rossi R, Milzani A, Colombo R, Dalle-Donne I. S-glutathionylation: from redox regulation of protein functions to human diseases. *J Cell Mol Med.* 2004;8 (2):201–212.
2. Biswas S, Chida AS, Rahman I. Redox modifications of protein-thiols: emerging roles in cell signaling. *Biochem Pharmacol.* 2006;71(5):551–564.
3. Tew KD, Manevich Y, Grek C, Xiong Y, Uys J, Townsend DM. The role of glutathione S-transferase P in signaling pathways and S-glutathionylation in cancer. *Free Radic Biol Med.* 2011;51(2):299–313.
4. Hsieh HL, Yang CM. Role of redox signaling in neuroinflammation and neurodegenerative diseases. *BioMed Res Int.* 2013;2013:484613.
5. Ghezzi P. Regulation of protein function by glutathionylation. *Free Radic Res.* 2005;39 (6):573–580.
6. Davis W, Ronai Z, Tew KD. Cellular thiols and reactive oxygen species in drug-induced apoptosis. *J Pharmacol Exp Ther.* 2001;296:1–6.
7. Hansen JM, Go YM, Jones DP. Nuclear and mitochondrial compartmentation of oxidative stress and redox signaling. *Annu Rev Pharmacol Toxicol.* 2006;46:215–234.
8. Sayre LM, Perry G, Smith MA. Oxidative stress and neurotoxicity. *Chem Res Toxicol.* 2008;21:172–188.
9. Schulz JB, Lindenau J, Seyfried J, Dichgans J. Glutathione, oxidative stress and neurodegeneration. *Eur J Biochem.* 2000;267:4904–4911.
10. Dringen R. Metabolism and functions of glutathione in brain. *Prog Neurobiol.* 2000;62: 649–671.
11. Aoyama K, Nakaki T. Impaired glutathione synthesis in neurodegeneration. *Int J Mol Sci.* 2013;14(10):21021–21044.
12. Dringen R, Gutterer JM, Hirrlinger J. Glutathione metabolism in brain: metabolic interactions between astrocytes and neurons in the defense against reactive oxygen species. *Eur J Biochem.* 2000;267:4912–4916.
13. Dringen R, Hirrlinger J. Glutathione pathways in the brain. *Biol Chem.* 2003;384: 505–516.
14. Meyer AJ. The integration of glutathione homeostasis and redox signaling. *J Plant Physiol.* 2008;165(13):1390–1403.
15. Flohe L. The fairytale of the GSSG/GSH redox potential. *Biochim Biophys Acta.* 2013;1830(5):3139–3142.
16. Popov D. Protein S-glutathionylation: from current basics to targeted modifications. *Arch Physiol Biochem.* 2014;120(4):123–130.
17. Hurd TR, Costa NJ, Dahm CC, et al. Glutathionylation of mitochondrial proteins. *Antioxid Redox Signal.* 2005;7(7&8):999–1010.
18. Ghezzi P. Protein glutathionylation in health and disease. *Biochim Biophys Acta.* 2013;1830(5):3165–3172.
19. Aesif SW, Janssen-Heininger YMW, Reynaert NL. Protocols for the detection of S-glutathionylated and s-nitrosylated proteins *in situ*. *Methods Enzymol.* 2010;474: 289–296.
20. Klatt P, Lamas S. Regulation of protein function by S-glutathiolation in response to oxidative and nitrosative stress. *Eur J Biochem.* 2000;267:4928–4944.
21. Dalle-Donne I, Rossi R, Giustarini D, Colombo R, Milzani A. S-glutathionylation in protein redox regulation. *Free Radic Biol Med.* 2007;43(6):883–898.

22. Grek CL, Zhang J, Manevich Y, Townsend DM, Tew KD. Causes and consequences of cysteine S-glutathionylation. *J Biol Chem.* 2013;288(37):26497–26504.
23. Adachi T, Schöneich C, Cohen RA. S-glutathiolation in redox-sensitive signaling. *Drug Discov Today Dis Mech.* 2005;2(1):39–46.
24. Groitl B, Jakob U. Thiol-based redox switches. *Biochim Biophys Acta.* 2014;1844(8): 1335–1343.
25. Hill BG, Bhatnagar A. Protein S-glutathiolation: redox-sensitive regulation of protein function. *J Mol Cell Cardiol.* 2012;52(3):559–567.
26. Ghezzi P, Di Simplicio P. Glutathionylation pathways in drug response. *Curr Opin Pharmacol.* 2007;7(4):398–403.
27. Fernandes AP, Holmgren A. Glutaredoxins: glutathione-dependent redox enzymes with functions far beyond a simple thioredoxin backup system. *Antioxid Redox Signal.* 2004;6(1):63–74.
28. Townsend DM, Manevich Y, He L, Hutchens S, Pazoles CJ, Tew KD. Novel role for glutathione S-transferase pi. Regulator of protein S-glutathionylation following oxidative and nitrosative stress. *J Biol Chem.* 2009;284(1):436–445.
29. Findlay VJ, Townsend DM, Morris TE, Fraser JP, He L, Tew KD. A novel role for human sulfiredoxin in the reversal of glutathionylation. *Cancer Res.* 2006;66 (13):6800–6806.
30. Gallogly MM, Mieyal JJ. Mechanisms of reversible protein glutathionylation in redox signaling and oxidative stress. *Curr Opin Pharmacol.* 2007;7(4):381–391.
31. Ghezzi P, Bonetto V, Fratelli M. Thiol–disulfide balance: from the concept of oxidative stress to that of redox regulation. *Antioxid Redox Signal.* 2005;7(7&8):964–972.
32. Mailloux RJ, Willmore WG. S-glutathionylation reactions in mitochondrial function and disease. *Front Cell Dev Biol.* 2014;2:68.
33. Mailloux RJ, McBride SL, Harper ME. Unearthing the secrets of mitochondrial ROS and glutathione in bioenergetics. *Trends Biochem Sci.* 2013;38(12):592–602.
34. O'Brian CA, Chu F. Post-translational disulfide modifications in cell signaling—role of inter-protein, intra-protein, S-glutathionyl, and S-cysteaminyl disulfide modifications in signal transmission. *Free Radic Res.* 2005;39(5):471–480.
35. Ghezzi P. Oxidoreduction of protein thiols in redox regulation. *Biochem Soc Trans.* 2005;33(6):1378–1381.
36. Go YM, Jones DP. Thiol/disulfide redox states in signaling and sensing. *Crit Rev Biochem Mol Biol.* 2013;48(2):173–181.
37. Jones DP, Go YM. Mapping the cysteine proteome: analysis of redox-sensing thiols. *Curr Opin Chem Biol.* 2011;15(1):103–112.
38. Jones DP, Go Y-M, Anderson CL, Ziegler TR, Kinkade Jr JM, Kirlin WG. Cysteine/cystine couple is a newly recognized node in the circuitry for biologic redox signaling and control. *FASEB J.* 2004;18(11):1246–1258.
39. Tew KD, Townsend DM. Glutathione-s-transferases as determinants of cell survival and death. *Antioxid Redox Signal.* 2012;17(12):1728–1737.
40. Shelton MD, Chock PB, Mieyal JJ. Glutaredoxin: role in reversible protein S-glutathionylation and regulation of redox signal transduction and protein translocation. *Antioxid Redox Signal.* 2005;7(3&4):348–366.
41. Lind C, Gerdes R, Hamnell Y, et al. Identification of S-glutathionylated cellular proteins during oxidative stress and constitutive metabolism by affinity purification and proteomic analysis. *Arch Biochem Biophys.* 2002;406(2):229–240.
42. Tew KD. Redox in redux: emergent roles for glutathione S-transferase P (GSTP) in regulation of cell signaling and S-glutathionylation. *Biochem Pharmacol.* 2007;73(9): 1257–1269.
43. Filomeni G, Rotilio G, Ciriolo MR. Cell signalling and the glutathione redox system. *Biochem Pharmacol.* 2002;64:1057–1064.

44. Mieyal JJ, Chock PB. Post-translational modification of cysteine in redox signaling and oxidative stress: focus on s-glutathionylation. *Antioxid Redox Signal.* 2012;16 (6):471–475.
45. Koob GF, Volkow ND. Neurocircuitry of addiction. *Neuropsychopharmacology.* 2010;35 (1):217–238.
46. Koob GF. Theoretical frameworks and mechanistic aspects of alcohol addiction: alcohol addiction as a reward deficit disorder. *Curr Top Behav Neurosci.* 2013;13:3–30.
47. Oliveira MT, Rego AC, Macedo TRA, Oliveira CR. Drugs of abuse induce apoptotic features in PC12 cells. *Ann NY Acad Sci.* 2003;1010:667–670.
48. Wise R. Dopamine and reward: the anhedonia hypothesis 30 years on. *Neurotox Res.* 2008;14(2–3):169–183.
49. Hyman S, Malenka R, Nestler E. Neural mechanisms of addiction: the role of reward-related learning and memory. *Annu Rev Neurosci.* 2006;29:565–598.
50. Hirrlinger J, Schulz JB, Dringen R. Effects of dopamine on the glutathione metabolism of cultured astroglial cells: implications for Parkinson's disease. *J Neurochem.* 2002;82:458–467.
51. Rabinovic AD, Hastings T. Role of endogenous glutathione in the oxidation of dopamine. *J Neurochem.* 1998;71:2071–2078.
52. Moszczynska A, Turenne S, Kish SJ. Rat striatal levels of the antioxidant glutathione are decreased following binge administration of methamphetamine. *Neurosci Lett.* 1998;255:49–52.
53. Hastings TG, Lewis DA, Zigmond MJ. Role of oxidation in the neurotoxic effects of intrastriatal dopamine injections. *Proc Natl Acad Sci USA.* 1996;93:1956–1961.
54. Stokes AH, Lewis DY, Lash LH, et al. Dopamine toxicity in neuroblastoma cells: role of glutathione depletion by L-BSO and apoptosis. *Brain Res.* 2000;858:1–8.
55. Kishida KT, Klann E. Sources and targets of reactive oxygen species in synaptic plasticity and memory. *Antioxid Redox Signal.* 2007;9(2):233–244.
56. Guindalini C, O'Gara C, Laranjeirad R, et al. A GSTP1 functional variant associated with cocaine dependence in a Brazilian population. *Pharmacogenet Genomics.* 2005;15:891–893.
57. Hashimoto T, Hashimoto K, Matsuzawa D, et al. A functional glutathione S-transferase P1 gene polymorphism is associated with methamphetamine-induced psychosis in Japanese population. *Am J Med Genet B Neuropsychiatr Genet.* 2005;135B(1):5–9.
58. Rabinovic AD, Lewis DA, Hastings TG. Role of oxidative changes in the degeneration of dopamine terminal after injection of neurotoxic levels of dopamine. *Neuroscience.* 2000;101(1):67–76.
59. Bosch M, Hayashi Y. Structural plasticity of dendritic spines. *Curr Opin Neurobiol.* 2012;22(3):383–388.
60. Meredith G. The synaptic framework for chemical signaling in the nucleus accumbens. *Ann NY Acad Sci.* 1999;29(877):140–156.
61. Wang J, Boja E, Tan W, et al. Reversible glutathionylation regulates actin polymerization in A431 cells. *J Biol Chem.* 2001;276(51):47763–47766.
62. Tanaka F, Shiratori Y, Yokosuka O, Imazeki F, Omata YT. Polymorphism of alcohol-metabolizing genes affects drinking behavior and alcoholic liver disease in Japanese men. *Alcoholism Clin Exp Res.* 1997;21(4):596–601.
63. Klatt P, Molina EP, Perez-Sala D, Lamas S. Novel application of S-nitrosoglutathione–Sepharose to identify proteins that are potential targets for S-nitrosoglutathione-induced mixed-disulphide formation. *Biochem J.* 2000;349:567–578.
64. Pastore A, Piemonte F. S-glutathionylation signaling in cell biology: progress and prospects. *Eur J Pharm Sci.* 2012;46(5):279–292.
65. Mlewski EC, Krapacher FA, Ferreras S, Paglini G. Transient enhanced expression of Cdk5 activator p25 after acute and chronic d-amphetamine administration. *Ann NY Acad Sci.* 2008;1139:89–102.

66. Bibb J, Chen J, Taylor J, et al. Effects of chronic exposure to cocaine are regulated by the neuronal protein Cdk5. *Nature.* 2001;410(6826):376–380.
67. Sun KH, Chang KH, Clawson S, et al. Glutathione-S-transferase P1 is a critical regulator of Cdk5 kinase activity. *J Neurochem.* 2011;118(5):902–914.
68. Calabrese B, Saffin JM, Halpain S. Activity-dependent dendritic spine shrinkage and growth involve downregulation of cofilin via distinct mechanisms. *PloS One.* 2014;9(4): e94787.
69. Spiga S, Mulas G, Piras F, Diana M. The "addicted" spine. *Front Neuroanat.* 2014;8:110.
70. Fratelli M, Demol H, Puype M, et al. Identification by redox proteomics of glutathionylated proteins in oxidatively stressed human T lymphocytes. *Proc Natl Acad Sci USA.* 2002;99(6):3505–3510.
71. Chen CK, Lin SK, Chiang SC, Su LW, Wang LJ. Polymorphisms of COMT Val158Met and DAT1 3′-UTR VNTR in illicit drug use and drug-related psychiatric disorders. *Subst Use Misuse.* 2014;49(11):1385–1391.
72. Vandenbergh DJ, Rodriguez LA, Miller IT, Uhl GR, Lachman HM. High-activity catechol-O-methyltransferase allele is more prevalent in polysubstance abusers. *Am J Med Genet.* 1997;74:439–442.
73. Cotton NJ, Stoddard B, Parson WW. Oxidative inhibition of human soluble catechol-O-methyltransferase. *J Biol Chem.* 2004;279(22):23710–23718.
74. Tew KD, Townsend DM. Regulatory functions of glutathione S-transferase P1-1 unrelated to detoxification. *Drug Metab Rev.* 2011;43(2):179–193.
75. Koskinen M, Bertling E, Hotulainen R, Tanhuanpaa K, Hotulainen P. Myosin IIb controls actin dynamics underlying the dendritic spine maturation. *Mol Cell Neurosci.* 2014;61:56–64.
76. Moen RJ, Cornea S, Oseid DE, Binder BP, Klein JC, Thomas DD. Redox-sensitive residue in the actin-binding interface of myosin. *Biochem Biophys Res Commun.* 2014;453 (3):345–349.
77. Wang H, Zhuo M. Group I metabotropic glutamate receptor-mediated gene transcription and implications for synaptic plasticity and diseases. *Front Pharmacol.* 2012;3:189.
78. Pineda-Molina E, Klatt P, Vázquez J, Marina A, Garcia de Lacoba M, Perez-Sala D, Lamas S. Glutathionylation of the p50 subunit of NF-kappaB: a mechanism for redox-induced inhibition of DNA binding. *Biochemistry.* 2001;40:14134–14142.
79. Chen J, Chen P, Chiang Y. Molecular mechanisms of psychostimulant addiction. *Chang Gung Med J.* 2009;32(2):148–154.
80. Cahill E, Salery M, Vanhoutte P, Caboche J. Convergence of dopamine and glutamate signaling onto striatal ERK activation in response to drugs of abuse. *Front Pharmacol.* 2014;4:172.
81. Lynch WJ, Taylor JR. Persistent changes in motivation to self-administer cocaine following modulation of cyclic AMP-dependent protein kinase A (PKA) activity in the nucleus accumbens. *Eur J Neurosci.* 2005;22(5):1214–1220.
82. Humphries KM, Deal MS, Taylor SS. Enhanced dephosphorylation of cAMP-dependent protein kinase by oxidation and thiol modification. *J Biol Chem.* 2005;280(4): 2750–2758.
83. Olive MF, Messing RO. Protein kanse C isozymes and addiction. *Mol Neurobiol.* 2004;29(2): 139–154.
84. Howell KK, Monk BR, Carmack SA, Mrowczynski OD, Clark RE, Anagnostaras SG. Inhibition of PKC disrupts addiction-related memory. *Front Behav Neurosci.* 2014;8:70.
85. Anselmo AN, Cobb MH. Protein kinase function and glutathionylation. *Biochem J.* 2004;381(Pt 3):e1–e2.
86. Penzes P, Rafalovich I. Regulation of the actin cytoskeleton in dendritic spines. *Adv Exp Med Biol.* 2012;970:81–95.

87. Daubner SC, Le T, Wang S. Tyrosine hydroxylase and regulation of dopamine synthesis. *Arch Biochem Biophys.* 2011;508(1):1–12.
88. Diana M. The dopamine hypothesis of drug addiction and its potential therapeutic value. *Front Psychiatry.* 2011;2:64.
89. Borges CR, Geddes T, Watson JT, Kuhn DM. Dopamine biosynthesis is regulated by S-glutathionylation. Potential mechanism of tyrosine hydroxylast inhibition during oxidative stress. *J Biol Chem.* 2002;277(50):48295–48302.
90. Self DW. Regulation of drug-taking and -seeking behaviors by neuroadaptations in the mesolimbic dopamine system. *Neuropharmacology.* 2004;47(suppl 1):242–255.
91. White F, Cooper D. The vicious cyclin of addiction. *Nat Med.* 2001;7(4):416–417.
92. Chergui K, Svenningsson P, Greengard P. Cyclin-dependent kinase 5 regulates dopaminergic and glutamatergic transmission in the striatum. *Proc Natl Acad Sci USA.* 2004;101(7):2191–2196.
93. Nakano T, Doi T, Yoshimoto J, Doya K. A kinetic model of dopamine- and calcium-dependent striatal synaptic plasticity. *PLoS Comput Biol.* 2010;6(2):e1000670.
94. Sanderson JL, Dell'Acqua ML. AKAP signaling complexes in regulation of excitatory synaptic plasticity. *Neuroscientist.* 2011;17(3):321–336.
95. Wang JQ, Guo ML, Jin DZ, Xue B, Fibuch EE, Mao LM. Roles of subunit phosphorylation in regulating glutamate receptor function. *Eur J Pharmacol.* 2014;728: 183–187.
96. Morabito MA, Sheng M, Tsai LH. Cyclin-dependent kinase 5 phosphorylates the N-terminal domain of the postsynaptic density protein PSD-95 in neurons. *J Neurosci.* 2004;24(4):865–876.
97. Boudreau AC, Ferrario CR, Glucksman MJ, Wolf ME. Signaling pathway adaptations and novel protein kinase A substrates related to behavioral sensitization to cocaine. *J Neurochem.* 2009;110(1):363–377.
98. Yao L, McFarland K, Fan P, Jiang Z, Inoue Y, Diamond I. Activator of G protein signaling 3 regulates opiate activation of protein kinase A signaling and relapse of heroin-seeking behavior. *Proc Natl Acad Sci USA.* 2005;102(24):8746–8751.
99. Lee AM, Messing RO. Protein kinases and addiction. *Ann NY Acad Sci.* 2008;1141: 22–57.
100. Zhang Y, Wang N, Su P, Lu J, Wang Y. Disruption of dopamine D1 receptor phosphorylation at serine 421 attenuates cocaine-induced behaviors in mice. *Neurosci Bull.* 2014;30(6):1025–1035.
101. Liu J-G, Anand KJS. Protein kinases modulate the cellular adaptations associated with opioid tolerance and dependence. *Brain Res Rev.* 2001;38:1–19.
102. Self DW, Genova LM, Spencer JJ, Hope BT, Nestler aEJ. Involvement of cAMP-dependent protein kinase in the nucleus accumbens in cocaine self-administration and relapse of cocaine-seeking behavior. *J Neurosci.* 1998;18(5):1848–1859.
103. Self DW, Genova LM, Hope BT, Barnhart WJ, Spencer JJ, Nestler EJ. Involvement of cAMP-dependent protein kinase in the nucleus accumbens in cocaine self-administration and relapse of cocaine-seeking behavior. *J Neurosci.* 1998;18(5):1848–1859.
104. Humphries KM, Juliano C, Taylor SS. Regulation of cAMP-dependent protein kinase activity by glutathionylation. *J Biol Chem.* 2002;277(45):43505–43511.
105. Escribá P, García-Sevilla J. Parallel modulation of receptor for activated C kinase 1 and protein kinase C-alpha and beta isoforms in brains of morphine-treated rats. *Br J Pharmacol.* 1999;127(2):343–348.
106. García-Sevilla J, Ventayol P, Busquets X, Harpe RL, Walzer C, Guimón J. Regulation of immunolabelled mu-opioid receptors and protein kinase C-alpha and zeta isoforms in the frontal cortex of human opiate addicts. *Neurosci Lett.* 1997;226(1):29–32.
107. Jung ME, Metzger DB. Alcohol withdrawal and brain injuries: beyond classical mechanisms. *Molecules.* 2010;15(7):4984–5011.

108. Ortinski PI, Briand LA, Pierce RC, Schmidt HD. Cocaine-seeking is associated with PKC-dependent reduction of excitatory signaling in accumbens shell D2 dopamine receptor-expressing neurons. *Neuropharmacology.* 2015;92:80–89.
109. Schmidt HD, Kimmey BA, Arreola AC, Pierce RC. Group I metabotropic glutamate receptor-mediated activation of PKC gamma in the nucleus accumbens core promotes the reinstatement of cocaine seeking. *Addict Biol.* 2015;20(2):285–296.
110. Taylor JR, Lynch WJ, Sanchez H, Olausson P, Nestler EJ, Bibb JA. Inhibition of Cdk5 in the nucleus accumbens enhances the locomotor-activating and incentive-motivational effects of cocaine. *Proc Natl Acad Sci USA.* 2007;104(10):4147–4152.
111. Robinson TE, Kolb B. Structural plasticity associated with exposure to drugs of abuse. *Neuropharmacology.* 2004;47(suppl 1):33–46.
112. Collo G, Cavalleri L, Spano P. Structural plasticity in mesencephalic dopaminergic neurons produced by drugs of abuse: critical role of BDNF and dopamine. *Front Pharmacol.* 2014;5:259.
113. Alvaro-Bartolome M, La Harpe R, Callado LF, Meana JJ, Garcia-Sevilla JA. Molecular adaptations of apoptotic pathways and signaling partners in the cerebral cortex of human cocaine addicts and cocaine-treated rats. *Neuroscience.* 2011;196:1–15.
114. Trivedi MS, Deth R. Redox-based epigenetic status in drug addiction: a potential contributor to gene priming and a mechanistic rationale for metabolic intervention. *Front Neurosci.* 2014;8:444.
115. Villalba RM, Smith Y. Differential striatal spine pathology in Parkinson's disease and cocaine addiction: a key role of dopamine? *Neuroscience.* 2013;251:2–20.
116. Uys JD, McGuier NS, Gass JT, Griffin III WC, Ball LE, Mulholland PJ. Chronic intermittent ethanol exposure and withdrawal leads to adaptations in nucleus accumbens core postsynaptic density proteome and dendritic spines. *Addict Biol.* 2015z. doi: 10.1111/adb.12238.
117. Lai KO, Liang Z, Fei E, Huang H, Ip NY. Cdk5-dependent phosphorylation of p70 ribosomal S6 kinase (S6K) is required for dendritic spine morphogenesis. *J Biol Chem.* 2015;290:14637–14646.
118. Drug Facts: Nationwide Trends. http://www.drugabuse.gov/publications/drugfacts/nationwide-trends; 2014. Accessed 28.05.2015
119. Koob GF. Alcoholism: allostasis and beyond. *Alcohol Clin Exp Res.* 2003;27(2):232–243.
120. Wrona MZ, Waskiewicz J, Han Q-P, Han J, Li H, Dryhurst G. Putative oxidative metabolites of 1-methyl-6-hydroxy-1,2,3,4-tetrahydro-β-carboline of potential relevance to the addictive and neurodegenerative consequences of ethanol abuse. *Alcohol.* 1997;14(3):213–223.
121. Erdozain AM, Callado LF. Neurobiological alterations in alcohol addiction: a review. *Adicciones.* 2014;26(3):360–370.
122. Jaatinen P, Rintala J. Mechanisms of ethanol-induced degeneration in the developing, mature, and aging cerebellum. *Cerebellum.* 2008;7(3):332–347.
123. Elibol-Can B, Jakubowska-Dogru E, Severcan M, Severcan F. The effects of short-term chronic ethanol intoxication and ethanol withdrawal on the molecular composition of the rat hippocampus by FT-IR spectroscopy. *Alcohol Clin Exp Res.* 2011;35(11):2050–2062.
124. Gonzalez A, Pariente JA, Salido GM. Ethanol stimulates ROS generation by mitochondria through Ca^{2+} mobilization and increases GFAP content in rat hippocampal astrocytes. *Brain Res.* 2007;1178:28–37.
125. Bleich S, Spilker K, Kurth C, et al. Oxidative stress and altered methionine metabolism in alcoholism. *Neurosci Lett.* 2000;286:171–174.
126. Thome J, Zhang J, Davids E, et al. Evidence for increased oxidative stress in alcohol-dependent patients provided by quantification of *in vivo* salicylate hydroxylation products. *Alcohol Clin Exp Res.* 1997;21(1):82–85.

127. Thome J, Foley P, Gsell W, et al. Increased concentrations of manganese superoxide dismutase in serum of alcohol-dependent patients. *Alcohol Alcohol.* 1997;32(1):65–69.
128. Ikegami Y, Goodenough S, Inoue Y, Dodd PR, Wilce PA, Matsumoto I. Increased TUNEL positive cells in human alcoholic brains. *Neurosci Lett.* 2003;349(3):201–205.
129. Heaton MB, Paiva M, Madorsky I, Shaw G. Ethanol effects on neonatal rat cortex: comparative analyses of neurotrophic factors, apoptosis-related proteins, and oxidative processes during vulnerable and resistant periods. *Dev Brain Res.* 2003;145(2):249–262.
130. Han Q-P, Dryhurst G. Influence of glutathione on the oxidation of 1-methyl-6-hydroxy-1,2,3,4-teterahydro-β-carboline: chemistry of potential relevance to the addictive and neurodegenerative consequences of ethanol abuse. *J Med Chem.* 1996;39: 1491–1508.
131. Lee DW, Nam YK, Kim TK, et al. Dose-dependent influence of short-term intermittent ethanol intoxication on cerebral neurochemical changes in rats detected by *ex vivo* proton nuclear magnetic resonance spectroscopy. *Neuroscience.* 2014;262: 107–117.
132. Yeligar SM, Harris FL, Hart CM, Brown LA. Glutathione attenuates ethanol-induced alveolar macrophage oxidative stress and dysfunction by downregulating NADPH oxidases. *Am J Physiol Lung Cell Mol Physiol.* 2014;306(5):L429–L441.
133. Götz ME, Janetzky B, Pohli S, et al. Chronic alcohol consumption and cerebral indices of oxidative stress: is there a link? *Alcohol Clin Exp Res.* 2001;25(5):717–725.
134. Conigrave K, Davies P, Haber P, Whitfield J. Traditional markers of excessive alcohol use. *Addiction.* 2003;98(suppl 2):31–43.
135. Balldin J, Berggren U, Berglund K, Blennow K, Fahlke C, Zetterberg H. Gamma-glutamyltransferase in alcohol use disorders: modification of decision limits in relation to treatment goals? *Scand J Clin Lab Invest.* 2010;70(2):71–74.
136. Kovacic P, Cooksy AL. Role of diacetyl metabolite in alcohol toxicity and addiction via electron transfer and oxidative stress. *Arch Toxicol.* 2005;79(3):123–128.
137. Brocardo PS, Gil-Mohapel J, Christie BR. The role of oxidative stress in fetal alcohol spectrum disorders. *Brain Res Rev.* 2011;67(1–2):209–225.
138. Huang M, Liu W, Li Q, Wu CF. Endogenous released ascorbic acid suppresses ethanol-induced hydroxyl radical production in rat striatum. *Brain Res.* 2002;944: 90–96.
139. Ledesma JC, Font L, Aragon CM. The H_2O_2 scavenger ebselen decreases ethanol-induced locomotor stimulation in mice. *Drug Alcohol Depend.* 2012;124(1–2):42–49.
140. Manrique HM, Miquel M, Aragon CM. Acute administration of 3-nitropropionic acid, a reactive oxygen species generator, boosts ethanol-induced locomotor stimulation. New support for the role of brain catalase in the behavioural effects of ethanol. *Neuropharmacology.* 2006;51(7–8):1137–1145.
141. Pastor R, Sanchis-Segura C, Aragon CM. Ethanol-stimulated behaviour in mice is modulated by brain catalase activity and H_2O_2 rate of production. *Psychopharmacology.* 2002;165(1):51–59.
142. Rouach H, Houze P, Gentil M, Orfanelli M-T, Nordmann R. Changes in some pro- and antioxidants in rat cerebellum after chronic alcohol intake. *Biochem Pharmacol.* 1997;53:539–545.
143. Pirlich M, Kiok K, Sandig G, Lochs H, Gruneb T. Alpha-lipoic acid prevents ethanol-induced protein oxidation in mouse hippocampal HT22 cells. *Neurosci Lett.* 2002;328:93–96.
144. Narasimhan M, Rathinam M, Patel D, Henderson G, Mahimainathan L. Astrocytes prevent ethanol induced apoptosis of Nrf2 depleted neurons by maintaining GSH homeostasis. *Open J Apoptosis.* 2012;1(2)z. doi: 10.4236/ojapo.2012.12002.
145. Kashem MA, Harper C, Matsumoto I. Differential protein expression in the corpus callosum (genu) of human alcoholics. *Neurochem Int.* 2008;53(1–2):1–11.

146. Kashem MA, Etages HD, Kopitar-Jerala N, McGregor IS, Matsumoto I. Differential protein expression in the corpus callosum (body) of human alcoholic brain. *J Neurochem.* 2009;110(2):486–495.
147. Henderson GI, Chen J, Schenker S. Ethanol, oxidative stress, reactive aldehydes and the fetus. *Front Biosci.* 1999;5:541–550.
148. Li Y, Yan GY, Zhou JQ, et al. ^{1}H NMR-based metabonomics in brain nucleus accumbens and striatum following repeated cocaine treatment in rats. *Neuroscience.* 2012;218:196–205.
149. Boess F, Ndikum-Moffor FM, Boelsterli UA, Roberts SM. Effects of cocaine and its oxidative metabolites on mitochondrial respiration and generation of reactive oxygen species. *Biochem Pharmacol.* 2000;60:615–623.
150. Ponsoda X, Bort R, Jover R, Gomez-Lechon MJ, Castell JV. Increased toxicity of cocaine on human hepatocytes induced by ethanol: role of glutathione. *Biochem Pharmacol.* 1999;58:1579–1585.
151. Labib R, Turkall R, Abdel-Rahman MS. Inhibition of cocaine oxidative metabolism attenuates endotoxin potentiation of cocaine mediated hepatotoxicity. *Toxicology.* 2002;179:9–19.
152. Cunha-Oliveira T, Rego AC, Cardoso SM, et al. Mitochondrial dysfunction and caspase activation in rat cortical neurons treated with cocaine or amphetamine. *Brain Res.* 2006;1089(1):44–54.
153. Cunha-Oliveira T, Silva L, Silva AM, Moreno AJ, Oliveira CR, Santos MS. Mitochondrial complex I dysfunction induced by cocaine and cocaine plus morphine in brain and liver mitochondria. *Toxicol Lett.* 2013;219(3):298–306.
154. Dietrich JB, Mangeol A, Revel MO, Burgun C, Aunis D, Zwiller J. Acute or repeated cocaine administration generates reactive oxygen species and induces antioxidant enzyme activity in dopaminergic rat brain structures. *Neuropharmacology.* 2005;48(7):965–974.
155. Dietrich JB, Poirier R, Aunis D, Zwiller J. Cocaine downregulates the expression of the mitochondrial genome in rat brain. *Ann NY Acad Sci.* 2004;1025:345–350.
156. Jang EY, Ryu YH, Lee BH, et al. Involvement of reactive oxygen species in cocaine-taking behaviors in rats. *Addict Biol.* 2014;20:663–675.
157. Klongpanichapak S, Govitrapong P, Sharma SK, Ebadi M. Attenuation of cocaine and methamphetamine neurotoxicity by coenzyme Q10. *Neurochem Res.* 2006;31(3):303–311.
158. Macedo DS, de Vasconcelos SM, dos Santos RS, et al. Cocaine alters catalase activity in prefrontal cortex and striatum of mice. *Neurosci Lett.* 2005;387(1):53–56.
159. Lee YW, Hennig B, Fiala M, Kim KS, Toborek M. Cocaine activates redox-regulated transcription factors and induces TNF-α expression in human brain endothelial cells. *Brain Res.* 2001;920:125–133.
160. Numa R, Baron M, Kohen R, Yaka R. Tempol attenuates cocaine-induced death of PC12 cells through decreased oxidative damage. *Eur J Pharmacol.* 2011;650(1):157–162.
161. Numa R, Kohen R, Poltyrev T, Yaka R. Tempol diminishes cocaine-induced oxidative damage and attenuates the development and expression of behavioral sensitization. *Neuroscience.* 2008;155(3):649–658.
162. Muriach M, Lopez-Pedrajas R, Barcia JM, Sanchez-Villarejo MV, Almansa I, Romero FJ. Cocaine causes memory and learning impairments in rats: involvement of nuclear factor kappa B and oxidative stress, and prevention by topiramate. *J Neurochem.* 2010;114(3):675–684.
163. Pomierny-Chamiolo L, Moniczewski A, Wydra K, Suder A, Filip M. Oxidative stress biomarkers in some rat brain structures and peripheral organs underwent cocaine. *Neurotox Res.* 2013;23(1):92–102.

164. Uys JD, Knackstedt L, Hurt P, et al. Cocaine-induced adaptations in cellular redox balance contributes to enduring behavioral plasticity. *Neuropsychopharmacology.* 2011;36 (12):2551–2560.
165. Bernstein HG, Trubner K, Krebs P, et al. Increased densities of nitric oxide synthase expressing neurons in the temporal cortex and the hypothalamic paraventricular nucleus of polytoxicomanic heroin overdose victims: possible implications for heroin neurotoxicity. *Acta Histochem.* 2014;116(1):182–190.
166. Cunha-Oliveira T, Rego AC, Garrido J, Borges F, Macedo T, Oliveira CR. Neurotoxicity of heroin-cocaine combinations in rat cortical neurons. *Toxicology.* 2010;276(1):11–17.
167. Pereska Z, Dejanova B, Bozinovska C, Petkovska L. Prooxidative/antioxidative homeostasis in heron addiction and detoxification. *Bratisl Lev Listy.* 2007;108(9): 393–398.
168. Kovatsi L, Njau S, Nikolaou K, Topouridou K, Papamitsou T, Koliakos G. Evaluation of prooxidant–antioxidant balance in chronic heroin users in a single assay: an identification criterion for antioxidant supplementation. *Am J Drug Alcohol Abuse.* 2010;36(4):228–232.
169. Kovatsi L, Njau S, Nikolaou K, Tsolakidou A, Karamouzis I, Thisiadou K. Isoprostane as a marker of oxidative stress in chronic heroin users: correlation with duration of heroin use or concomitant hepatitis C infection. *Am J Drug Alcohol Abuse.* 2010;36(1):13–17.
170. Gutowicz M, Sadurska B, Cholojczyk M, Pokorska-Lis M, Siwinska-Ziolkowska A, Baranczyk-Kuzma A. Antioxidant status in different regions of heroin addicts' brain. *Environ Toxicol Pharmacol.* 2006;21(1):80–85.
171. Xu B, Wang Z, Li G, et al. Heroin-administered mice involved in oxidative stress and exogenous antioxidant-alleviated withdrawal syndrome. *Basic Clin Pharmacol Toxicol.* 2006;99:153–161.
172. Cemek M, Buyukokuroglu ME, Hazman O, Konuk M, Bulut S, Birdane YO. The roles of melatonin and vitamin E plus selenium in prevention of oxidative stress induced by naloxone-precipitated withdrawal in heroin-addicted rats. *Biol Trace Elem Res.* 2011;142(1):55–66.
173. Qi C, Wang X, Ge F, et al. mGluR5 in the nucleus accumbens shell regulates morphine-associated contextual memory through reactive oxygen species signaling. *Addict Biol.* 2015;20:927–940.
174. Pinelli A, Cighetti G, Trivulzio S. Plasma malondialdehyde levels and opiate withdrawal signs observed in rats treated with morphine plus naloxone: effects of alphalipoic acid administration. *Fundam Clin Pharmacol.* 2008;22(4):439–445.
175. Jiang LH, Luo X, He W, Huang XX, Cheng TT. Effects of exogenous hydrogen sulfide on apoptosis proteins and oxidative stress in the hippocampus of rats undergoing heroin withdrawal. *Arch Pharm Res.* 2011;34(12):2155–2162.
176. Pan J, Zhang Q, Zhang Y, Ouyang Z, Zheng Q, Zheng R. Oxidative stress in heroin administered mice and natural antioxidants protection. *Life Sci.* 2005;77(2):183–193.
177. Govindaraj J, Pillai SS. Rosmarinic acid modulates the antioxidant status and protects pancreatic tissues from glucolipotoxicity mediated oxidative stress in high-fat diet: streptozotocin-induced diabetic rats. *Mol Cell Biochem.* 2015;404(1–2):143–159.
178. Debersac P, Vernevaut M, Amiot M, Suschetet M, Siess M-H. Effects of a water-soluble extract of rosemary and its purified component rosmarinic acid on xenobiotic-metabolizing enzymes in rat liver. *Food Chem Toxicol.* 2001;39(2):109–117.
179. Zhou W, Kalivas PW. *N*-acetylcysteine reduces extinction responding and induces enduring reductions in cue- and heroin-induced drug-seeking. *Biol Psychiatry.* 2008;63 (3):338–340.

180. Schneider Jr R, Santos CF, Clarimundo V, Dalmaz C, Elisabetsky E, Gomez R. *N*-acetylcysteine prevents behavioral and biochemical changes induced by alcohol cessation in rats. *Alcohol.* 2015;49(3):259–263.
181. Reichel CM, Moussawi K, Do PH, Kalivas PW, See RE. Chronic *N*-acetylcysteine during abstinence or extinction after cocaine self-administration produces enduring reductions in drug seeking. *J Pharmacol Exp Ther.* 2011;337(2):487–493.
182. Lu SC. Glutathione synthesis. *Biochim Biophys Acta.* 2013;1830(5):3143–3153.
183. Fernandez-Fernandez S, Almeida A, Bolanos JP. Antioxidant and bioenergetic coupling between neurons and astrocytes. *Biochem J.* 2012;443(1):3–11.
184. Albano R, Raddatz NJ, Hjelmhaug J, Baker DA, Lobner D. Regulation of system xc(−) by pharmacological manipulation of cellular thiols. *Oxid Med Cell Longev.* 2015;2015: 269371.
185. Seib TM, Patel SA, Bridges RJ. Regulation of the system x(C)-cystine/glutamate exchanger by intracellular glutathione levels in rat astrocyte primary cultures. *Glia.* 2011;59(10):1387–1401.
186. Badisa RB, Goodman CB, Fitch-Pye CA. Attenuating effect of *N*-acetyl-L-cysteine against acute cocaine toxicity in rat C6 astroglial cells. *Int J Mol Med.* 2013;32 (2):497–502.
187. Badisa RB, Kumar SS, Mazzio E, et al. *N*-acetyl cysteine mitigates the acute effects of cocaine-induced toxicity in astroglia-like cells. *PloS One.* 2015;10(1):e0114285.
188. McClure EA, Gipson CD, Malcolm RJ, Kalivas PW, Gray KM. Potential role of *N*-acetylcysteine in the management of substance use disorders. *CNS Drugs.* 2014;28 (2):95–106.
189. Shi M, Bradner J, Bammler TK, et al. Identification of glutathione S-transferase pi as a protein involved in Parkinson disease progression. *Am J Pathol.* 2009;175(1):54–65.
190. Ehrhart J, Gluck M, Mieyal J, Zeevalk GD. Functional glutaredoxin (thioltransferase) activity in rat brain and liver mitochondria. *Parkinsonism Relat Disord.* 2006;8(6): 395–400.

CHAPTER FOUR

Molecular Mechanisms of *Cannabis* Signaling in the Brain

Patrick J. Ronan[*,†,‡,1], **Narin Wongngamnit**[*,§,¶], **Thomas P. Beresford**[*,§]

[*] Department of Veterans Affairs Medical Center, Laboratory for Clinical and Translational Research in Psychiatry, Denver, Colorado, USA
[†] Research Service, Sioux Falls VA Health Care System, Sioux Falls, South Dakota, USA
[‡] Department of Psychiatry and Division of Basic Biomedical Sciences, Sanford School of Medicine at the University of South Dakota, Sioux Falls, South Dakota, USA
[§] Department of Psychiatry, School of Medicine, University of Colorado, Denver, Colorado, USA
[¶] Substance Abuse Treatment Program, Department of Veterans Affairs, Denver, Colorado, USA
[1] Corresponding author: e-mail address: Patrick.Ronan@VA.gov.

Contents

Abstract

Cannabis has been cultivated and used by humans for thousands of years. Research for decades was focused on understanding the mechanisms an illegal/addictive drug. This led to the discovery of the vast endocannabinoid system. Research has now

Progress in Molecular Biology and Translational Science, Volume 137
ISSN 1877-1173
http://dx.doi.org/10.1016/bs.pmbts.2015.10.002

This work was supported in part by an Advanced VA Research Fellowship (P.J.R.) from the United States (U.S.) Department of Veterans Affairs Office of Academic Affairs. The contents of this publication do not represent the views of the U.S. Department of Veterans Affairs or the United States Government.

shifted to understanding fundamental biological questions related to one of the most widespread signaling systems in both the brain and the body. Our understanding of cannabinoid signaling has advanced significantly in the last two decades. In this review, we discuss the state of knowledge on mechanisms of *Cannabis* signaling in the brain and the modulation of key brain neurotransmitter systems involved in both brain reward/addiction and psychiatric disorders. It is highly probable that various cannabinoids will be found to be efficacious in the treatment of a number of psychiatric disorders. However, while there is clearly much potential, marijuana has not been properly vetted by the medical-scientific evaluation process and there are clearly a range of potentially adverse side-effects—including addiction. We are at crossroads for research on endocannabinoid function and therapeutics (including the use of exogenous treatments such as *Cannabis*). With over 100 cannabinoid constituents, the majority of which have not been studied, there is much *Cannabis* research yet to be done. With more states legalizing both the medicinal and recreational use of marijuana the rigorous scientific investigation into cannabinoid signaling is imperative.

1. INTRODUCTION

Cannabis has been used medicinally and as an intoxicant for recreational and religious purposes for thousands of years.[1] *Cannabis* research for decades was focused on understanding mechanisms of an illegal drug. This led to the discovery of the vast endocannabinoid systems and research has shifted to understanding fundamental biological questions related to one of the most widespread signaling systems in both the brain and the body. There has been increased interest in the therapeutic use of *Cannabis* and cannabinoids in an extremely wide range of disorders. Our understanding of cannabinoid function in the brain has advanced significantly in the last two decades. In this review, we will discuss the state of knowledge on mechanisms of *Cannabis* signaling in the brain and the modulation of key brain neurotransmitter systems involved in both brain reward and psychiatric disorders.

2. CHEMICAL COMPONENTS OF *CANNABIS*

Most studies on *Cannabis* have focused on the primary psychoactive component $(-)$-Δ^9-*trans*-(6aR,10aR)-tetrahydrocannabinol (Δ^9-THC)[2] but recent discoveries highlight the potential therapeutic benefits of many other components—either alone or in combination. The chemistry of *Cannabis* is very complex and the biological activity of many constituents is still unknown (for review see Ref. [3]). Analysis of *Cannabis sativa* L. has identified over 500 natural compounds, many of which interact with each other. To date, over 100 cannabinoids have been identified in *Cannabis*.[3–7] The term phytocannabinoids is used to distinguish the many cannabinoid components of *Cannabis* from endocannabinoids and synthetic cannabinoids.

Other identified cannabinoid types include cannabidiol (CBD), cannabigerol (CBG), cannabichromene (CBC), cannabicyclol (CBL), cannabielsoin (CBE), cannabinol (CBN), cannabinodiol (CBND), cannabitriol (CBT) along with 14 other miscellaneous types (for review see Ref. [3]). CBD is also enriched in marijuana and has known bioactive and neurologic effects. It does not produce a euphoria or "high" like Δ^9-THC and has garnered much attention as a potential pharmacotherapy for numerous disorders.[8] The remaining 400 plus chemical components of *Cannabis* consist of a wide range of flavonoids, nitrogenous compounds, amino acids, proteins, glycoproteins, sugars, hydrocarbons, simple alcohols, ketones and acids, fatty acids, esters, lactones, steroids, terpenes, noncannabinoid phenols, vitamins, and pigments.[4,5] Research driven by the discovery that Δ^9-THC is largely responsible for the psychotropic effects of *Cannabis* led to the monumental discovery of the first "cannabinoid" receptor[9] and the vast endocannabinoid system.[10,11]

3. ENDOCANNABINOID SYSTEM

The endocannabinoid system is now recognized as a major signaling system not only in the brain but in peripheral tissues as well. It has been called one of the most widespread and versatile signaling systems ever discovered (for review see Refs [11–13]). This system plays a significant role in a wide range of physiologic processes and behaviors including neurogenesis, neural development, immune function, metabolism and energy homeostasis, synaptic plasticity and learning, pain, emotional state, arousal and sleep, stress reactivity, and reward processing/addiction to other drugs of abuse.[12,14–16]

In the brain, the endocannabinoid system makes up an extensive interconnected network of neuromodulators that control (usually inhibit) synaptic neurotransmitter release (for review see Refs [12,17]). Endocannabinoids are arachidonic acid derivatives that are released by postsynaptic neurons to regulate presynaptic activity.[17–21] The two best-characterized endocannabinoids are *N*-arachidonoylethanolamine (anandamide; AEA—"bliss" in Sanskrit) and 2-arachidonoylglycerol (2-AG). Unlike classical neurotransmitters they are not stored in vesicles but are synthesized on demand *de novo*. Other cannabinoid-like compounds have been identified but their presence and activity in intact tissues is still under some debate.[16] Along with the endocannabinoids this system is comprised of the G-coupled receptors, CB_1 and CB_2 (and potentially others), regulatory proteins, a membrane transporter, and an array of enzymes that control the bioavailability of the endocannabinoids through their synthesis and degradation.[11,13,17,22]

4. SIGNAL TRANSDUCTION

Both CB_1 and CB_2 belong to the superfamily of guanine-binding regulatory protein (G protein) coupled receptors of $G_{i/o}$ type. CB_1 and CB_2 have 48% sequence homology[23] and both inhibit adenylyl cyclase and activate mitogen-activated protein kinase (MAPK) cascades.[23–25] Agonism of CB_1 receptors also releases G-proteins that interact with multiple ion channels. They increase A-type and inwardly rectify K^+ currents and inhibit N- and P/Q type Ca^{2+}currents.[26,27] Importantly, CB_1 regulation of these ion channels is a primary mechanism mediating cannabinoid-induced synaptic plasticity (will be discussed later). Activation of $G_{i/o}$ leads to $G\beta\gamma$-mediated inhibition of Ca^{2+}influx and inhibition of cAMP/PKA phosphorylation of ion channels.[28,29] Interestingly, CB_1 can also stimulate Gs-coupled proteins and is modulated by accessory proteins.[30] There is a general agreement based on a body of evidence which suggests that the modulatory effects of cannabinoids cannot be fully explained by their actions at CB_1 and CB_2 alone and that other G-coupled receptors will be discovered.[31] One candidate is the orphan G-protein-coupled receptor 55 (GPR55)[32] or, as it has been called, "CB_3." Both Δ^9-THC and CBD bind this receptor but other groups have had discrepant results leaving its status as a true cannabinoid receptor in doubt. Other non-G-coupled receptors that may play a direct role in cannabinoid signaling include the transient receptor potential vanilloid type 1 Ca^{2+}channels (TRPV1). Endocannabinoids also bind nuclear peroxisome proliferator-activated receptors α and γ that play a role in the regulation of lipid and glucose metabolism along with inflammatory responses (see Ref. [30] for review). The net effect is a growing body of evidence suggesting that cannabinoids dampen immune responses.

Functional multimeric forms of CB_1 receptors are common in the brain. Multimeric homodimer forms of CB_1 exist throughout the brain and it is thought that this could be the native state of CB_1 receptors. CB_1 also forms heteromers with CB_2 receptors.[21] There is also evidence that CB_1 forms heteromers with a number of other G-coupled receptors including D1 and D2, orexin, μ-opioid, and adenosine$_{2A}$. Little is known about the physiologic role of these G-protein receptor heteromers but there is evidence that they play a role in disease states such as Parkinson's and addiction.[33] CB_1 activation also causes an interaction and transduction via receptor and non-receptor tyrosine kinases (reviewed in Ref. [20]). The functionality of these receptor/receptor interactions needs to be further described but they provide intriguing insight into the complexity of cannabinoid signaling and perhaps provide clues as to potential targets for investigating cannabinoid pharmacotherapy.

Another finding with important implications for pharmacotherapy is the seminal discovery that G-protein-coupled receptors such as CB_1 can activate multiple and even opposing signaling cascades (functional selectivity) depending on binding. Different agonists lead to alternate stable receptor conformations that engage selective transduction mechanisms.[34] For CB_1 receptors different cannabinoids demonstrated this biased agonism. It is more complicated as even a single cannabinoid ligand can show biased agonism depending on concentration (reviewed in Ref. [35]). One proposed mechanism for this phenomenon for CB_1 receptors has been described. Different ligands of CB_1 receptors produced varying endocytic dwell times (the time during which receptors are clustered into clathrin pits together with β-arrestins before endocytosis) that controlled the functional selectivity of this receptor.[36]

Regulatory proteins also play a role in controlling cannabinoid signaling via CB_1 receptors (reviewed in Ref. [37]). The cannabinoid receptor-interacting proteins 1a and 1b (CRIP1a and CRIP1b) are only known to interact with CB_1 receptors. Both CRIP1a and CRIP1b directly interact with the C-terminus of the CB_1 receptor but not CB_2.[37,38] Not much is known about the regulation and function of these interacting proteins. Evidence suggests that $CRIP_{1a}$ attenuates CB_1 receptor-induced inhibition of Ca^{2+}efflux.[37,38] $CRIP_{1a}$, like CB_1, is widely expressed in the brain at high levels.[38] It has been suggested that these regulatory proteins play a critical role in modulating CB_1 receptor function in the pathogenesis or neuroadaptive response to epilepsy.[37] There are a number of other regulatory proteins that interact with CB_1 to affect signal transduction. Some of these are nonspecific G-protein receptors regulators such as G-protein-coupled receptor kinase-3 (GRK3) and β-arrestin2. These proteins facilitate desensitization and intracellular trafficking of CB_1 receptors.[39,40] Other regulatory proteins including AP3 and G protein receptor associated sorting protein-1 (GASP1) associate with CB_1 and limited number of other G-coupled receptors. These proteins are less promiscuous and associate with only a limited subset of G-coupled receptors. They direct CB_1 targeting to lysosomes,[41,42] a mechanism of downregulation of response.[43]

Another mechanism that downregulates CB_1 receptor activity is trafficking and internalization. CB_1 receptors are rapidly internalized following agonist binding. This leads to the termination of CB_1 signaling. Brief exposure to cannabinoid agonists leads to rapid recycling of the functional receptor to the cell membrane. Prolonged exposure on the other hand leads to degradation of these receptors and requires new receptor synthesis. Studies have demonstrated that after chronic exposure new protein synthesis is required to regain functional cell surface receptors. From this evidence one can deduce that the receptors that had undergone endocytosis had been degraded by lysosomes instead of recycled to the cell membrane as is the case in acute administration. This is thought to be a mechanism that explains, at least in part, the rapid tolerance seen due to repeated *Cannabis* use.[41,44,45] Desensitization, uncoupling a receptor from its G protein and transduction mechanisms, is also a cause of tolerance. Cannabinoid agonists, especially Δ^9-THC, cause a rapid desensitization of CB_1 receptors.[46] Additionally, CB_1 receptor expression is downregulated by chronic *Cannabis* use in key limbic and cortical regions of both humans and rodents—an effect reversed by abstinence.[47–49]

5. CB_1 DISTRIBUTION

CB_1 receptors were first thought to be expressed exclusively in the brain but are expressed both in the central nervous system (CNS) and to a lesser degree in peripheral organs. They are among the most abundant G-coupled receptors in the CNS.[50] Their distribution corresponds to the behavioral effects of *Cannabis*—pleasure, anxiety, fear, panic, learning and memory, thinking, concentration, movement, coordination, increased appetite, and sensory and time perception.[51,52] High levels of expression of CB_1 receptors are found in the amygdala, hippocampus, cortex, basal ganglia outflow tracts, and cerebellum. They are principally found on axons and presynaptic terminals of neurons. A majority appear to be on GABAergic cells where they inhibit GABA release but functional receptors are also found on glutamatergic cells and inhibit glutamate release.[50,53] As pointed out by

Mackie,[52] careful analysis of the details of this distribution and cellular location allows us to generate hypotheses regarding potential neurologic and psychiatric disorders for which these receptors may be involved and which might provide fruitful targets for pharmacotherapy.

The fact that the CB_1 receptor is located on presynaptic terminals of both GABA and glutamatergic neurons has tremendous implications for therapeutic use for phyto- and endocannabinoids. Systemic delivery of most medicinal and recreational pharmacotherapies, which in the case of cannabinoids may result in contraindicated effects. Global activation of CB_1 receptors likely results in a complex web of inhibition and excitation of various neural pathways. As we are discovering with so many signaling systems in the brain, therapeutic targeting of receptors may rely on our ability to specifically activate particular circuitry to the exclusion of others.

6. CB_2 RECEPTORS

Much less is known about the function of CB_2 receptors in the brain. These receptors were first identified on immune cells and were not thought to play a significant role in signaling in CNS. Although the anatomic data is not as clear as it is for CB_1, CB_2 receptors have now been found to be expressed on neurons, glia, and epithelial cells in the brain, albeit at much lower levels than CB_1. The majority of CNS CB_2 receptors are found in the cerebellum and brainstem. Different groups using a variety of methods (*in situ* hybridization, RT-PCR, immunohistochemistry, western blot, autoradiography) have found CB_2 mRNA or protein in other regions such as striatum, hippocampus, thalamus, and hypothalamus. Immunohistochemical findings of CB_2 in midbrain, olfactory bulb, and cortex are controversial due to issues with proper immunohistochemical controls.[11,16,54]

Recently, the idea that CB_2 plays only a minor role in brain cannabinoid signaling has been vigorously challenged by experimental evidence. The results from Callen *et al.*,[21] clearly suggest that CB_2 receptors play a key role in CB_1-mediated signaling through the formation of functional heteromers. Their results suggest that even at low expression levels, CB_2 receptors could have a significant impact on CB_1 signaling by reducing the cellular response through CB_1 receptors. Another finding that points to an important functional role for central CB_2 receptors is that their expression increases in response to certain pathologic conditions.

It has been proposed that CB_2 receptors play a critical role as part of a general protective system, including both its immune and neuroprotective roles, involved in "preventing, attenuating, or repairing inflicted damage" (for review see Ref. [55]). Glia have often been thought of as CNS cells that provide support and protection for neurons—an immune system of sorts (though we now understand that this family of cells do much more). Microglia are the macrophages of the brain and play a negative role in many neurodegenerative diseases. These cells express CB_2 receptors and disease pathology is associated with their activation. CB_2 receptor activation appears to attenuate microglial activation and prevent neurodegeneration in animal models of Huntington's.[56]

7. *CANNABIS* AND ADDICTION

Cannabis use has the hallmark pathologic features shared by all substance use disorders. Like other drugs of abuse, *Cannabis* causes compulsive drug-seeking behaviors, loss of self-control, withdrawal symptoms, and the propensity to relapse.[57] *Cannabis* use disorder has been recognized as a bona fide disorder and has been included in the most recent version of the *Diagnostic and Statistical Manual of Mental Disorders* (fifth edition).[58] Long-term clinical outcomes for *Cannabis* use disorder may be less severe than for other drugs of abuse but the addictive processes appears to be very similar. Animal models of addiction have clearly supported the notion that *Cannabis* has addictive properties.

8. ANIMAL MODELS

Classic preclinical models of rewarding/reinforcing properties of compounds such as intracranial self-stimulation (ICSS), conditioned place preference (CPP), self-administration, and others have all provided evidence that cannabinoids act on brain reward substrates just like other drugs of abuse. Self-administration of drugs of abuse by humans and animals is an indicator of the addictive or rewarding properties of a drug. Briefly, this paradigm has many variations but is based on operant conditioning where a behavior such as lever pressing or touching a lighted button causes self-administration, usually IV, of a drug. For decades researchers were not able to reliably demonstrate self-administration of cannabinoids, mainly Δ^9-THC. Altering the experimental parameters (e.g., employing food deprivation in rodents) has resulted in a number of studies demonstrating clear self-administration of Δ^9-THC or other synthetic cannabinoids in both rodents and primates (for reviews see Refs [59–61]). An interaction between CB_1 and opioids is seen in this model. Naloxone blocks cannabinoid self-administration in rodents[62] and naltrexone blocks it in primates.[63] As Maldonado and Fonseca[59] point out, this cannabinoid/opioid interaction in self-administration is bidirectional. Researchers were unable to achieve self-administration of morphine in CB_1 knockout mice.[64] Consistent with these findings is anatomic data demonstrating co-expression of $CB_1$and μ-opioid receptors in regions such as the periaqueductal gray.[65] Opioid signaling is also implicated in Δ^9-THC effects in another model.

Another measure of the rewarding and potentially addictive properties of a compound is ICSS. This is an operant behavioral procedure where animals learn to press a lever to deliver brief excitatory electrical pulses via implanted electrodes to discreet brain regions that mediate reward processing.[60,66] Most drugs of abuse lower the threshold for ICSS. Cannabinoids have presented inconsistent results. Initially, Gardner's and coworkers reported that systemic administration of Δ^9-THC (1.5 mg/kg) reliably decreased the ICSS threshold in rats.[67] This finding was repeated.[68] A number of other studies, however, showed that Δ^9-THC either attenuated or had no effect on ICSS.[69–72] A recent review of the literature regarding cannabinoid effects on ICSS concluded that these disparate findings are due to methodological variables including animal strain, cannabinoid compound, and dose.[73] The Δ^9-THC-induced decrease in ICSS can be blocked by naloxone, suggesting that Δ^9-THC is acting through an opioid mechanism to induce its rewarding effects.[74] This cannabinoid interaction with opioid signaling has been extensively studied in relation to pain, an indication for which there is mounting evidence that *Cannabis* has clinical effects.[75]

Finally, behavioral evidence employing the CPP model demonstrates both addictive and potentially aversive properties of cannabinoids. CPP is a paradigm based on Pavlovian conditioning in which rodents tend to have a preference for a contextually distinct chamber that had been previously paired with a rewarding drug, or conversely have a negative preference for a chamber the compound has aversive properties. The first CPP study showed that Δ^9-THC induced a strong place preference.[76] Interestingly, higher doses of 2 and 4 mg/kg produced CPP but when a washout period between training sessions was used (48 h) these same doses actually had an aversive effect whereas a lower dose (1.0 mg/kg) had a positive effect. This dichotomy dependent on dosing and scheduling was the forerunner of other complicated findings with Δ^9-THC and cannabinoid agonists/antagonists in the CPP. Cannabinoids produce both positive and negative results in the CPP depending on details of the experimental variables and procedures (for review see Ref. [73]). One interesting study showed that while Δ^9-THC alone had no effect on CPP, combining Δ^9-THC with CBD did induce CPP.[77] There is some evidence that this interaction can also mitigate negative effects. In one study, CBD was able to prevent the transient impairment of memory caused by Δ9-THC.[78]

This suggests that there is indeed an interactive or synergistic effect of combining multiple constituents of *Cannabis*. This observation may be the foundation for the current popular notion that use of the whole *Cannabis* plant is functionally different than using its isolated components. Together, these models provide evidence that *Cannabis* and cannabinoids share rewarding properties with other drugs of abuse and presumably work, at least in part, through common neural pathways and molecular mechanisms.

One common effect of drugs of abuse is the acute activation of the ventral tegmental area (VTA)–nucleus accumbens (NAc)/prefrontal cortex (PFC) dopamine pathway.[57] Dopamine (DA) release in the NAc has long been considered to play a critical role in the acute rewarding effects of drugs of abuse and the long-term neuroadaptations leading to addiction.[79–84] Recently, phasic DA activity (a specific firing pattern) in this pathway has been shown to play the key causal role in the instrumental learning underlying the development of addiction,[85] reviewed in Ref. [57].

A variety of techniques in animals and humans have demonstrated that cannabinoids increase both tonic and phasic DA activity (reviewed in Ref. [86]). *In vivo* microdialysis and electrophysiologic studies in rodents have confirmed that Δ^9-THC activates tonic DA release in the NAc.[87–94] Using fast-scan cyclic voltammetry that measures DA concentrations in real time, the CB_1 agonist WIN55,212-2 dose-dependently increased phasic DA events in the NAc.[95,96] Importantly, increased phasic DA events arising from high-frequency bursts of neural activity[97] are a mechanism reinforcing conditioned reward-related addiction behaviors.[85,98–100] Consistent with these effects on terminal release of DA, cannabinoid agonists including Δ^9-THC dose-dependently increase the firing rate and burst activity of VTA DA neurons.[86,90,91,93,101] These increases in DA activity are blocked by rimonabant, a CB_1 receptor-selective antagonist indicating that the observed effects of Δ^9-THC and other cannabinoid agonists act through a CB_1 mechanism. This enhanced DA signaling is thought to underlie the long-term molecular adaptations that cause addiction.

9. MOLECULAR ADAPTATIONS TO CHRONIC USE

Common molecular adaptations and mechanisms within the brain are shared by drugs of abuse, including *Cannabis*.[83] Substantial evidence has been accumulated over the years indicating that chronic use of *Cannabis* leads to long-term molecular adaptations in brain. One such common well-documented adaptation to chronic drugs of abuse is long-term induction of the transcription factor ΔFosB, an extremely stable truncated splice variant of the *fosB* gene.[102–107] Prolonged induction of ΔFosB in the striatum, in particular the ventral striatum or NAc, is involved in mediating long-term aspects of addiction by promoting reward and motivation and serving as a mechanism of drug sensitization and drug self-administration.[107] Like cocaine, nicotine, morphine, and ethanol, chronic administration of Δ^9-THC causes a robust increase in the expression of ΔFosB in the NAc.[108] This study found that Δ^9-THC activates ΔFosB in a wider extended reward system that includes the amygdala, bed nucleus of the stria terminalis (BNST), hippocampus, and medial prefrontal cortex (mPFC).

Long-lasting Δ^9-THC-induced expression of ΔFosB in brain structures involved in stress responses has implications for the mediation of aversive aspects of addiction and may hint to a mechanism driving the protracted propensity for stress-induced relapse. These regions provide the functional connections that mediate the increased DA signaling in the NAc seen in humans in response to drug conditioned and aversive cues.[109] Studies have confirmed that in regular marijuana users, exposure to marijuana cues elicit enhanced activity in a number of limbic structures including the amygdala and hippocampus.[110,111] *Cannabis* has major effects on stress-related behaviors in humans and the endocannabinoid system plays a significant role in their modulation via stress circuitry.[112,113] Another piece of evidence suggesting that ΔFosB may serve as a molecular mechanism connecting stress and addiction is the fact that chronic stress alone, in the absence of drugs, also induces sustained ΔFosB expression in reward circuitry.[114]

10. SYNAPTIC PLASTICITY

Chronic *Cannabis* also impairs synaptic plasticity and cognitive function. The development of addiction involves not only the acute rewarding effects of a drug but long-term molecular adaptations in neural circuits.[83]

Retrograde endocannabinoid signaling has emerged as a major player in long-term synaptic plasticity (reviewed in Refs [11,20,115,116]). CB_1 receptors located on presynaptic GABA and glutamate terminals inhibit release of the transmitters. Depending on the cell type on which they are located their activation induces a depolarization-induced suppression of inhibition (GABA release) and excitation (glutamate release). Whereas this effect for endocannabinoids is relatively transient (<1 min) with Δ^9-THC, the effect can last hours or even weeks.[116]

This persistent Δ^9-THC effect can lead to neuroadaptations that have lasting effects. Endocannabinoid signaling is a key mechanism driving the neural processes of long-term depression (LTD) and long-term potentiation (LTP). These cause alterations in synaptic strength that last for extended periods and are thought to be the mechanisms underlying learning and encoding experience.[117] Long-term *Cannabis* use may lead to molecular adaptations causing system effects. Imaging studies in adolescents reveal that chronic *Cannabis* use negatively alters connectivity in neural networks that control a wide range of critical functions including learning and memory and impulse control.[118]

11. *CANNABIS* WITHDRAWAL

There are well-documented clinically significant withdrawal symptoms caused by cessation or significant reduction from chronic (daily use for months) cannabinoid use (reviewed in Ref. [86]). These symptoms include irritability/anger or aggression, nervousness or anxiety, depressed mood, decreased appetite, weight loss, restlessness, sleep disturbances (insomnia, strange/disturbing dreams), chills, stomach pain, discomfort, shakiness, and sweating.[58,86,119] These symptoms follow the classic negative-affect-withdrawal component of the addiction cycle, first described by Koob and Le Moal, that drives addictive behaviors.[120,121] This syndrome is often not severe enough to require medical attention but users do report that these symptoms contribute to an inability to quit and recurrence of use.[122]

Key changes to neurotransmitter systems caused by withdrawal from other drugs of abuse also occur during cannabinoid withdrawal.[86] One primary and consistent shared change is decreased DA activity in the VTA–NAc. DA release is tonically decreased during withdrawal.[123] This is thought to be partially responsible for the anhedonia and depressed mood that occurs during withdrawal. Another consistent effect of withdrawal on brain systems that is thought to drive use and relapse is the acute increase in the stress neuropeptide CRF in the central nucleus of the amygdala (CeA; reviewed in Ref. [124]).

Interestingly, acute administration of cannabinoid agonists decrease CRF release in the CeA while precipitated withdrawal causes an increase in both release[125] and CRF mRNA expression (and presumably CRF peptide).[126] Chronically enhanced amygdalar CRF signaling is thought to be one of the mechanisms underlying a number of stress-induced psychiatric disorders including depression, anxiety, and posttraumatic stress disorder (PTSD).[124,127] Enhanced mRNA (and presumably CRF peptide) expression is significant because along with local circuits CRF produced in cells of the CeA project to many other brain regions, including the dopaminergic VTA.[128] It is intriguing that cannabinoid treatment is capable of attenuating this increased CRF signaling, albeit only in cannabinoid naive animals. We recently showed that silencing expression of CRF in the CeA attenuates anxiety-like behaviors and endocrine responses to stressors.[127] The fact that chronic use seems to augment this anxiogenic CRF expression is troubling for the prospect of long-term *Cannabis* treatment of anxiety disorders. Cannabinoids are known to have a bidirectional effect on other stress systems and responses as well, capable of both reducing acutely and enhancing them with chronic use (for review see Ref. [129]).

12. MONOAMINERGIC SYSTEMS

Monoaminergic signaling (DA, norepinephrine, serotonin) is central to wide ranging behavioral and physiologic functions including reward-related behaviors already described, stress responses and mental health. It is critically involved in the pathophysiology of depression, anxiety, and PTSD.[130,131] Cannabinoids interact with monoaminergic systems to influence metabolism,[132–134] direct inhibition of monoamine release or inhibition of release of other neurotransmitter systems that interact with monoamine pathways such as GABA, glutamate, acetylcholine, adenosine, numerous neuropeptides, and other systems.[134]

12.1 Serotonin

A number of studies have provided evidence suggesting that the modulation of serotonergic signaling by cannabinoids underlies many of their behavioral effects—including stress responses. There is significant overlap in the effects of both the endocannabinoid and serotonergic systems. Both regulate processes and behaviors such as sleep and arousal, feeding, body temperature, and emotional processes, however, cannabinoid regulation requires 5-HT suggesting that it has an upstream role in modulating 5-HT.[113]

Cannabinoids, both endo and phyto, control the release of serotonin (5-HT) in terminal regions and activity of serotonergic neurons. Cannabinoids have both direct and indirect effects on 5-HT activity and can be both inhibitory and excitatory. *In vivo* microdialysis studies have shown that agonism of CB_1 receptors decreases 5-HT release in hippocampus and PFC.[135] Conversely, release of serotonin in the NAc is increased by systemic administration of a CB_1 agonist while direct application to terminals in NAc decreases in release. This enhanced 5-HT release in the NAc likely plays a role in the rewarding effects of Δ^9-THC. Direct application in the NAc likely causes activation of presynaptic CB_1 receptors on 5-HT terminals and an inhibition of release. The increased release was blocked with a GABA antagonist suggesting cannabinoid removal of the strong tonic GABAergic inhibitory inputs to dorsal raphe (DR) projections to NAc.[136–138] The circuitry controlling cannabinoid-mediated excitability of DR neurons is complex. CB_1 receptors are located on both GABA and glutamate terminals in the brain.[139,140] There is evidence for CB_1-mediated inhibition of excitatory glutamate inputs in the DR.[141,142] Other studies have demonstrated that systemic antagonism of CB_1 receptors enhances 5-HT release in medial PFC.[143–145]

Cannabinoid modulation of serotonergic transmission has generated interest in targeting this system as first-line antidepressant therapies also target 5-HT signaling.[146,147] Recently this idea was tested in a preclinical animal model of antidepressant efficacy—the forced swim test (FST). Δ^9-THC had antidepressant-like effects in this test but only when administered (i.p.) for 5 days, as is the case with classic serotonin reuptake inhibitors such as citalopram. Also, repeated but not single doses of Δ^9-THC increased spontaneous DR 5-HT neuron activity and postsynaptic 5-HT_{1a} receptor activity in dorsal hippocampus, also effects common to antidepressants.[148,149] Clinical evidence suggests that long-term *Cannabis* use does possess antidepressant properties. A review of the available literature provides evidence that *Cannabis* use, and especially heavy *Cannabis* use, may be associated with an increased risk for developing depressive disorders.[150]

12.2 Norepinephrine

Another common pharmacotherapy target of psychiatric disorders that is influenced by cannabinoid signaling is the noradrenergic system (see review by Ref. [131]). Dysregulation of noradrenergic signaling has been demonstrated in mood disorders, such as depression and anxiety disorders including PTSD. As with serotonin, increasing both noradrenergic and cannabinoid activity improves mood. Also, excessive activation of both systems also can induce mania. Together, these findings suggest that these systems interact or share common pathways.[131]

Although, the majority of CB_1 receptors are found on postsynaptic norepinephrine (NE) terminals and presynaptic glutamate terminals in the locus coeruleus (LC), CB_1 receptors located on presynaptic GABAergic terminals[151,152] appear to play a significant role in the functional response to exogenous cannabinoid agonist application. CB_1 activation results in increased activity of noradrenergic neurons presumably via release of inhibitory tone. Microdialysis and tissue studies show cannabinoid agonists including Δ^9-THC increase NE release and turnover (used as a static measure of release) in a variety of brain regions including PFC, NAc, LC, hippocampus, hypothalamus, and cerebellum. Electrophysiologic studies increase spontaneous firing and firing rate of LC neurons. Overall, cannabinoid interaction with noradrenergic system has important functions in both physiologic and behavioral responses.[131]

13. CANNABIDIOL

CBD has many diverse functions in the brain and there has been much interest in it as a pharmacotherapy for numerous disorders (reviewed in Refs [8,153]). As of 2015, over 75 studies using CBD were registered on clinicaltrials.gov. Studies in preclinical animal models and humans have provided evidence suggesting that CBD has neuroprotective, anti-inflammatory, anxiolytic, and antipsychotic properties. CBD, along with Δ^9-THC, is found in relatively high levels in *Cannabis*.[154] Some strains of *Cannabis* are particularly rich in CBD (CBD:Δ^9-THC content greater than 20:1).

CBD is considered a multitarget drug and interacts with a diverse array of signaling systems. Unlike Δ^9-THC, CBD is not a direct ligand for CB_1 or CB_2 receptors. Depending on the interaction, CBD can have both enhance and inhibit signaling. CBD increases serotonergic activity via enhanced 5-HT_{1a} signaling. It also has a facilitatory effect on $\alpha 3$ and $\alpha 1$ glycine receptors, transient receptor potential of ankyrin type 1 channel (TRPA1) and can both enhance and inhibit intracellular calcium flux. At higher concentrations CBD can activate TRPV1 and TRPV2 channels. CBD can inhibit multiple targets at low concentrations. It reduces signaling at the putative but poorly characterized cannabinoid receptor GPR55. It can also act as an antagonist of the transient receptor potential of melastatin type 8 channel (TRPM8) and the equilibrative nucleoside transporter (ENT).[8,32,153,155–158]

One critical mechanism of CBD action that has been described is its ability to enhance intrinsic signaling of the endocannabinoid anandamide by decreasing cellular uptake and FAAH-mediated catabolism.[159] This could play a significant role in the therapeutic effects of CBD. Clinical studies have shown that endogenous anandamide levels are inversely correlated with psychotic symptoms[160,161] and administration of CBD in patients with schizophrenia enhances serum anandamide levels that correlate with significant clinical improvement.[162] These results are in contrast to findings correlating *Cannabis* use with the development of psychosis.[163,164]

14. THE CASE OF CHARLOTTE'S WEB

The case for using *Cannabis* therapeutically for a number of disorders is largely based on anecdotal reports. There has been particular interest in CBD as a therapy for epilepsy.[8,165] Due to the lack of effective treatments for his child epileptic patients, Dr Edward Maa of Denver Health and Hospitals, researched medical marijuana after hearing the story of a California mother who successfully treated her child's epilepsy with *Cannabis*. Synthesizing all available information from sources ranging from published articles, doctors, patients, to *Cannabis* growers and dispensary operators, Dr Maa believed that CBD had potential as a treatment of a particularly devastating form of early onset and severe epilepsy called Dravet syndrome.[165] His patient, Charlotte, was only 5 years old at the time and there were many concerns with using medical marijuana to treat anyone let alone what would be the youngest patient ever. Dr Maa received approval from a team of epileptologists, pediatricians, and the state of Colorado and started Charlotte on low doses of a sublingual treatment of an extract from a strain of *Cannabis* that would become known as "Charlotte's Web." This is a rare strain of *Cannabis* specifically bred by the Stanley brothers in Colorado to be extremely rich in CBD with virtually no Δ^9-THC. In Charlotte's case the results were tremendous, or as Dr Maa called it "too good to be true." Charlotte had no seizures at all during the first week of treatment for the first time since her seizures started and by the third month had over 90% reduction in seizures. This case and the publicity it received has led to many families from around the country with children suffering from epilepsy syndromes to move to Colorado for its liberal medical marijuana laws in order to obtain this treatment. Dr Maa is the first to point out that what is needed is "thoughtful and thorough pharmacologic and clinical research into *Cannabis* and its many constituent compounds to confirm or disprove its safety and antiepileptic potential." Researchers are excited given this published case report and the many other reports of its efficacy.[166] There are now randomized placebo-controlled studies planned or underway both in Colorado and elsewhere to determine if CBD is effective in the treatment of Dravets and other forms of epilepsy such as the devastating early onset Lennox–Gastaut syndrome.

15. CONCLUSIONS

We are at a crossroad for *Cannabis* research. With the liberalization of marijuana laws in many states, the stigma surrounding *Cannabis* research is being lifted—and importantly, more funding is being directed to conduct this research. Currently, Δ^9-THC and CBD, the most concentrated phytocannabinoids, are by far the most studied. Both have been shown to play a significant role in signaling in the brain. There are over 100 cannabinoid compounds identified in *Cannabis* and the majority of them have not even been studied. Given the science to date and the fact that the endocannabinoid system is one of the most widespread modulatory systems in the brain, it is rational to believe that various cannabinoids will be efficacious in a number of psychiatric disorders. While there is clearly much potential for *Cannabis* components to be used medicinally, marijuana has not been properly vetted by the medical-scientific evaluation process[167] and there are clearly a range of potentially adverse side effects—including addiction. Findings such as those cited earlier with Dravet's syndrome highlight the need for more thorough study of the wide range of phytocannabinoids. We need to fully understand their mechanisms of action and their complex influence on brain function and behavior.

REFERENCES

1. Russo EB. History of cannabis and its preparations in saga, science, and sobriquet. *Chem Biodivers.* 2007;4(8):1614–1648.
2. Mechoulam R, Gaoni Y. The absolute configuration of delta-1-tetrahydrocannabinol, the major active constituent of hashish. *Tetrahedron Lett.* 1967;12:1109–1111.

3. Elsohly MA, Slade D. Chemical constituents of marijuana: the complex mixture of natural cannabinoids. *Life Sci.* 2005;78(5):539–548.
4. Turner CE, Elsohly MA, Boeren EG. Constituents of *Cannabis sativa* L. XVII. A review of the natural constituents. *J Nat Prod.* 1980;43(2):169–234.
5. Ross SA, ElSohly MA. The volatile oil composition of fresh and air-dried buds of *Cannabis sativa*. *J Nat Prod.* 1996;59(1):49–51.
6. Mehmedic Z, Chandra S, Slade D, et al. Potency trends of Delta9-THC and other cannabinoids in confiscated cannabis preparations from 1993 to 2008. *J Forensic Sci.* 2010;55(5):1209–1217.
7. Sharma P, Murthy P, Bharath MM. Chemistry, metabolism, and toxicology of cannabis: clinical implications. *Iran J Psychiatry.* 2012;7(4):149–156.
8. Devinsky O, Cilio MR, Cross H, et al. Cannabidiol: pharmacology and potential therapeutic role in epilepsy and other neuropsychiatric disorders. *Epilepsia.* 2014;55(6):791–802.
9. Devane WA, Dysarz III FA, Johnson MR, Melvin LS, Howlett AC. Determination and characterization of a cannabinoid receptor in rat brain. *Mol Pharmacol.* 1988;34(5):605–613.
10. Pertwee RG. Ligands that target cannabinoid receptors in the brain: from THC to anandamide and beyond. *Addict Biol.* 2008;13(2):147–159.
11. Mechoulam R, Parker LA. The endocannabinoid system and the brain. *Annu Rev Psychol.* 2013;64:21–47.
12. Castillo PE, Younts TJ, Chavez AE, Hashimotodani Y. Endocannabinoid signaling and synaptic function. *Neuron.* 2012;76(1):70–81.
13. Maccarrone M, Bab I, Biro T, et al. Endocannabinoid signaling at the periphery: 50 years after THC. *Trends Pharmacol Sci.* 2015;36(5):277–296.
14. Fernandez-Ruiz J, Berrendero F, Hernandez ML, Ramos JA. The endogenous cannabinoid system and brain development. *Trends Neurosci.* 2000;23(1):14–20.
15. Murillo-Rodriguez E, Blanco-Centurion C, Sanchez C, Piomelli D, Shiromani PJ. Anandamide enhances extracellular levels of adenosine and induces sleep: an *in vivo* microdialysis study. *Sleep.* 2003;26(8):943–947.
16. Serrano A, Parsons LH. Endocannabinoid influence in drug reinforcement, dependence and addiction-related behaviors. *Pharmacol Ther.* 2011;132(3):215–241.
17. Piomelli D. The molecular logic of endocannabinoid signalling. *Nat Rev Neurosci.* 2003;4(11):873–884.
18. Freund TF, Katona I, Piomelli D. Role of endogenous cannabinoids in synaptic signaling. *Physiol Rev.* 2003;83(3):1017–1066.
19. Katona I, Sperlagh B, Sik A, et al. Presynaptically located CB_1 cannabinoid receptors regulate GABA release from axon terminals of specific hippocampal interneurons. *J Neurosci.* 1999;19(11):4544–4558.
20. Dalton GD, Bass CE, Van Horn CG, Howlett AC. Signal transduction via cannabinoid receptors. *CNS Neurol Disord Drug Targets.* 2009;8(6):422–431.
21. Callen L, Moreno E, Barroso-Chinea P, et al. Cannabinoid receptors CB_1 and CB_2 form functional heteromers in brain. *J Biol Chem.* 2012;287(25):20851–20865.
22. Di Marzo V, Ligresti A, Morera E, Nalli M, Ortar G. The anandamide membrane transporter. Structure–activity relationships of anandamide and oleoylethanolamine analogs with phenyl rings in the polar head group region. *Bioorg Med Chem.* 2004;12(19):5161–5169.
23. Howlett AC. The cannabinoid receptors. *Prostaglandins Other Lipid Mediat.* 2002;68–69:619–631.
24. Klee WA, Sharma SK, Nirenberg M. Opiate receptors as regulators of adenylate cyclase. *Life Sci.* 1975;16(12):1869–1874.
25. Nathanson NM, Klein WL, Nirenberg M. Regulation of adenylate cyclase activity mediated by muscarinic acetylcholine receptors. *Proc Natl Acad Sci USA.* 1978;75(4):1788–1791.

26. Mackie K, Hille B. Cannabinoids inhibit N-type calcium channels in neuroblastoma-glioma cells. *Proc Natl Acad Sci USA.* 1992;89(9):3825–3829.
27. Mackie K, Lai Y, Westenbroek R, Mitchell R. Cannabinoids activate an inwardly rectifying potassium conductance and inhibit Q-type calcium currents in AtT20 cells transfected with rat brain cannabinoid receptor. *J Neurosci.* 1995;15(10):6552–6561.
28. Deadwyler SA, Hampson RE, Mu J, Whyte A, Childers S. Cannabinoids modulate voltage sensitive potassium A-current in hippocampal neurons via a cAMP-dependent process. *J Pharmacol Exp Ther.* 1995;273(2):734–743.
29. Dalton GD, Peterson LJ, Howlett AC. CB(1) cannabinoid receptors promote maximal FAK catalytic activity by stimulating cooperative signaling between receptor tyrosine kinases and integrins in neuronal cells. *Cell Signal.* 2013;25(8):1665–1677.
30. Tantimonaco M, Ceci R, Sabatini S, et al. Physical activity and the endocannabinoid system: an overview. *Cell Mol Life Sci.* 2014;71(14):2681–2698.
31. Console-Bram L, Marcu J, Abood ME. Cannabinoid receptors: nomenclature and pharmacological principles. *Prog Neuropsychopharmacol Biol Psychiatry.* 2012;38(1): 4–15.
32. Ryberg E, Larsson N, Sjogren S, et al. The orphan receptor GPR55 is a novel cannabinoid receptor. *Br J Pharmacol.* 2007;152(7):1092–1101.
33. Gomes I, Gupta A, Devi LA. G-protein-coupled heteromers: regulation in disease. *Methods Enzymol.* 2013;521:219–238.
34. Urban JD, Clarke WP, von Zastrow M, et al. Functional selectivity and classical concepts of quantitative pharmacology. *J Pharmacol Exp Ther.* 2007;320(1):1–13.
35. Turu G, Hunyady L. Signal transduction of the CB_1 cannabinoid receptor. *J Mol Endocrinol.* 2010;44(2):75–85.
36. Flores-Otero J, Ahn KH, Delgado-Peraza F, Mackie K, Kendall DA, Yudowski GA. Ligand-specific endocytic dwell times control functional selectivity of the cannabinoid receptor 1. *Nat Commun.* 2014;5:4589.
37. Smith TH, Blume LC, Straiker A, et al. Cannabinoid receptor-interacting protein 1a modulates CB_1 receptor signaling and regulation. *Mol Pharmacol.* 2015;87(4):747–765.
38. Niehaus JL, Liu Y, Wallis KT, et al. CB_1 cannabinoid receptor activity is modulated by the cannabinoid receptor interacting protein CRIP 1a. *Mol Pharmacol.* 2007;72 (6):1557–1566.
39. Jin W, Brown S, Roche JP, et al. Distinct domains of the CB_1 cannabinoid receptor mediate desensitization and internalization. *J Neurosci.* 1999;19(10):3773–3780.
40. Nguyen PT, Schmid CL, Raehal KM, Selley DE, Bohn LM, Sim-Selley LJ. β-arrestin2 regulates cannabinoid CB_1 receptor signaling and adaptation in a central nervous system region-dependent manner. *Biol Psychiatry.* 2012;71(8):714–724.
41. Martini L, Waldhoer M, Pusch M, et al. Ligand-induced down-regulation of the cannabinoid 1 receptor is mediated by the G-protein-coupled receptor-associated sorting protein GASP1. *FASEB J.* 2007;21(3):802–811.
42. Rozenfeld R, Devi LA. Regulation of CB_1 cannabinoid receptor trafficking by the adaptor protein AP-3. *FASEB J.* 2008;22(7):2311–2322.
43. Beguinot L, Lyall RM, Willingham MC, Pastan I. Down-regulation of the epidermal growth factor receptor in KB cells is due to receptor internalization and subsequent degradation in lysosomes. *Proc Natl Acad Sci USA.* 1984;81(8):2384–2388.
44. Hsieh C, Brown S, Derleth C, Mackie K. Internalization and recycling of the CB_1 cannabinoid receptor. *J Neurochem.* 1999;73(2):493–501.
45. Leterrier C, Bonnard D, Carrel D, Rossier J, Lenkei Z. Constitutive endocytic cycle of the CB_1 cannabinoid receptor. *J Biol Chem.* 2004;279(34):36013–36021.
46. Wu DF, Yang LQ, Goschke A, et al. Role of receptor internalization in the agonist-induced desensitization of cannabinoid type 1 receptors. *J Neurochem.* 2008;104(4): 1132–1143.

47. Fattore L, Fadda P, Fratta W. Endocannabinoid regulation of relapse mechanisms. *Pharmacol Res.* 2007;56(5):418–427.
48. Villares J. Chronic use of marijuana decreases cannabinoid receptor binding and mRNA expression in the human brain. *Neuroscience.* 2007;145(1):323–334.
49. Hirvonen J, Goodwin RS, Li CT, et al. Reversible and regionally selective downregulation of brain cannabinoid CB_1 receptors in chronic daily cannabis smokers. *Mol Psychiatry.* 2012;17(6):642–649.
50. Marsicano G, Lutz B. Expression of the cannabinoid receptor CB_1 in distinct neuronal subpopulations in the adult mouse forebrain. *Eur J Neurosci.* 1999;11(12):4213–4225.
51. Tsou K, Brown S, Sanudo-Pena MC, Mackie K, Walker JM. Immunohistochemical distribution of cannabinoid CB_1 receptors in the rat central nervous system. *Neuroscience.* 1998;83(2):393–411.
52. Mackie K. Distribution of cannabinoid receptors in the central and peripheral nervous system. *Handb Exp Pharmacol.* 2005;168:299–325.
53. Kano M, Ohno-Shosaku T, Hashimotodani Y, Uchigashima M, Watanabe M. Endocannabinoid-mediated control of synaptic transmission. *Physiol Rev.* 2009;89 (1):309–380.
54. Atwood BK, Mackie K. CB_2: a cannabinoid receptor with an identity crisis. *Br J Pharmacol.* 2010;160(3):467–479.
55. Pacher P, Mechoulam R. Is lipid signaling through cannabinoid 2 receptors part of a protective system? *Prog Lipid Res.* 2011;50(2):193–211.
56. Palazuelos J, Aguado T, Pazos MR, et al. Microglial CB_2 cannabinoid receptors are neuroprotective in Huntington's disease excitotoxicity. *Brain.* 2009;132(Pt 11): 3152–3164.
57. Britt JP, Bonci A. Optogenetic interrogations of the neural circuits underlying addiction. *Curr Opin Neurobiol.* 2013;23(4):539–545.
58. American Psychiatric Association.*DSM-5 Task Force. Diagnostic and Statistical Manual of Mental Disorders: DSM-5.* 5th ed Washington, DC: American Psychiatric Association; 2013.
59. Maldonado R, Rodriguez de Fonseca F. Cannabinoid addiction: behavioral models and neural correlates. *J Neurosci.* 2002;22(9):3326–3331.
60. Vlachou S, Panagis G. Regulation of brain reward by the endocannabinoid system: a critical review of behavioral studies in animals. *Curr Pharm Des.* 2014;20(13): 2072–2088.
61. Panagis G, Vlachou S, Nomikos GG. Behavioral pharmacology of cannabinoids with a focus on preclinical models for studying reinforcing and dependence-producing properties. *Curr Drug Abuse Rev.* 2008;1(3):350–374.
62. Navarro M, Carrera MR, Fratta W, et al. Functional interaction between opioid and cannabinoid receptors in drug self-administration. *J Neurosci.* 2001;21(14):5344–5350.
63. Justinova Z, Tanda G, Munzar P, Goldberg SR. The opioid antagonist naltrexone reduces the reinforcing effects of delta 9 tetrahydrocannabinol (THC) in squirrel monkeys. *Psychopharmacology.* 2004;173(1–2):186–194.
64. Ledent C, Valverde O, Cossu G, et al. Unresponsiveness to cannabinoids and reduced addictive effects of opiates in CB_1 receptor knockout mice. *Science.* 1999;283(5400): 401–404.
65. Wilson-Poe AR, Morgan MM, Aicher SA, Hegarty DM. Distribution of CB_1 cannabinoid receptors and their relationship with mu-opioid receptors in the rat periaqueductal gray. *Neuroscience.* 2012;213:191–200.
66. Olds J, Milner P. Positive reinforcement produced by electrical stimulation of septal area and other regions of rat brain. *J Comp Physiol Psychol.* 1954;47(6):419–427.
67. Gardner EL, Paredes W, Smith D, et al. Facilitation of brain stimulation reward by delta 9-tetrahydrocannabinol. *Psychopharmacology.* 1988;96(1):142–144.

68. Lepore M, Liu X, Savage V, Matalon D, Gardner EL. Genetic differences in delta 9-tetrahydrocannabinol-induced facilitation of brain stimulation reward as measured by a rate-frequency curve-shift electrical brain stimulation paradigm in three different rat strains. *Life Sci.* 1996;58(25):PL365–PL372.
69. Stark P, Dews PB. Cannabinoids. I. Behavioral effects. *J Pharmacol Exp Ther.* 1980;214 (1):124–130.
70. Vlachou S, Nomikos GG, Stephens DN, Panagis G. Lack of evidence for appetitive effects of Delta 9-tetrahydrocannabinol in the intracranial self-stimulation and conditioned place preference procedures in rodents. *Behav Pharmacol.* 2007;18(4):311–319.
71. Kwilasz AJ, Negus SS. Dissociable effects of the cannabinoid receptor agonists Delta9-tetrahydrocannabinol and CP55940 on pain-stimulated versus pain-depressed behavior in rats. *J Pharmacol Exp Ther.* 2012;343(2):389–400.
72. Wiebelhaus JM, Grim TW, Owens RA, et al. Delta9-tetrahydrocannabinol and endocannabinoid degradative enzyme inhibitors attenuate intracranial self-stimulation in mice. *J Pharmacol Exp Ther.* 2015;352(2):195–207.
73. Panagis G, Mackey B, Vlachou S. Cannabinoid regulation of brain reward processing with an emphasis on the role of CB_1 receptors: a step back into the future. *Front Psychiatry.* 2014;5:92.
74. Gardner EL, Vorel SR. Cannabinoid transmission and reward-related events. *Neurobiol Dis.* 1998;5(6 Pt B):502–533.
75. Whiting PF, Wolff RF, Deshpande S, et al. Cannabinoids for medical use: a systematic review and meta-analysis. *JAMA.* 2015;313(24):2456–2473.
76. Lepore M, Vorel SR, Lowinson J, Gardner EL. Conditioned place preference induced by delta 9-tetrahydrocannabinol: comparison with cocaine, morphine, and food reward. *Life Sci.* 1995;56(23–24):2073–2080.
77. Klein C, Karanges E, Spiro A, et al. Cannabidiol potentiates delta(9)-tetrahydrocannabinol (THC) behavioural effects and alters THC pharmacokinetics during acute and chronic treatment in adolescent rats. *Psychopharmacology.* 2011;218(2):443–457.
78. Morgan CJ, Schafer G, Freeman TP, Curran HV. Impact of cannabidiol on the acute memory and psychotomimetic effects of smoked cannabis: naturalistic study: naturalistic study [corrected]. *Br J Psychiatry.* 2010;197(4):285–290.
79. Wise RA, Bozarth MA. Brain reward circuitry: four circuit elements "wired" in apparent series. *Brain Res Bull.* 1984;12(2):203–208.
80. Wise RA. Action of drugs of abuse on brain reward systems. *Pharmacol Biochem Behav.* 1980;13(suppl 1):213–223.
81. Koob GF, Bloom FE. Cellular and molecular mechanisms of drug dependence. *Science.* 1988;242(4879):715–723.
82. Di Chiara G, Imperato A. Drugs abused by humans preferentially increase synaptic dopamine concentrations in the mesolimbic system of freely moving rats. *Proc Natl Acad Sci USA.* 1988;85(14):5274–5278.
83. Nestler EJ. Is there a common molecular pathway for addiction? *Nat Neurosci.* 2005;8 (11):1445–1449.
84. Volkow ND, Wang GJ, Fowler JS, Tomasi D. Addiction circuitry in the human brain. *Annu Rev Pharmacol Toxicol.* 2012;52:321–336.
85. Tsai HC, Zhang F, Adamantidis A, et al. Phasic firing in dopaminergic neurons is sufficient for behavioral conditioning. *Science.* 2009;324(5930):1080–1084.
86. Oleson EB, Cheer JF. A brain on cannabinoids: the role of dopamine release in reward seeking. *Cold Spring Harb Perspect Med.* 2012;2(8):1–13.
87. Ng Cheong Ton JM, Gerhardt GA, Friedemann M, et al. The effects of delta 9-tetrahydrocannabinol on potassium-evoked release of dopamine in the rat caudate nucleus: an *in vivo* electrochemical and *in vivo* microdialysis study. *Brain Res.* 1988;451 (1–2):59–68.

88. Chen J, Paredes W, Lowinson JH, Gardner EL. Delta 9-tetrahydrocannabinol enhances presynaptic dopamine efflux in medial prefrontal cortex. *Eur J Pharmacol.* 1990;190 (1–2):259–262.
89. Chen JP, Paredes W, Li J, Smith D, Lowinson J, Gardner EL. Delta 9-tetrahydrocannabinol produces naloxone-blockable enhancement of presynaptic basal dopamine efflux in nucleus accumbens of conscious, freely-moving rats as measured by intracerebral microdialysis. *Psychopharmacology.* 1990;102(2):156–162.
90. French ED. delta9-Tetrahydrocannabinol excites rat VTA dopamine neurons through activation of cannabinoid CB_1 but not opioid receptors. *Neurosci Lett.* 1997;226 (3):159–162.
91. French ED, Dillon K, Wu X. Cannabinoids excite dopamine neurons in the ventral tegmentum and substantia nigra. *Neuroreport.* 1997;8(3):649–652.
92. Tanda G, Pontieri FE, Di Chiara G. Cannabinoid and heroin activation of mesolimbic dopamine transmission by a common mu1 opioid receptor mechanism. *Science.* 1997;276(5321):2048–2050.
93. Gessa GL, Melis M, Muntoni AL, Diana M. Cannabinoids activate mesolimbic dopamine neurons by an action on cannabinoid CB_1 receptors. *Eur J Pharmacol.* 1998;341 (1):39–44.
94. Malone DT, Taylor DA. Modulation by fluoxetine of striatal dopamine release following Delta9-tetrahydrocannabinol: a microdialysis study in conscious rats. *Br J Pharmacol.* 1999;128(1):21–26.
95. Cheer JF, Wassum KM, Heien ML, Phillips PE, Wightman RM. Cannabinoids enhance subsecond dopamine release in the nucleus accumbens of awake rats. *J Neurosci.* 2004;24 (18):4393–4400.
96. Cheer JF, Wassum KM, Sombers LA, et al. Phasic dopamine release evoked by abused substances requires cannabinoid receptor activation. *J Neurosci.* 2007;27(4): 791–795.
97. Sombers LA, Beyene M, Carelli RM, Wightman RM. Synaptic overflow of dopamine in the nucleus accumbens arises from neuronal activity in the ventral tegmental area. *J Neurosci.* 2009;29(6):1735–1742.
98. Grace AA. The tonic/phasic model of dopamine system regulation and its implications for understanding alcohol and psychostimulant craving. *Addiction.* 2000;95(suppl 2): S119–S128.
99. Schultz W. Reward signaling by dopamine neurons. *Neuroscientist.* 2001;7(4):293–302.
100. Adamantidis AR, Tsai HC, Boutrel B, et al. Optogenetic interrogation of dopaminergic modulation of the multiple phases of reward-seeking behavior. *J Neurosci.* 2011;31 (30):10829–10835.
101. Wu X, French ED. Effects of chronic delta9-tetrahydrocannabinol on rat midbrain dopamine neurons: an electrophysiological assessment. *Neuropharmacology.* 2000;39 (3):391–398.
102. Hiroi N, Brown JR, Haile CN, Ye H, Greenberg ME, Nestler EJ. FosB mutant mice: loss of chronic cocaine induction of Fos-related proteins and heightened sensitivity to cocaine's psychomotor and rewarding effects. *Proc Natl Acad Sci USA.* 1997;94 (19):10397–10402.
103. Hope BT, Nye HE, Kelz MB, et al. Induction of a long-lasting AP-1 complex composed of altered Fos-like proteins in brain by chronic cocaine and other chronic treatments. *Neuron.* 1994;13(5):1235–1244.
104. Young ST, Porrino LJ, Iadarola MJ. Cocaine induces striatal c-fos-immunoreactive proteins via dopaminergic D1 receptors. *Proc Natl Acad Sci USA.* 1991;88 (4):1291–1295.
105. Ulery PG, Nestler EJ. Regulation of DeltaFosB transcriptional activity by Ser27 phosphorylation. *Eur J Neurosci.* 2007;25(1):224–230.

106. Nestler EJ. Molecular basis of long-term plasticity underlying addiction. *Nat Rev Neurosci.* 2001;2(2):119–128.
107. Nestler EJ. FosB: a transcriptional regulator of stress and antidepressant responses. *Eur J Pharmacol.* 2015;753:66–72.
108. Perrotti LI, Weaver RR, Robison B, et al. Distinct patterns of DeltaFosB induction in brain by drugs of abuse. *Synapse.* 2008;62(5):358–369.
109. Volkow ND, Wang GJ, Fowler JS, Tomasi D, Telang F. Addiction: beyond dopamine reward circuitry. *Proc Natl Acad Sci USA.* 2011;108(37):15037–15042.
110. Goldman M, Szucs-Reed RP, Jagannathan K, et al. Reward-related brain response and craving correlates of marijuana cue exposure: a preliminary study in treatment-seeking marijuana-dependent subjects. *J Addict Med.* 2013;7(1):8–16.
111. Filbey FM, Schacht JP, Myers US, Chavez RS, Hutchison KE. Marijuana craving in the brain. *Proc Natl Acad Sci USA.* 2009;106(31):13016–13021.
112. Williamson EM, Evans FJ. Cannabinoids in clinical practice. *Drugs.* 2000;60(6): 1303–1314.
113. Haj-Dahmane S, Shen RY. Modulation of the serotonin system by endocannabinoid signaling. *Neuropharmacology.* 2011;61(3):414–420.
114. Perrotti LI, Hadeishi Y, Ulery PG, et al. Induction of deltaFosB in reward-related brain structures after chronic stress. *J Neurosci.* 2004;24(47):10594–10602.
115. Heifets BD, Castillo PE. Endocannabinoid signaling and long-term synaptic plasticity. *Annu Rev Physiol.* 2009;71:283–306.
116. Fratta W, Fattore L. Molecular mechanisms of cannabinoid addiction. *Current Opin Neurobiol.* 2013;23(4):487–492.
117. Malenka RC. Synaptic plasticity in the hippocampus: LTP and LTD. *Cell.* 1994;78 (4):535–538.
118. Batalla A, Bhattacharyya S, Yucel M, et al. Structural and functional imaging studies in chronic cannabis users: a systematic review of adolescent and adult findings. *PloS One.* 2013;8(2):e55821.
119. Budney AJ, Hughes JR. The cannabis withdrawal syndrome. *Curr Opin Psychiatry.* 2006;19(3):233–238.
120. Koob GF, Le Moal M. Drug abuse: hedonic homeostatic dysregulation. *Science.* 1997;278(5335):52–58.
121. Koob GF, Le Moal M. Drug addiction, dysregulation of reward, and allostasis. *Neuropsychopharmacology.* 2001;24(2):97–129.
122. Elkashef A, Vocci F, Huestis M, et al. Marijuana neurobiology and treatment. *Subst Abus.* 2008;29(3):17–29.
123. Tanda G, Loddo P, Di Chiara G. Dependence of mesolimbic dopamine transmission on delta9-tetrahydrocannabinol. *Eur J Pharmacol.* 1999;376(1–2):23–26.
124. Ronan PJ, Summers CH. Molecular signaling and translational significance of the corticotropin releasing factor system. *Prog Mol Biol Transl Sci.* 2011;98:235–292.
125. Rodriguez de Fonseca F, Carrera MR, Navarro M, Koob GF, Weiss F. Activation of corticotropin-releasing factor in the limbic system during cannabinoid withdrawal. *Science.* 1997;276(5321):2050–2054.
126. Caberlotto L, Rimondini R, Hansson A, Eriksson S, Heilig M. Corticotropin-releasing hormone (CRH) mRNA expression in rat central amygdala in cannabinoid tolerance and withdrawal: evidence for an allostatic shift? *Neuropsychopharmacology.* 2004;29(1):15–22.
127. Callahan LB, Tschetter KE, Ronan PJ. Inhibition of corticotropin releasing factor expression in the central nucleus of the amygdala attenuates stress-induced behavioral and endocrine responses. *Front Neurosci.* 2013;7:195.
128. Rodaros D, Caruana DA, Amir S, Stewart J. Corticotropin-releasing factor projections from limbic forebrain and paraventricular nucleus of the hypothalamus to the region of the ventral tegmental area. *Neuroscience.* 2007;150(1):8–13.

129. Hyman SM, Sinha R. Stress-related factors in cannabis use and misuse: implications for prevention and treatment. *J Subst Abuse Treat.* 2009;36(4):400–413.
130. Van Bockstaele EJ. Cannabinoid receptor signaling and modulation of monoamines: implications for psychiatric and neurological disorders. *Neuropsychopharmacol Biol Psychiatry* 2012;38(1):1–3.
131. Carvalho AF, Van Bockstaele EJ. Cannabinoid modulation of noradrenergic circuits: implications for psychiatric disorders. *Prog Neuropsychopharmacol Biol Psychiatry.* 2012;38 (1):59–67.
132. Moranta D, Esteban S, Garcia-Sevilla JA. Differential effects of acute cannabinoid drug treatment, mediated by CB_1 receptors, on the *invivo* activity of tyrosine and tryptophan hydroxylase in the rat brain. *Naunyn Schmiedebergs Arch Pharmacol.* 2004;369 (5):516–524.
133. Fisar Z. Inhibition of monoamine oxidase activity by cannabinoids. *Naunyn Schmiedebergs Arch Pharmacol.* 2010;381(6):563–572.
134. Fisar Z. Cannabinoids and monoamine neurotransmission with focus on monoamine oxidase. *Prog Neuropsychopharmacol Biol Psychiatry.* 2012;38(1):68–77.
135. Egashira N, Mishima K, Katsurabayashi S, et al. Involvement of 5-hydroxytryptamine neuronal system in Delta(9)-tetrahydrocannabinol-induced impairment of spatial memory. *Eur J Pharmacol.* 2002;445(3):221–229.
136. Tao R, Ma Z. Neural circuit in the dorsal raphe nucleus responsible for cannabinoid-mediated increases in 5-HT efflux in the nucleus accumbens of the rat brain. *ISRN Pharmacol.* 2012;2012:276902.
137. Tao R, Auerbach SB. Regulation of serotonin release by GABA and excitatory amino acids. *J Psychopharmacol.* 2000;14(2):100–113.
138. Tao R, Auerbach SB. Influence of inhibitory and excitatory inputs on serotonin efflux differs in the dorsal and median raphe nuclei. *Brain Res.* 2003;961(1):109–120.
139. Katona I, Freund TF. Multiple functions of endocannabinoid signaling in the brain. *Annu Rev Neurosci.* 2012;35:529–558.
140. Chiarlone A, Bellocchio L, Blazquez C, et al. A restricted population of CB_1 cannabinoid receptors with neuroprotective activity. *Pro Natl Acad Sci USA.* 2014;111 (22):8257–8262.
141. Haj-Dahmane S, Shen RY. Endocannabinoids suppress excitatory synaptic transmission to dorsal raphe serotonin neurons through the activation of presynaptic CB_1 receptors. *J Pharmacol Exp Ther.* 2009;331(1):186–196.
142. Haj-Dahmane S, Shen RY. Regulation of plasticity of glutamate synapses by endocannabinoids and the cyclic-AMP/protein kinase A pathway in midbrain dopamine neurons. *J Physiol.* 2010;588(Pt 14):2589–2604.
143. Darmani NA, Janoyan JJ, Kumar N, Crim JL. Behaviorally active doses of the CB_1 receptor antagonist SR 141716A increase brain serotonin and dopamine levels and turnover. *Pharmacol Biochem Behav.* 2003;75(4):777–787.
144. Tzavara ET, Davis RJ, Perry KW, et al. The CB_1 receptor antagonist SR 141716A selectively increases monoaminergic neurotransmission in the medial prefrontal cortex: implications for therapeutic actions. *Br J Pharmacol.* 2003;138(4):544–553.
145. Aso E, Renoir T, Mengod G, et al. Lack of CB_1 receptor activity impairs serotonergic negative feedback. *J Neurochem.* 2009;109(3):935–944.
146. Vaswani M, Linda FK, Ramesh S. Role of selective serotonin reuptake inhibitors in psychiatric disorders: a comprehensive review. *Prog Neuropsychopharmacol Biol Psychiatry.* 2003;27(1):85–102.
147. Micale V, Di Marzo V, Sulcova A, Wotjak CT, Drago F. Endocannabinoid system and mood disorders: priming a target for new therapies. *Pharmacol Ther.* 2013;138(1):18–37.
148. Haddjeri N, Blier P, de Montigny C. Long-term antidepressant treatments result in a tonic activation of forebrain 5-HT1A receptors. *J Neurosci.* 1998;18(23):10150–10156.

149. Bambico FR, Hattan PR, Garant JP, Gobbi G. Effect of delta-9-tetrahydrocannabinol on behavioral despair and on pre- and postsynaptic serotonergic transmission. *Prog Neuropsychopharmacol Biol Psychiatry.* 2012;38(1):88–96.
150. Lev-Ran S, Roerecke M, Le Foll B, George TP, McKenzie K, Rehm J. The association between cannabis use and depression: a systematic review and meta-analysis of longitudinal studies. *Psychol Med.* 2014;44(4):797–810.
151. Scavone JL, Mackie K, Van Bockstaele EJ. Characterization of cannabinoid-1 receptors in the locus coeruleus: relationship with mu-opioid receptors. *Brain Res.* 2010;1312:18–31.
152. Carvalho AF, Mackie K, Van Bockstaele EJ. Cannabinoid modulation of limbic forebrain noradrenergic circuitry. *Eur J Neurosci.* 2010;31(2):286–301.
153. Mechoulam R, Peters M, Murillo-Rodriguez E, Hanus LO. Cannabidiol—recent advances. *Chem Biodivers.* 2007;4(8):1678–1692.
154. Mechoulam R. Marihuana chemistry. *Science.* 1970;168(3936):1159–1166.
155. Russo EB, Burnett A, Hall B, Parker KK. Agonistic properties of cannabidiol at 5-HT1a receptors. *Neurochem Res.* 2005;30(8):1037–1043.
156. Zuardi AW. Cannabidiol: from an inactive cannabinoid to a drug with wide spectrum of action. *Rev Bras Psiquiatr.* 2008;30(3):271–280.
157. Pertwee RG. The diverse CB_1 and CB_2 receptor pharmacology of three plant cannabinoids: delta9-tetrahydrocannabinol, cannabidiol and delta9-tetrahydrocannabivarin. *Br J Pharmacol.* 2008;153(2):199–215.
158. Ryan D, Drysdale AJ, Lafourcade C, Pertwee RG, Platt B. Cannabidiol targets mitochondria to regulate intracellular Ca^{2+} levels. *J Neurosci.* 2009;29(7):2053–2063.
159. Bisogno T, Hanus L, De Petrocellis L, et al. Molecular targets for cannabidiol and its synthetic analogues: effect on vanilloid VR1 receptors and on the cellular uptake and enzymatic hydrolysis of anandamide. *Br J Pharmacol.* 2001;134(4):845–852.
160. Leweke FM, Giuffrida A, Wurster U, Emrich HM, Piomelli D. Elevated endogenous cannabinoids in schizophrenia. *Neuroreport.* 1999;10(8):1665–1669.
161. Giuffrida A, Leweke FM, Gerth CW, et al. Cerebrospinal anandamide levels are elevated in acute schizophrenia and are inversely correlated with psychotic symptoms. *Neuropsychopharmacology.* 2004;29(11):2108–2114.
162. Leweke FM, Piomelli D, Pahlisch F, et al. Cannabidiol enhances anandamide signaling and alleviates psychotic symptoms of schizophrenia. *Transl Psychiatry.* 2012;2:e94.
163. Damjanovic A, Pantovic M, Damjanovic A, et al. Cannabis and psychosis revisited. *Psychiatr Danub.* 2015;27(1):97–100.
164. Hall W, Degenhardt L. High potency cannabis: a risk factor for dependence, poor psychosocial outcomes, and psychosis. *BMJ.* 2015;350:h1205.
165. Maa E, Figi P. The case for medical marijuana in epilepsy. *Epilepsia.* 2014;55(6):783–786.
166. Porter BE, Jacobson C. Report of a parent survey of cannabidiol-enriched cannabis use in pediatric treatment-resistant epilepsy. *Epilepsy Behav.* 2013;29(3):574–577.
167. Wongngamnit NK, G. Marijuana—Medicine by Legislation. *NCADD Addiction Medicine Update*; 2014

CHAPTER FIVE

The Nociceptin Receptor as an Emerging Molecular Target for Cocaine Addiction

Kabirullah Lutfy*,[1], Nurulain T. Zaveri†

* Department of Pharmaceutical Sciences, College of Pharmacy, Western University of Health Sciences, Pomona, California, USA

† Astraea Therapeutics, LLC, Mountain View, California, USA

[1] Corresponding author: e-mail address: klutfy@westernu.edu.

Contents

Abstract

Cocaine addiction is a global public health and socioeconomic issue that requires pharmacological and cognitive therapies. Currently there are no FDA-approved medications to treat cocaine addiction. However, in preclinical studies, interventions ranging from herbal medicine to deep-brain stimulation have shown promise for the therapy of cocaine addiction. Recent developments in molecular biology, pharmacology, and medicinal chemistry have enabled scientists to identify novel molecular targets along the pathways involved in drug addiction. In 1994, a receptor that showed a great deal of homology to the traditional opioid receptors was characterized. However, endogenous and exogenous opioids failed to bind to this receptor,

Progress in Molecular Biology and Translational Science, Volume 137
ISSN 1877-1173
http://dx.doi.org/10.1016/bs.pmbts.2015.10.003

which led scientists to name it opioid receptor-like receptor, now referred to as the nociceptin receptor. The endogenous ligand of NOPr was identified a year later and named orphanin FQ/nociceptin. Nociceptin and NOPr are widely distributed throughout the CNS and are involved in many physiological responses, such as food intake, nociceptive processing, neurotransmitter release, etc. Furthermore, exogenous nociceptin has been shown to regulate the activity of mesolimbic dopaminergic neurons, glutamate, and opioid systems, and the stress circuit. Importantly, exogenous nociceptin has been shown to reduce the rewarding and addictive actions of a number of drugs of abuse, such as psychostimulants, alcohol, and opioids. This paper reviews the existing literature on the role of endogenous nociceptin in the rewarding and addictive actions of cocaine. The effect of exogenous nociceptin on these processes is also reviewed. Furthermore, the effects of novel small-molecule NOPr ligands on these actions of cocaine are discussed. Overall, a review of the literature suggests that NOPr could be an emerging target for cocaine addiction pharmacotherapy.

1. INTRODUCTION

Addiction to cocaine and other addictive substances is a chronic and relapsing brain disorder brought about by neuronal adaptive changes along numerous brain circuits and is characterized by uncontrollable drug-taking and drug-seeking behaviors in spite of adverse consequences (such as loss of family, friendship, productivity, and most importantly risk of sudden death due to cocaine overdose) associated with continuation of such behaviors. Cocaine addiction is a major public health issue and places a tremendous burden on the person, relationships, productivity and society as a whole. Estimates show that about 1.2 million people consumed cocaine for the very first time in the United States in 2001.[1] Furthermore, in 2007, it was estimated that more than 2 million people over the age of 12 were current cocaine users and females were found to be more sensitive to the addictive actions of cocaine than males.[2] Estimates show that the lifetime prevalence of cocaine use in 2013 was 14.3% (http://www.drugabuse.gov/drugs-abuse/cocaine) in adults aged 12 and older. Despite these alarming estimates and negative consequences, there is no FDA-approved pharmacotherapy to treat cocaine addiction. Thus, further research is needed to characterize novel targets to develop new pharmacotherapy and effectively curb this chronic and relapsing brain disorder. Most drugs of abuse, such as cocaine, nicotine, opioids, and amphetamines increase dopamine levels in the nucleus accumbens albeit through different mechanisms. Cocaine enhances accumbal dopamine levels by blocking the dopamine transporter. Therefore, the dopamine receptor and dopamine transporter have been the conventional targets for medication development for cocaine addiction. However, this strategy has failed to yield clinically useful cocaine dependence medications. Several other receptor systems that affect different phases of cocaine addiction are now being investigated. Among these, opioidergic, γ-amino butyric acid[3] -ergic,[4] endocannabinoid,[5] serotonergic, and cholinergic agonists have been investigated for their inhibition of cocaine reward. A growing body of evidence suggests that the nociceptin receptor[6] could be a potential target for novel medications to treat cocaine addiction. This review discusses the existing literature around this molecular target showing that NOPr agonists could potentially provide beneficial effects in curbing cocaine addiction.

2. ORPHANIN FQ/NOCICEPTIN/NOPr SYSTEM

In 1994, several laboratories[7–10] cloned a receptor that showed a great deal of homology to the classical opioid receptors, namely mu, delta, and kappa, in particular in the transmembrane domains, yet neither endogenous opioids nor exogenous opioid ligands (except a few) showed appreciable affinity toward this receptor. Therefore, this receptor was named the opioid receptor-like (ORL) receptor. Later on, the International Union of Pharmacology (IUPHAR) in agreement with the scientific community renamed it nociceptin/orphanin FQ receptor (NOR) and then nociceptin receptor (NOPr). In 1995, two laboratories independently identified a heptadecapeptide as the endogenous ligand of this receptor.[11,12] One group named this peptide nociceptin because intracerebroventricular administration of the peptide shortened hot-plate latency, indicative of pain in rodents;[11] while the other group called it orphanin FQ because the peptide was the endogenous ligand of an orphan receptor, and F and Q stand for the first (phenylalanine) and last (glutamine) amino acid of the peptide sequence.[12] Orphanin FQ/nociceptin (OFQ/N) is a heptadecapeptide that shows similarities to the endogenous opioid peptides and in particular to dynorphin 1–17.[11,13] Interestingly, the NOPr and classical opioid receptors also share common signaling mechanisms. Activation of the opioid receptors and the NOPr leads to inhibition of adenylyl cyclase,[11,12,14] decrease in calcium currents,[15,16] activation of potassium channels[17,18] and mitogen-activated protein kinase.[19–22]

Nociceptin and its receptor (NOPr) are widely distributed throughout the CNS[23,24] and have been implicated in many physiological responses ranging from modulation of pain to learning and memory, food intake, and drug reward (for review see Ref. [25]). Importantly, exogenously applied OFQ/N has been shown to reduce the activity of the mesolimbic dopaminergic neurons,[26–28] glutamatergic neurons,[29] opioidergic neurons,[30] and the stress circuit.[31–37] Interestingly, these neurons and circuits are involved in motivated behaviors, reward, learning and memory, anxiety and stress response, processes that play a major role in the development and maintenance of addictive behaviors. Therefore, the OFQ/N–NOPr system could be a potential target to develop small-molecule NOPr agonists to treat cocaine addiction because OFQ/N and other NOPr agonists have multiple effects that are opposite of cocaine's action at these molecular targets (for details see Section 3).

2.1 Effects of Exogenous OFQ/N on the Actions of Cocaine

The discovery of NOPr and its endogenous ligand, and the demonstration that OFQ/N exerts antiopioid effects in the brain, prompted Murphy and coworkers (1996) to assess the effect of this peptide on extracellular dopamine in the nucleus accumbens.[26] These authors found that intracerebroventricular administration of OFQ/N reduced accumbal dopamine in the microdialysate in anesthetized rats in a concentration-dependent manner,[26] suggesting that OFQ/N acts as a brake to regulate the function of the mesolimbic dopaminergic neurons. Subsequent studies demonstrated that local OFQ/N administration in the ventral tegmental area (VTA) reduced extracellular levels of dopamine in the nucleus accumbens (NAc),[27] suggesting that the VTA may be at least one of the sites where OFQ/N exerts its regulatory action on extracellular accumbal dopamine. Consistent with this notion, other investigators reported the presence of NOPr on dopaminergic neurons in the VTA.[38,39] Furthermore, exogenously administered OFQ/N was found to regulate the activity of dopaminergic neurons in the VTA.[40] The regulatory action of OFQ/N on extracellular dopamine in the NAc was later shown in freely moving mice.[41] These authors also showed the selectivity of OFQ/N's action in this regard, as OFQ/N failed to reduce the level of accumbal dopamine in mice lacking NOPr.[41] OFQ/N has also been shown to decrease synthesis of dopamine in synaptosomes obtained from rat NAc and striatum, as OFQ/N reduced phosphorylation of tyrosine hydroxylase, the rate-limiting enzyme in the synthesis of catecholamines.[42] The action of OFQ/N on extracellular dopamine is reminiscent of the effect of dynorphin, which is structurally similar to OFQ/N. However, unlike dynorphin and related κ-opioid receptor agonists, OFQ/N does not induce aversion in rats,[43] although Sakoori and Murphy have found a modest aversive response following OFQ/N administration in mice.[44] Nevertheless, it is not entirely clear why, even though the two peptides reduce extracellular accumbal dopamine, only dynorphin elicits aversion.[45,46] Thus, further research is needed in this area to delineate the underlying mechanism of this divergent effect of dynorphin and nociceptin on hedonic homeostasis.

While there is moderate to high density of NOPr expression in the mesolimbic areas, few studies have examined the effect of exogenous NOPr stimulation on cocaine-elevated extracellular dopamine in the NAc. Given that cocaine elevates extracellular dopamine by blocking the dopamine transporter and that this action of cocaine is impulse-dependent, OFQ/N might also reduce the ability of cocaine to elevate extracellular dopamine in the NAc. We tested this hypothesis first and showed that intracerebroventricular OFQ/N administration prior to systemic cocaine treatment reduced cocaine-elevated accumbal dopamine levels in anesthetized rats.[47] Recently, Vazquez-DeRose and coworkers demonstrated that this action of OFQ/N could also be mediated locally in the NAc, as they showed that retrodialysis of OFQ/N into the NAc shell abolished cocaine-elevated extracellular dopamine in the NAc.[48] Given that the ability of cocaine to elevate extracellular dopamine levels in the NAc plays an essential role in the motor stimulatory action of cocaine,[49] we tested whether OFQ/N could also reduce hyperlocomotion induced by cocaine. Our results revealed that intracerebroventricular administration of OFQ/N dose-dependently reduced cocaine-induced motor stimulation.[47] OFQ/N may exert its action, at least in part, in the VTA because local injection of OFQ/N in this brain region blunted cocaine-stimulated locomotor activity in rats.[50]

The effect of OFQ/N has also been studied on the rewarding actions of cocaine and other psychostimulants. In these studies, the conditioned place preference (CPP) paradigm has been used as an animal model of drug reward or drug-seeking behavior.[51] The first study reporting the effect of OFQ/N on the expression of CPP was by Kotlinska and coworkers, who showed that intracerebroventricular OFQ/N administration suppressed the expression of cocaine-induced CPP in rats.[52] This group also studied the effect of OFQ/N on the acquisition of amphetamine-induced CPP.[53] Rats treated with OFQ/N prior to each amphetamine injection during the acquisition phase of CPP were found to exhibit a significant reduction in amphetamine-induced CPP compared to their vehicle-treated controls. These authors also showed that while OFQ/N did not affect the acute stereotypic effect of amphetamine, it was able to reduce the sensitized response that developed to this effect of amphetamine following its repeated administration, suggesting that OFQ/N affects the neuronal adaptive changes that develop following repeated drug administration.[53] In a similar experimental paradigm, using methamphetamine instead of amphetamine, Zhao and coworkers reported that OFQ/N reduced the rewarding action of methamphetamine in rats.[54] The effect of OFQ/N on the rewarding action of cocaine was also studied.[44] These authors demonstrated that intracerebroventricular administration of OFQ/N prior to cocaine administration on each conditioning day attenuated the CPP response, showing that OFQ reduced the rewarding action of cocaine.[44]

Repeated cocaine treatment causes neuronal adaptive changes along the mesolimbic dopaminergic and glutamatergic neurons (for a review see Ref. [55]). Along with these neuroadaptive alterations, the motor stimulatory, rewarding, and other behavioral effects of cocaine are enhanced, a phenomenon known as behavioral sensitization. Bebawy and coworkers showed that repeated cocaine treatment enhanced the rewarding action of cocaine and this conditioned rewarding effect of cocaine was reduced in mice treated with OFQ/N before each cocaine injection.[56]

The effect of exogenous OFQ/N on cocaine-induced locomotor sensitization has been studied.[50,56,57] The phenomenon of locomotor sensitization is referred to as an enduring increase in the motor-stimulatory action of cocaine that develops following its repeated intermittent treatment. Neuronal adaptive changes develop along the mesolimbic dopaminergic and corticolimbic glutamatergic neurons in animals sensitized to cocaine and other addictive drugs. Interestingly, similar alterations have been reported in human addicts and these changes are thought to mediate the motivational behavioral changes observed in drug addicts, such as craving and relapse. Therefore, the phenomenon of behavioral sensitization in rodents has been proposed to mimic some aspects of addiction in humans.[55,58–65] An earlier report failed to show any effect of OFQ/N on cocaine sensitization.[57] These authors used the same concentration of OFQ/N during the induction of sensitization (i.e., on days 1–3) and found that the inhibitory effect of OFQ/N on cocaine-induced motor stimulation was evident only on day 1 but not on days 2 and 3 of the cocaine-sensitizing regimen, raising the possibility that the lack of effect of OFQ/N might have been due to the development of tolerance. Interestingly, a rapid tolerance develops toward the motor suppressant effect of OFQ/N.[66]

However, using escalating doses of OFQ/N during the induction of sensitization, Lutfy *et al.*, demonstrated that intracerebroventricular as well as intra-VTA injection of OFQ/N blocked the development of locomotor sensitization in rats.[50] These results were replicated in C57BL/6 mice,[56] showing the generality of the inhibitory action of OFQ/N on this phenomenon. These authors also demonstrated that OFQ/N failed to alter cocaine-induced locomotor sensitization in mice lacking NOPr, showing that OFQ/N exerts its inhibitory action on cocaine sensitization via NOPr. An intriguing observation from these studies was that OFQ/N not only blocked the development of locomotor sensitization but also reversed the time course of an established sensitized response.[56] The reversal of cocaine sensitization suggests that OFQ/N may facilitate the process leading to extinction of the conditioned response. Interestingly, we recently discovered that extinction of cocaine-induced CPP was facilitated in mice treated with a novel small-molecule NOPr agonist (Lutfy and Zaveri, unpublished data), whereas the opposite was evident in mice lacking nociceptin as well as in wild-type mice treated with a NOPr antagonist, J-113397 (Marquez and Lutfy, unpublished data). Thus, the OFQ/N–NOPr system appears to alter behaviors induced by cocaine in mice, and may be a potential target to develop medications to treat cocaine addiction.

While OFQ/N clearly reduces the rewarding effects of psychostimulant drugs as measured in the place-conditioning paradigm, it does not appear to affect cocaine-seeking behavior after extinction in rats.[67] However, OFQ/N is able to attenuate stress-induced ethanol-seeking behavior. These results are in line with the previous observations that OFQ/N failed to affect heroin self-administration in rats at any dose, even though the higher doses (10 μg) caused suppression of spontaneous locomotor activity.[68] It is possible that the reinforcing properties of a drug, measured by the operant self-administration paradigm are different than the rewarding, hedonic effects of a drug, as measured in the place-conditioning paradigm, and that OFQ/N affects the rewarding effects but has no effect on the reinforcing properties of an abused drug. Alternatively, the OFQ/N–NOPr system may become overwhelmed following repeated drug administration in the operant conditioning paradigm, which involves more dosing with the drugs of abuse. Clearly, further studies using more aggressive OFQ/N treatment would be needed to understand these differences. Furthermore, studies with small-molecule NOPr agonists, which are expected to mimic NOPr activation by the neuropeptide OFQ/N, would be useful to determine the potential of NOPr agonism as a novel pharmacologic approach to treat the various aspects of substance abuse disorders.

2.2 Endogenous OFQ/N as a Brake to the Rewarding and Addictive Actions of Cocaine

While there is a clear indication that exogenous OFQ/N regulates the activity of the mesolimbic dopaminergic neurons, the role of the endogenous OFQ/N is not entirely clear although some research has been performed in this area. Koizumi *et al.* showed that while intracerebroventricular OFQ/N administration suppressed extracellular dopamine in the NAc, the NOPr antagonist peptide, [Nphe1, Arg14, Lys15] nociceptin-NH(2), known as UFP-101, failed to alter the levels of dopamine, suggesting that endogenous OFQ/N may not regulate the activity of mesolimbic dopaminergic neurons.[41] However, in a different report, these authors showed that basal level of dopamine was mildly higher (but not significant) in mice lacking NOPr compared to their wild-type controls.[69] These authors also showed that administration of Compound B (1-[(3R, 4R)-1-cyclooctylmethyl-3-hydroxymethyl-4-piperidyl]-3-ethyl-1,3-dihydro-2H-benzimidazol-2-one), a small-molecule NOPr antagonist, increased extracellular dopamine in the NAc. However, this response was not affected by prior administration of OFQ/N. Also, the increase in dopamine induced by Compound B was not altered in mice lacking NOPr, suggesting that receptors other than NOPr may be involved in the action of Compound B. Thus, further research with selective NOP antagonists is needed to fully characterize whether endogenous OFQ/N regulates the activity of mesolimbic dopaminergic neurons.

The endogenous OFQ/N also plays a role in the rewarding and addictive actions of cocaine. Likewise, acute and chronic cocaine treatment each appears to modulate the level of endogenous OFQ/N. For instance, Romualdi and coworkers showed that the level of OFQ/N was reduced in the medial region of the caudate putamen as well as in the NAc shell and substantia nigra following chronic cocaine administration.[70] Furthermore, Lutfy *et al.* showed that the level of OFQ/N-immunoreactivity (OFQ/N-IR) was reduced in the midbrain area 1 h after a single cocaine (20 mg/kg) administration. Reductions were also seen in the striatum and prefrontal cortex but these changes were not significantly different from the saline-treated controls.[71] However, tissue content does not provide information about the release, metabolism, and/or synthesis of the peptide. Additionally, given that these changes were measured 1 h after cocaine administration, future studies using microdialysis in freely moving animals are needed to assess the dose--response and time course effect of single and repeated cocaine administration on the release of endogenous OFQ/N. Interestingly, however, the level of OFQ/N-IR was found to be increased in the hippocampus of rats that received repeated intermittent cocaine treatment and showed locomotor sensitization.[71] These results suggest that the endogenous OFQ/N may be involved in the acute and chronic actions of cocaine.

Consistent with these findings, Marquez and coworkers showed that the rewarding action of acute cocaine was enhanced in mice lacking NOPr compared to their wild-type littermates/controls.[72] In addition, cocaine-induced locomotor sensitization induced by repeated intermittent cocaine administration was enhanced in these mice compared to their wild-type controls.[73] This response appeared to be selective for the action of cocaine since locomotor sensitization induced by repeated amphetamine treatment was not altered in mice lacking NOPr.

2.3 Effect of Small-Molecule NOPr Agonists on Actions of Cocaine

Relatively little is known about the effect of small-molecule NOPr ligands on the rewarding and addictive effects of cocaine and other psychostimulants such as amphetamines. A number of laboratories have developed NOPr ligands that are relatively selective and have shown promise to reduce the rewarding action of morphine[74,75] and alcohol.[76,77] However, small-molecule NOPr agonists have not been extensively evaluated for their effects on cocaine-induced reward or other effects of cocaine. The few studies published with nonpeptide small-molecule NOPr agonists showed equivocal results regarding the effects of these compounds on cocaine-induced reward. Ro 65-6570, an NOPr agonist that belongs to the same class of NOPr agonists as Ro 64-6198[78] but has higher affinity at the μ-opioid receptor than Ro 64-6198, showed no effect on the expression of cocaine-induced place preference in rats[52] or on the acquisition of cocaine-induced place preference in rats,[79] although Ro 64-6198 itself was not investigated for these effects. It is possible that lower selectivity of this compound for NOPr versus the μ-opioid receptor (MOPr) may be responsible for the lack of any effect of this compound. Thus, a combination of a mu antagonist with such nonselective NOPr agonists may reduce the rewarding effect of cocaine more strongly than the nonselective compounds administered alone. The potential of such an approach is highlighted in a recent report by Cordery and coworkers, who used a combination of low doses of buprenorphine (which has higher affinity for MOPr than NOPr) and naltrexone, a nonselective opioid receptor antagonist, that elicited neither CPP nor aversion but completely blocked cocaine-primed reinstatement of an extinguished CPP in rats.[80]

Given the inhibitory effect of exogenous OFQ/N on cocaine-induced elevation of dopamine in the NAc and on the expression and acquisition of cocaine CPP, it is possible that selective small-molecule NOP agonists may decrease the rewarding effects of cocaine in these paradigms. Indeed, we found that AT-202, a selective NOP agonist which has a much higher affinity and selectivity for the NOPr over MOPr compared to buprenorphine, blocked drug- and stress-induced reinstatement of extinguished CPP in ICR mice.[81] However, AT-202 did not reduce the acquisition of cocaine CPP in these mice,[81] raising the possibility that following repeated administration of these drugs, for example, during the acquisition phase of CPP, as opposed to one time application of these compounds, that is, prior to reinstatement of CPP, the mu-mediated effect of these compounds may interfere with their NOPr-mediated actions, thereby leading to lack of a significant effect of these drugs on the development of cocaine CPP. Clearly, further studies with newer and more selective small-molecule NOPr agonists are warranted to establish the potential of NOPr agonism as an approach to treat cocaine (and possibly other psychostimulant) addiction.

3. MOLECULAR MECHANISMS OF COCAINE ADDICTION

Chronic cocaine use causes neuroadaptive changes along a number of brain circuits, leading to uncontrollable drug-taking and drug-seeking behaviors. However, there is no FDA-approved pharmacotherapy to treat cocaine addiction. The lack of approved treatment may be due to the fact that different neurochemicals contribute to different stages of cocaine addiction. The mesolimbic dopaminergic neurons and the corticolimbic glutamatergic neurons have been implicated in the initiation and maintenance of addictive behaviors to cocaine and other addictive drugs, respectively.[55,82–91] Furthermore, the endogenous opioid system has been implicated in cocaine addiction although the role of each opioid peptide and its receptor is distinct. For example, activation of the μ- and δ-opioid receptors is known to be proaddictive while stimulation of the κ-opioid receptor appears antiaddictive.[77,92–110] Recent evidence suggests that the dynorphin/κ-opioid receptor system may also mediate the negative-affective states associated with drug withdrawal.[93,111–117] Stress and anxiety have been reported to be among the leading causes of craving, relapse, and compulsive cocaine use and abuse.[114,118–128] Therefore, a drug that reduces anxiety and has antistress and antiopioid effects on the brain, decreases cocaine-mediated elevation of dopamine in the NAc, and attenuates neuronal plasticity mediated by glutamate would have a broader spectrum of activity to reduce cocaine reward and reinforcement, and be a more effective pharmacotherapeutic agent to treat cocaine addiction. A growing body of literature suggests that OFQ/N possesses all of the previously mentioned characteristics (for review see Refs [5,35]) and the OFQ/N–NOPr system may be a potential target for medication development to treat cocaine addiction. The possible mechanism(s) through which OFQ/N could reduce the rewarding and addictive actions of cocaine have been discussed later in the chapter.

3.1 OFQ/N Alters the Positive Reinforcing Action of Cocaine

The primary mechanism of action of cocaine is to block the dopamine transporter[129–131] and elevate extracellular dopamine in the NAc.[132] The cocaine-elevated extracellular dopamine in the NAc is thought to mediate the motor stimulatory, rewarding, and motivational effects of cocaine. A number of studies using pharmacologic tools[130,131,133–135] or neurotoxins to cause lesion of the dopaminergic neurons[133,136–138] showed that dopamine plays an important role in the motor-stimulatory action of cocaine. The observation that mice lacking the dopamine transporter were hyperactive compared to their wild-type counterparts and that cocaine-induced hyperlocomotion was blunted in these mice, confirms that the site of action of cocaine to induce its motor-stimulatory action is the dopamine transporter. Interestingly, OFQ/N has been shown to decrease cocaine-elevated extracellular dopamine in the NAc and also reduce the motor-stimulatory action of cocaine.[47] Although a causal relationship has not been established between the two effects, they appeared to correlate with each other. The rise in extracellular dopamine in the NAc is also thought to mediate the acute pleasurable effects of cocaine as well as is proposed to be important in the initiation of cocaine addiction.[132] However, recent evidence suggests that dopamine may be involved in drug "wanting" than "liking."[139] Dopamine is also implicated in long-term potentiation (LTP) in the hippocampus and amygdala,[3,140–142] error prediction and aversion.[143–147] Nevertheless, dopamine-deficient mice exhibit a robust aversive response following administration of a κ-opioid receptor agonist,[148] raising concern about the importance of dopamine in the aversive effect of kappa agonists. Similarly, the role of dopamine in the rewarding and addictive actions of cocaine has been debated. For example, mice lacking the dopamine transporter and their wild-type controls exhibit a comparable cocaine-induced CPP, showing that cocaine retains its rewarding action in mice lacking the dopamine transporter.[149] Similarly, cocaine was able to induce CPP in dopamine-deficient mice although the response was attenuated in dopamine-deficient mice compared to their wild-type controls. Nevertheless, cocaine CPP was blunted in wild-type mice treated with SCH-23390, a dopamine D1 receptor antagonist, suggesting that dopamine D1 receptor activation is necessary for the CPP response in wild-type mice.[150] Additionally, neither dopamine transporter knockout nor dopamine-deficient mice have been tested in the CPP paradigm following a single conditioning with cocaine, a measure of acute rewarding action of cocaine.[72,151] As an example, we have shown that the rewarding action of acute morphine was blunted in mice lacking the pituitary adenylyl cyclase activating polypeptide (PACAP) compared to their wild-type controls. However, following repeated conditioning, a comparable morphine-induced CPP was observed in mice of the two genotypes.[152] Thus, it is possible that dopamine is still important in the acute rewarding action of cocaine but the CPP paradigm used is unable to capture that. Clearly, in the orthodox CPP paradigm, where repeated conditioning is used, other mechanisms become engaged in the rewarding actions of cocaine, as we observed regarding the role of PACAP in morphine-induced CPP.[152] Indeed, Sora and coworkers used mice lacking not only the dopamine transporter but also the serotonin transporter and their wild-type controls and discovered that cocaine was no longer able to induce CPP in these mice, suggesting that in the absence of the dopamine transporter, the serotonin transporter compensates and, therefore, the absence of both transporters was necessary for the abolishment of the rewarding action of cocaine.[153] Thus, drugs that target both transporters, such as some of the currently used antidepressants with dual inhibitory action on dopamine and serotonin transporters may also be potential candidates to treat cocaine addiction.

The dopamine system has been implicated in the early phases of cocaine addiction, as shown in preclinical studies.[154–156] However, as stated earlier, targeting only the dopamine system has not been a fruitful approach to treat cocaine addiction since dopamine may be involved in the initiation phase of addiction. The OFQ/N–NOPr system, on the other hand, is an attractive target for medication development to treat cocaine addiction because OFQ/N has been shown to alter the activity of the mesolimbic dopaminergic neurons and the effect of addictive drugs on extracellular dopamine in the NAc[26,47,48,157] via a broader effect on multiple neurotransmitter systems involved in drug reward. As discussed later in the chapter, the beneficial effects of OFQ/N on cocaine reward do not appear to be solely due to its ability to reduce extracellular dopamine in the NAc. OFQ/N has other actions that make it a unique peptide with the potential to reduce the rewarding and addictive actions of cocaine and other addictive drugs. Interestingly, OFQ/N has also been shown to alter the activity of serotonergic neurons.[33,158–163] Notably, OFQ/N may be devoid of any rewarding or aversive effects, as injection of OFQ/N into the lateral ventricles failed to induce CPP or aversion[43] although the same treatment reduced basal motor activity.[66] Thus, the OFQ/N–NOPr system may be a potentially better target to develop pharmacotherapy for cocaine addiction without inducing aversion. This also has a potential advantage for treatment of substance abuse in general, since OFQ/N has also been shown to reduce the rewarding action of other addictive drugs.[164–167] However, further research is needed to define the underlying mechanism of OFQ/N's action in reducing the rewarding action of cocaine and other addictive drugs. OFQ/N not only affects the dopamine and serotonin systems but also other mechanisms that could contribute to its antiaddictive properties. One such mechanism is the ability of OFQ/N to regulate the action of endogenous opioid peptides to exert its antirewarding actions (see below).

3.2 OFQ/N as an Antiopioid Peptide in the Brain to Reduce Actions of Cocaine

The endogenous opioid system has long been known as a major regulator of the mesolimbic dopaminergic neuronal activity. In particular, beta-endorphin and enkephalins have been shown to increase the activity of the mesolimbic dopaminergic neurons and stimulate the release of dopamine in the NAc.[94,100] The endogenous opioid system has also been implicated in the rewarding and addictive actions of cocaine and other addictive drugs. For example, naloxone, a nonselective opioid receptor antagonist, has been shown to reduce the motor stimulatory and rewarding actions of cocaine.[101] Similar observations have been reported using selective μ-opioid receptor antagonists[168] and antisense oligonucleotides against the μ-opioid receptor,[103] suggesting that cocaine may exert some of its action via the endogenous opioid system. The observation that beta-endorphin induces reward[4] and that it has higher affinity for the mu compared to the δ- and κ-opioid receptors prompted us to hypothesize that cocaine may exert its effects via the release of beta-endorphin. Indeed, using microdialysis in freely moving rats, Olive and coworkers showed that the level of beta-endorphin increases following systemic administration of cocaine.[169] Later studies showed that the increase in the level of beta-endorphin is also observed in animals engaged in cocaine self-administration[170–172] as well as in the incubation of cocaine-seeking behaviors.[173] Using the CPP paradigm as an animal model of reward,[51] we demonstrated that beta-endorphin may be involved in the rewarding action of acute cocaine, as mice lacking beta-endorphin exhibited blunted CPP compared to their wild-type littermates/controls.[151,174] Changes in the level of opioid peptides and their mRNA have also been reported following chronic cocaine treatment.[175–181] Given that OFQ/N is considered an antiopioid peptide in the brain[25,182] and that it regulates the activity of beta-endorphin-containing neurons in the hypothalamic arcuate nucleus,[30] which provide opioidergic inputs to the VTA and NAc,[183,184] we propose that OFQ/N may exert some of its inhibitory action on cocaine reward by blocking the action of beta-endorphin. Interestingly, we have observed that the rewarding action of acute cocaine was enhanced in mice lacking NOPr,[72] which is opposite of what is observed in mice lacking beta-endorphin, that is, reduced rewarding action of acute cocaine.[151] The two peptides (beta-endorphin and OFQ/N) exert opposite effects on the mesolimbic dopaminergic neurons causing, respectively, an increase[185] or decrease[26] in extracellular accumbal dopamine. Further investigation is needed to examine if OFQ/N exerts its action on cocaine reward via blocking the effect of endogenous opioid peptide(s), which may yield other potential targets and approaches to treat opioid and other forms of addiction.

3.3 OFQ/N as an Antistress Peptide in the Brain

Stress is known to increase the risk of drug use, abuse, and relapse (for review see Ref. [186]). Indeed, a positive correlation has been reported between negative-affective life events and drug abuse. For example, child abuse in the form of sexual or physical abuse has been linked to an earlier onset of substance use and abuse and other forms of mental illnesses.[187–190] In such individuals, a heightened responsiveness of the hypothalamic–pituitary–adrenal (HPA) axis may be induced by chronic stress or drug use, resulting in a greater sensitivity of these individuals to the reinforcing properties of drugs of abuse, thereby contributing to compulsive drug use (for review see Refs [114,191]). The acquisition of cocaine self-administration has been facilitated by different forms of stress, such as isolation[192] or social defeat[123] (for reviews see Refs [114,120–122,191,193]).

The initial step in activation of the HPA axis is the release of corticotropin-releasing hormone (CRH) or corticotropin-releasing factor (CRF) from the paraventricular nucleus of the hypothalamus. Interestingly, CRH is one of the major neuropeptides that play a prominent role in the development of addiction by inducing negative-affective states, such as anxiety and reward deficit via an action in the amygdala (for a review see Ref. [194]). Interestingly, OFQ/N has been shown to act as an antistress peptide and a physiologic CRF antagonist in the amygdala.[31,35,195] OFQ/N has been shown to reduce stress-mediated reinstatement of alcohol- (but not that of cocaine-) seeking behaviors.[67] Furthermore, AT-202, a small-molecule NOPr agonist has been shown to reduce stress-induced reinstatement of cocaine-induced CPP in mice.[81]

The negative reinforcing actions of cocaine also contribute to the maintenance of cocaine addiction. Indeed, the theory of the "dark side of addiction" suggests that occasional drug use leads to neuroadaptive changes along numerous neuronal circuits, in particular the mesolimbic dopaminergic neurons, the corticolimbic glutamatergic neurons, the stress and fear circuits involving CRH, dynorphin, and other peptides, leading to negative-affective states. For example, research in the past decade has shown that the level of CRH and dynorphin is altered following repeated cocaine administration, making the drug users experience negative-affective states in the absence of the drug, and forcing them to compulsively seek and take cocaine to reduce dysphoria, anxiety, and depression elicited by drug withdrawal. Thus, drugs that reduce these negative-affective states that develop following drug withdrawal may be useful to treat addiction to cocaine and other addictive drugs. It is important to note that OFQ/N has been shown to possess anti-CRF properties.[31,196,197] OFQ/N has also been shown to exert anxiolytic effects in a number of animal models of anxiety. For example, low concentration of OFQ/N as well as small-molecule NOPr agonists elicited anxiolytic effects in rats.[34,196–199] Consistent with these observations, mice lacking NOPr exhibit greater anxiety-like phenotype compared to their wild-type control in the elevated plus maze test.[200] Thus, activation of the OFQ/N–NOPr system has the potential to reduce negative-affective states following cocaine withdrawal and reduce craving, and this may be an additional mechanism through which small-molecule NOPr agonists may exert their beneficial effects on actions of cocaine and other addictive drugs. Interestingly, OFQ/N has been found to exert greater anxiolytic effects in animals with prior alcohol experience compared to their alcohol-naive controls.[201] However, in the same study, OFQ/N failed to reduce anxiety in alcohol-naive rats.[201] Thus, further studies are needed to determine if

OFQ/N would exert an anxiolytic effect in subjects with prior cocaine experience. This is particularly important because there are some reports showing that OFQ/N could be anxiogenic under certain experimental conditions.[196,197] Nevertheless, these studies highlight another potentially advantageous molecular mechanism and pharmacologic profile of NOPr agonists as drug addiction pharmacotherapy.

3.4 OFQ/N Alters Glutamate-Mediated Neuronal Plasticity

Earlier studies showed that glutamate neurotransmission is involved in behavioral changes observed following repeated cocaine administration. For example, Karler *et al.* showed that MK-801 (dizocilpine), which blocks the *N*-methyl-D-aspartate (NMDA) subtype of glutamate receptors, reduced cocaine sensitization.[202] Neurotransmission at the excitatory synapses through the release of glutamate is now implicated in a number of neuropsychiatric disorders, including drug addiction.[85,203,204] In particular, neuronal plastic changes have been shown to occur in the NAc core following cocaine withdrawal. For example, calcium permeable α-amino-3-hydroxy-5-methyl-4-isoxazolepropionic acid (AMPA) receptor accumulates in the NAc, and mediates the intensified cue-induced craving following withdrawal from long-access cocaine self-administration.[205] Additionally, glutamate may mediate the effect of neutral environmental stimuli that gain saliency once they are associated with drug administration and can serve as cues in the absence of the drug to elicit craving. Indeed, previous studies have shown that relapse to drug-seeking behavior occurs following exposure to drug-related cues and the motivation to self-administer drugs of abuse increases over the course of abstinence.[206–211] One of the major problems in the pharmacotherapy of drug addiction is the high frequency of relapse.[212,213] However, if the drug is unavailable (cue but no drug) during this critical period, these repeated cue-no-drug associations could lead to a decrease in the strength of the conditioned response or consolidation of new memory, thereby resulting in extinction of the conditioned response.

Research has shown that the level of basal glutamate is reduced in the NAc following repeated cocaine administration, raising the possibility that the activity of glutamate transporters is altered following cocaine treatment.[214] Indeed, the expression of glutamate transporter 1 is decreased in animals that underwent withdrawal, leading to increased glutamatergic neurotransmission in this brain region.[215] Therefore, drugs that reduce glutamatergic neurotransmission in this brain region may have potential to reduce cue-induced cocaine-seeking behaviors. Interestingly, Li and Markou have recently proposed that a viable approach to reduce glutamatergic neurotransmission is via activation of the metabotropic glutamate 7 receptors (mGluR7), which are located presynaptically.[216] These authors showed that systemic or local administration of an allosteric modulator of mGluR7, AMN082, into the NAc reduced cocaine self-administration and cue-induced cocaine-seeking behaviors.[216] Administration of an mGluR1 receptor allosteric modulator is another useful approach since this treatment has been shown to reduce the accumulation of calcium permeable AMPA receptors in the NAc that occur following cocaine withdrawal.[217,218] These studies show that a drug that can reduce glutamate release or reverse neuronal adaptive changes induced by repeated cocaine administration may have potential to reduce cocaine-seeking behaviors. Interestingly, OFQ/N has been shown to reduce potassium-evoked glutamate release in cortical slices.[29] OFQ/N has also been shown to reverse cocaine-induced behavioral sensitization.[56] The glutamate system has been implicated in mnemonic processes and OFQ/N exerts inhibitory effects on learning and memory and these mnemonic effects of OFQ/N may involve the NMDA receptor.[219–223] Thus, this may be another mechanism through which OFQ/N and related NOPr agonists reduce the rewarding and addictive actions of cocaine and other addictive drugs (see later in the chapter). However, studies are needed to assess whether OFQ/N can affect changes in glutamate neurotransmission in the NAc that occur in response to cocaine treatment and withdrawal.

3.5 Mnemonic Effects of OFQ/N and Cocaine Addiction

The transition from occasional drug use to compulsive drug intake or abuse involves instrumental and classical conditioning processes (for reviews see Refs [63,82,91,224,225]). Thus, when administration of cocaine or another drug of abuse is repeatedly associated with a distinct environment, the context serves as a cue and this cue-drug association can elicit positive subjective feelings even in the absence of the drug. Once this conditioned response is acquired and consolidated, it can last for months and even years, due to its persistent nature.[226] The reason for this may be that compulsive drug taking can further lead to overlearning and "automatic" compulsive behavior (for reviews see Refs [91,224,226]). Importantly, subsequent exposure to the same surroundings, even in the absence of drugs, brings up old memories in abstinent addicts, thereby leading to drug-seeking behaviors and relapse.[63,224,225,227,228]

A growing body of evidence suggests that OFQ/N and other NOPr agonists impair learning and memory. The first electrophysiologic study by Yu and coworkers showed that bath application of OFQ/N reduced LTP in the CA1 region of the hippocampus.[229] Subsequently, these authors demonstrated a similar inhibitory effect of OFQ/N in the dentate gyrus.[223] Behavioral studies also demonstrated that local injection of OFQ/N in the CA3 region of the hippocampus caused memory impairment in the water maze test.[230,231] This effect of OFQ/N was mediated via NOPr because it was blocked by UFP-101, a NOPr antagonist.[230] Exogenous OFQ/N has also been shown to reduce learning and memory in the passive avoidance and contextual fear tests.[219,232–234] The inhibitory effect of OFQ/N on mnemonic processes has been replicated using small-molecule NOPr agonists, as Ro64-6198 reduced contextual but not cued fear learning in C57BL/6N mice.[219] Considering that behavioral sensitization and place preference each involves conditioned learning, that is, association between subjective effects of addictive drugs and contextual cues, that LTP and other forms of neuronal plasticity, such as long-term depression (LTD), are the basis of memory formation, and that OFQ/N reduces LTP[223,229] and impairs spatial and other forms of memory,[231,235,236] it is possible that OFQ/N may selectively reduce the learning and remembrance of reward-related behaviors because OFQ/N failed to reduce naloxone-induced conditioned place aversion.[44] However, as stated earlier, OFQ/N has been shown to regulate potassium-evoked glutamate release in cortical slices,[29] suggesting that OFQ/N may exert its effects on the rewarding and addictive actions of cocaine via more than regulation of learning and memory in the hippocampus. Further studies are definitely needed to assess the impact of OFQ/N, locally injected into the hippocampal regions compared to its administration in the lateral ventricles, on the rewarding and addictive effects of cocaine.

Not only the exogenous but also endogenous OFQ/N regulates mnemonic processes, as mice lacking NOPr exhibit enhanced memory.[221,233,235,236] Additionally, these mice have also been reported to show heightened cocaine-induced CPP[72] and behavioral sensitization,[73] raising the possibility that the amplified response observed in NOPr knockout mice was due to enhanced memory in these mice. However, morphine-induced CPP[72] and amphetamine-induced locomotor sensitization[73] were comparable between mice lacking NOPr and their wild-type littermates/controls. Thus, further studies are needed to delineate the exact mechanisms of regulatory action of endogenous OFQ/N on process of memory, as related to drug reward.

3.6 The OFQ/N–NOPr System and Gene Expression

A number of studies have shown that epigenetic mechanisms play an important role in the development and maintenance of addictive behaviors. It has been well documented that the effect of addictive drugs, contextual cues, and other environmental stimuli are translated through gene expression/repression and involve chromatin remodeling.[237] Chromatin is made of DNA tightly wrapped around by a protein called histone. Different forms of histone proteins exist, such as H2A, H2B, H3, and H4. Chemical modifications of the N-terminus of histones make the DNA sequence accessible or inaccessible for translation. Acetylation of histones is shown to increase gene expression.[238] On the other hand, deacetylation or methylation (but not in all cases) of histone proteins leads to tight wrapping of DNA around the histone protein and unavailability of the DNA sequence for translation.[239]

An overwhelming body of evidence suggests that drugs of abuse increase or decrease plasticity at different synapses and induce structural and behavioral changes through epigenetic mechanisms in different brain areas (for reviews see Refs [237,240]). Chronic cocaine treatment, for example, has been shown to increase histone (H3) acetylation of cyclin-dependent kinase 5 (CDK5) and brain-derived neurotrophic factor (BDNF) at their promoters.[241] Interestingly, inhibitors of histone deacetylase 5 (HDAC5), an enzyme responsible for deacetylation of histone proteins has been shown to enhance the rewarding action of cocaine,[241] suggesting that HDAC5-derived genes are involved in the rewarding actions of cocaine. Indeed, cocaine has been shown to alter the activity of HDAC5 in the NAc. Additionally, overexpression of HDAC5 in the NAc has been shown to reduce the rewarding action of cocaine.[242] HDAC has also been implicated in the reinstatement process, as an HDAC inhibitor has been shown to reduce reinstatement of cocaine-induced CPP.[243] However, the role of HDAC5 in cocaine self-administration is not yet clear (for a review see Ref. [240]).

While there is a great deal of interest in epigenetic mechanisms, a survey of the literature revealed only a single report, which studied the effect of cocaine on gene expression at the promoter regions of OFQ/N and NOPr.[6] The authors of this study showed that continuous subcutaneous cocaine infusion reduced pOFQ/N gene expression in the NAc and lateral, but not medial, caudate. Concomitantly, they showed increased NOPr gene expression in the NAc and reduced expression in the lateral, but not medial, caudate. These changes were associated with methylation of tyrosine 4 and 27 of H3 protein.[6] However, further studies are needed to delineate the effect of cocaine on expression of genes encoding for OFQ/N and NOPr, because the previously mentioned study utilized subcutaneous cocaine infusion, which does not mimic cocaine regimens of cocaine addicts. In addition, further studies are needed to understand the role of endogenous OFQ/N on gene expression induced by cocaine treatment. Similarly, the effect of exogenous OFQ/N on cocaine-induced gene expression needs to be studied. OFQ/N reduces the level of phosphorylated CREB (pCREB, cAMP response element-binding protein), a transcriptional factor involved in gene expression. We propose that OFQ/N may possibly alter chromatin remodeling via reducing the level of pCREB. Interestingly, mice lacking the CREB-binding protein showed reduced cocaine-induced locomotor activity and attenuated locomotor sensitization.[244] Given that OFQ/N not only blocked the development of behavioral sensitization but also reversed its time course, we hypothesize that exogenous OFQ/N would also reverse chromatin remodeling induced by repeated cocaine treatment. Future studies are thus needed to enhance our understanding of the role of endogenous OFQ/N and the effect of exogenous OFQ/N on changes in chromatin remodeling induced by cocaine treatment.

4. CONCLUSIONS

Molecular targets of cocaine addiction involve a number of neurotransmitter and neuropeptide systems and therefore it is difficult, if not impossible, to treat this chronic and relapsing brain disorder. Drugs that affect more than one of these molecular targets and neurotransmitter systems may be useful to reduce the rewarding and addictive actions of cocaine. OFQ/N has been shown to possess antistress, anxiolytic, and antiopioid effects as well as to reduce the activity of dopaminergic and glutamatergic neurons. Thus, the OFQ/N–NOPr system is a promising target to develop medications to treat cocaine addiction. The development of nonpeptide small-molecule NOPr agonists and investigation of their efficacies on the various aspects of cocaine addiction will provide further validation of this approach for cocaine addiction treatment. Indeed, this approach has been proven useful for other drug addictions, as small-molecule NOPr agonists have been shown to reduce the rewarding action of morphine and alcohol

(Lutfy and Zaveri, unpublished data). We therefore expect that small-molecule NOPr agonists would also attenuate the rewarding and possibly addictive effects of cocaine.

REFERENCES

1. O'Brien MS, Anthony JC. Risk of becoming cocaine dependent: epidemiological estimates for the United States, 2000–2001. *Neuropsychopharmacology.* 2005;30(5): 1006–1018.
2. Kasperski SJ, Vincent KB, Caldeira KM, Garnier-Dykstra LM, O'Grady KE, Arria AM. College students' use of cocaine: results from a longitudinal study. *AddictBehav.* 2011;36 (4):408–411.
3. Bissiere S, Humeau Y, Luthi A. Dopamine gates LTP induction in lateral amygdala by suppressing feedforward inhibition. *Nat Neurosci.* 2003;6(6):587–592.
4. Amalric M, Cline EJ, Martinez Jr JL, Bloom FE, Koob GF. Rewarding properties of beta-endorphin as measured by conditioned place preference. *Psychopharmacology (Berl).* 1987;91(1):14–19.
5. Heilig M, Egli M. Pharmacological treatment of alcohol dependence: target symptoms and target mechanisms. *PharmacolTher.* 2006;111(3):855–876.
6. Caputi FF, Di Benedetto M, Carretta D, et al. Dynorphin/KOP and nociceptin/NOP gene expression and epigenetic changes by cocaine in rat striatum and nucleus accumbens. *Prog Neuropsychopharmacol Biol Psychiatry.* 2014;49:36–46.
7. Bunzow JR, Saez C, Mortrud M, et al. Molecular cloning and tissue distribution of a putative member of the rat opioid receptor gene family that is not a mu, delta or kappa opioid receptor type. *FEBS Lett.* 1994;347(2–3):284–288.
8. Chen Y, Fan Y, Liu J, et al. Molecular cloning, tissue distribution and chromosomal localization of a novel member of the opioid receptor gene family. *FEBSLett.* 1994;347 (2–3):279–283.
9. Fukuda K, Kato S, Mori K, et al. cDNA cloning and regional distribution of a novel member of the opioid receptor family. *FEBS Lett.* 1994;343(1):42–46.
10. Mollereau C, Parmentier M, Mailleux P, et al. ORL1, a novel member of the opioid receptor family. Cloning, functional expression and localization. *FEBSLett.* 1994;341(1): 33–38.
11. Meunier JC, Mollereau C, Toll L, et al. Isolation and structure of the endogenous agonist of opioid receptor-like ORL1 receptor. *Nature.* 1995;377(6549):532–535.
12. Reinscheid RK, Nothacker HP, Bourson A, et al. Orphanin FQ: a neuropeptide that activates an opioidlike G protein-coupled receptor. *Science.* 1995;270(5237): 792–794.
13. Julius D. Pharmacology. Home for an orphan endorphin. *Nature.* 1995;377(6549):476.
14. Meis S. Nociceptin/orphanin FQ: actions within the brain. *Neuroscientist.* 2003;9 (2):158–168.
15. Connor M, Yeo A, Henderson G. The effect of nociceptin on Ca^{2+} channel current and intracellular Ca^{2+} in the SH-SY5Y human neuroblastoma cell line. *Br J Pharmacol.* 1996;118(2):205–207.
16. Knoflach F, Reinscheid RK, Civelli O, Kemp JA. Modulation of voltage-gated calcium channels by orphanin FQ in freshly dissociated hippocampal neurons. *J Neurosci.* 1996;16(21):6657–6664.
17. Matthes H, Seward EP, Kieffer B, North RA. Functional selectivity of orphanin FQ for its receptor coexpressed with potassium channel subunits in *Xenopus laevis* oocytes. *Mol Pharmacol.* 1996;50(3):447–450.

18. Vaughan CW, Christie MJ. Increase by the ORL1 receptor (opioid receptor-like1) ligand, nociceptin, of inwardly rectifying K conductance in dorsal raphe nucleus neurones. *Br J Pharmacol.* 1996;117(8):1609–1611.
19. Fawzi AB, Zhang H, Weig B, Hawes B, Graziano MP. Nociceptin activation of the human ORL1 receptor expressed in Chinese hamster ovary cells: functional homology with opioid receptors. *Eur J Pharmacol.* 1997;336(2–3):233–242.
20. Fukuda K, Shoda T, Morikawa H, Kato S, Mori K. Activation of mitogen-activated protein kinase by the nociceptin receptor expressed in Chinese hamster ovary cells. *FEBS Lett.* 1997;412(2):290–294.
21. Huang P, Kehner GB, Cowan A, Liu-Chen LY. Comparison of pharmacological activities of buprenorphine and norbuprenorphine: norbuprenorphine is a potent opioid agonist. *J Pharmacol Exp Ther.* 2001;297(2):688–695.
22. Lutfy K, Eitan S, Bryant CD, et al. Buprenorphine-induced antinociception is mediated by mu-opioid receptors and compromised by concomitant activation of opioid receptor-like receptors. *J Neurosci.* 2003;23(32):10331–10337.
23. Neal Jr CR, Mansour A, Reinscheid R, et al. Opioid receptor-like (ORL1) receptor distribution in the rat central nervous system: comparison of ORL1 receptor mRNA expression with (125)I-[(14)Tyr]-orphanin FQ binding. *J Comp Neurol.* 1999;412(4): 563–605.
24. Neal Jr CR, Mansour A, Reinscheid R, Nothacker HP, Civelli O, Watson Jr SJ. Localization of orphanin FQ (nociceptin) peptide and messenger RNA in the central nervous system of the rat. *J Comp Neurol.* 1999;406(4):503–547.
25. Mogil JS, Pasternak GW. The molecular and behavioral pharmacology of the orphanin FQ/nociceptin peptide and receptor family. *Pharmacol Rev.* 2001;53(3):381–415.
26. Murphy NP, Ly HT, Maidment NT. Intracerebroventricular orphanin FQ/nociceptin suppresses dopamine release in the nucleus accumbens of anaesthetized rats. *Neuroscience.* 1996;75(1):1–4.
27. Murphy NP, Maidment NT. Orphanin FQ/nociceptin modulation of mesolimbic dopamine transmission determined by microdialysis. *J Neurochem.* 1999;73(1):179–186.
28. Murphy NP, Tan AM, Lam HA, Maidment NT. Nociceptin/orphanin FQ modulation of rat midbrain dopamine neurons in primary culture. *Neuroscience.* 2004;127(4): 929–940.
29. Nicol B, Lambert DG, Rowbotham DJ, Smart D, McKnight AT. Nociceptin induced inhibition of K+ evoked glutamate release from rat cerebrocortical slices. *Br J Pharmacol.* 1996;119(6):1081–1083.
30. Wagner EJ, Ronnekleiv OK, Grandy DK, Kelly MJ. The peptide orphanin FQ inhibits beta-endorphin neurons and neurosecretory cells in the hypothalamic arcuate nucleus by activating an inwardly-rectifying K+ conductance. *Neuroendocrinology.* 1998;67(2): 73–82.
31. Ciccocioppo R, Biondini M, Antonelli L, Wichmann J, Jenck F, Massi M. Reversal of stress- and CRF-induced anorexia in rats by the synthetic nociceptin/orphanin FQ receptor agonist, Ro 64-6198. *Psychopharmacology (Berl).* 2002;161(2):113–119.
32. Ciccocioppo R, Cippitelli A, Economidou D, Fedeli A, Massi M. Nociceptin/orphanin FQ acts as a functional antagonist of corticotropin-releasing factor to inhibit its anorectic effect. *Physiol Behav.* 2004;82(1):63–68.
33. Filaferro M, Ruggieri V, Novi C, et al. Functional antagonism between nociceptin/orphanin FQ and corticotropin-releasing factor in rat anxiety-related behaviors: involvement of the serotonergic system. *Neuropeptides.* 2014;48(4): 189–197.
34. Jenck F, Moreau JL, Martin JR, et al. Orphanin FQ acts as an anxiolytic to attenuate behavioral responses to stress. *Proc Natl Acad Sci USA.* 1997;94(26):14854–14858.

35. Witkin JM, Statnick MA, Rorick-Kehn LM, et al. The biology of Nociceptin/Orphanin FQ (N/OFQ) related to obesity, stress, anxiety, mood, and drug dependence. *PharmacolTher.* 2014;141(3):283–299.
36. Devine DP, Hoversten MT, Ueda Y, Akil H. Nociceptin/orphanin FQ content is decreased in forebrain neurones during acute stress. *J Neuroendocrinol.* 2003;15 (1):69–74.
37. Devine DP, Watson SJ, Akil H. Nociceptin/orphanin FQ regulates neuroendocrine function of the limbic-hypothalamic-pituitary-adrenal axis. *Neuroscience.* 2001;102 (3):541–553.
38. Maidment NT, Chen Y, Tan AM, Murphy NP, Leslie FM. Rat ventral midbrain dopamine neurons express the orphanin FQ/nociceptin receptor ORL-1. *Neuroreport.* 2002;13(9):1137–1140.
39. Norton CS, Neal CR, Kumar S, Akil H, Watson SJ. Nociceptin/orphanin FQ and opioid receptor-like receptor mRNA expression in dopamine systems. *J Comp Neurol.* 2002;444(4):358–368.
40. Zheng F, Grandy DK, Johnson SW. Actions of orphanin FQ/nociceptin on rat ventral tegmental area neurons *in vitro*. *BrJ Pharmacol.* 2002;136(7):1065–1071.
41. Koizumi M, Midorikawa N, Takeshima H, Murphy NP. Exogenous, but not endogenous nociceptin modulates mesolimbic dopamine release in mice. *J Neurochem.* 2004;89(1):257–263.
42. Olianas MC, Dedoni S, Boi M, Onali P. Activation of nociceptin/orphanin FQ-NOP receptor system inhibits tyrosine hydroxylase phosphorylation, dopamine synthesis, and dopamine D(1) receptor signaling in rat nucleus accumbens and dorsal striatum. *J Neurochem.* 2008;107(2):544–556.
43. Devine DP, Reinscheid RK, Monsma Jr FJ, Civelli O, Akil H. The novel neuropeptide orphanin FQ fails to produce conditioned place preference or aversion. *Brain Res.* 1996;727(1–2):225–229.
44. Sakoori K, Murphy NP. Central administration of nociceptin/orphanin FQ blocks the acquisition of conditioned place preference to morphine and cocaine, but not conditioned place aversion to naloxone in mice. *Psychopharmacology (Berl).* 2004;172(2): 129–136.
45. Chefer VI, Backman CM, Gigante ED, Shippenberg TS. Kappa opioid receptors on dopaminergic neurons are necessary for kappa-mediated place aversion. *Neuropsychopharmacology.* 2013;38(13):2623–2631.
46. Spanagel R, Almeida OF, Bartl C, Shippenberg TS. Endogenous kappa-opioid systems in opiate withdrawal: role in aversion and accompanying changes in mesolimbic dopamine release. *Psychopharmacology (Berl).* 1994;115(1–2):121–127.
47. Lutfy K, Do T, Maidment NT. Orphanin FQ/nociceptin attenuates motor stimulation and changes in nucleus accumbens extracellular dopamine induced by cocaine in rats. *Psychopharmacology (Berl).* 2001;154(1):1–7.
48. Vazquez-DeRose J, Stauber G, Khroyan TV, Xie XS, Zaveri NT, Toll L. Retrodialysis of N/OFQ into the nucleus accumbens shell blocks cocaine-induced increases in extracellular dopamine and locomotor activity. *EurJ Pharmacol.* 2013;699(1–3):200–206.
49. Giros B, Jaber M, Jones SR, Wightman RM, Caron MG. Hyperlocomotion and indifference to cocaine and amphetamine in mice lacking the dopamine transporter. *Nature.* 1996;379(6566):606–612.
50. Lutfy K, Khaliq I, Carroll FI, Maidment NT. Orphanin FQ/nociceptin blocks cocaine-induced behavioral sensitization in rats. *Psychopharmacology (Berl).* 2002;164(2): 168–176.
51. Bardo MT, Bevins RA. Conditioned place preference: what does it add to our preclinical understanding of drug reward? *Psychopharmacology (Berl).* 2000;153(1):31–43.

52. Kotlinska J, Wichmann J, Legowska A, Rolka K, Silberring J. Orphanin FQ/nociceptin but not Ro 65-6570 inhibits the expression of cocaine-induced conditioned place preference. *Behav Pharmacol.* 2002;13(3):229–235.
53. Kotlinska J, Rafalski P, Biala G, Dylag T, Rolka K, Silberring J. Nociceptin inhibits acquisition of amphetamine-induced place preference and sensitization to stereotypy in rats. *Eur J Pharmacol.* 2003;474(2–3):233–239.
54. Zhao RJ, Woo RS, Jeong MS, Shin BS, Kim DG, Kim KW. Orphanin FQ/nociceptin blocks methamphetamine place preference in rats. *Neuroreport.* 2003;14 (18):2383–2385.
55. Thomas MJ, Kalivas PW, Shaham Y. Neuroplasticity in the mesolimbic dopamine system and cocaine addiction. *Br J Pharmacol.* 2008;154(2):327–342.
56. Bebawy D, Marquez P, Samboul S, Parikh D, Hamid A, Lutfy K. Orphanin FQ/ nociceptin not only blocks but also reverses behavioral adaptive changes induced by repeated cocaine in mice. *Biol Psychiatry.* 2010;68(3):223–230.
57. Narayanan S, Maidment NT. Orphanin FQ and behavioral sensitization to cocaine. *Pharmacol Biochem Behav.* 1999;63(2):271–277.
58. Kalivas PW, Lalumiere RT, Knackstedt L, Shen H. Glutamate transmission in addiction. *Neuropharmacology.* 2009;56(suppl 1):169–173.
59. Kalivas PW, O'Brien C. Drug addiction as a pathology of staged neuroplasticity. *Neuropsychopharmacology.* 2008;33(1):166–180.
60. Kalivas PW, Pierce RC, Cornish J, Sorg BA. A role for sensitization in craving and relapse in cocaine addiction. *J Psychopharmacol.* 1998;12(1):49–53.
61. Kalivas PW, Sorg BA, Hooks MS. The pharmacology and neural circuitry of sensitization to psychostimulants. *Behav Pharmacol.* 1993;4(4):315–334.
62. Kalivas PW, Stewart J. Dopamine transmission in the initiation and expression of drug- and stress-induced sensitization of motor activity. *Brain Res Brain Res Rev.* 1991;16(3): 223–244.
63. Kelley AE. Memory and addiction: shared neural circuitry and molecular mechanisms. *Neuron.* 2004;44(1):161–179.
64. Robinson TE, Berridge KC. The neural basis of drug craving: an incentive-sensitization theory of addiction. *Brain Res Brain Res Rev.* 1993;18(3):247–291.
65. Robinson TE, Berridge KC. The psychology and neurobiology of addiction: an incentive-sensitization view. *Addiction.* 2000;95(suppl 2):S91–S117.
66. Devine DP, Taylor L, Reinscheid RK, Monsma Jr FJ, Civelli O, Akil H. Rats rapidly develop tolerance to the locomotor-inhibiting effects of the novel neuropeptide orphanin FQ. *Neurochem Res.* 1996;21(11):1387–1396.
67. Martin-Fardon R, Ciccocioppo R, Massi M, Weiss F. Nociceptin prevents stress-induced ethanol—but not cocaine-seeking behavior in rats. *Neuroreport.* 2000;11(9): 1939–1943.
68. Walker JR, Spina M, Terenius L, Koob GF. Nociceptin fails to affect heroin self-administration in the rat. *Neuroreport.* 1998;9(10):2243–2247.
69. Koizumi M, Sakoori K, Midorikawa N, Murphy NP. The NOP (ORL1) receptor antagonist Compound B stimulates mesolimbic dopamine release and is rewarding in mice by a non-NOP-receptor-mediated mechanism. *Br J Pharmacol.* 2004;143(1): 53–62.
70. Romualdi P, Di Benedetto M, D'Addario C, et al. Chronic cocaine produces decreases in N/OFQ peptide levels in select rat brain regions. *J Mol Neurosci.* 2007;31(2):159–164.
71. Lutfy K, Lam H, Narayanan S. Alterations in the level of OFQ/N-IR in rat brain regions by cocaine. *Neuropharmacology.* 2008;55(2):198–203.
72. Marquez P, Nguyen AT, Hamid A, Lutfy K. The endogenous OFQ/N/ORL-1 receptor system regulates the rewarding effects of acute cocaine. *Neuropharmacology.* 2008;54 (3):564–568.

73. Marquez P, Hamid A, Lutfy K. The role of NOP receptors in psychomotor stimulation and locomotor sensitization induced by cocaine and amphetamine in mice. *Eur J Pharmacol.* 2013;707(1–3):41–45.
74. Shoblock JR, Wichmann J, Maidment NT. The effect of a systemically active ORL-1 agonist, Ro 64-6198, on the acquisition, expression, extinction, and reinstatement of morphine conditioned place preference. *Neuropharmacology.* 2005;49(4): 439–446.
75. Toll L, Khroyan TV, Polgar WE, Jiang F, Olsen C, Zaveri NT. Comparison of the antinociceptive and antirewarding profiles of novel bifunctional nociceptin receptor/mu-opioid receptor ligands: implications for therapeutic applications. *J Pharmacol Exp Ther.* 2009;331(3):954–964.
76. Ciccocioppo R, Stopponi S, Economidou D, et al. Chronic treatment with novel brain-penetrating selective NOP receptor agonist MT-7716 reduces alcohol drinking and seeking in the rat. *Neuropsychopharmacology.* 2014;39(11):2601–2610.
77. Kuzmin AV, Semenova S, Gerrits MA, Zvartau EE, Van Ree JM. Kappa-opioid receptor agonist U50,488H modulates cocaine and morphine self-administration in drug-naive rats and mice. *Eur J Pharmacol.* 1997;321(3):265–271.
78. Zaveri N. Peptide and nonpeptide ligands for the nociceptin/orphanin FQ receptor ORL1: research tools and potential therapeutic agents. *Life Sci.* 2003;73(6): 663–678.
79. Rutten K, De Vry J, Bruckmann W, Tzschentke TM. Effects of the NOP receptor agonist Ro65-6570 on the acquisition of opiate- and psychostimulant-induced conditioned place preference in rats. *Eur J Pharmacol.* 2010;645(1–3):119–126.
80. Cordery SF, Taverner A, Ridzwan IE, et al. A non-rewarding, non-aversive buprenorphine/naltrexone combination attenuates drug-primed reinstatement to cocaine and morphine in rats in a conditioned place preference paradigm. *Addict Biol.* 2014;19(4): 575–586.
81. Khroyan TV, Polgar W, Toll L, Jiang F, Zaveri NT. AT-202, a nociceptin receptor agonist, blocks drug- and stress-induced reinstatement of extinguished cocaine conditioned place preference. *Society for Neuroscience*; 2013. Program No. 733.24. 2013 Neuroscience Meeting Planner
82. Koob G. Drug addiction. *Neurobiol Dis.* 2000;7(5):543–545.
83. Koob GF. Hedonic homeostatic dysregulation as a driver of drug-seeking behavior. *Drug Discov Today Dis Models.* 2008;5(4):207–215.
84. Koob GF, Volkow ND. Neurocircuitry of addiction. *Neuropsychopharmacology.* 2010;35 (1):217–238.
85. Tsai G. Glutamatergic neurotransmission in alcoholism. *J Biomed Sci.* 1998;5(5): 309–320.
86. Tsai G, Coyle JT. The role of glutamatergic neurotransmission in the pathophysiology of alcoholism. *Annu Rev Med.* 1998;49:173–184.
87. Tsai GE, Ragan P, Chang R, Chen S, Linnoila VM, Coyle JT. Increased glutamatergic neurotransmission and oxidative stress after alcohol withdrawal. *Am J Psychiatry.* 1998;155(6):726–732.
88. Vanderschuren LJ, Kalivas PW. Alterations in dopaminergic and glutamatergic transmission in the induction and expression of behavioral sensitization: a critical review of preclinical studies. *Psychopharmacology (Berl).* 2000;151(2–3):99–120.
89. Volkow ND, Wang GJ, Fowler JS, Tomasi D, Telang F, Baler R. Addiction: decreased reward sensitivity and increased expectation sensitivity conspire to overwhelm the brain's control circuit. *BioEssays.* 2010;32(9):748–755.
90. Wise RA. Dopamine, learning and motivation. *Nat Rev Neurosci.* 2004;5(6):483–494.
91. Wolf ME. Addiction: making the connection between behavioral changes and neuronal plasticity in specific pathways. *Mol Interv.* 2002;2(3):146–157.

92. Chartoff EH, Potter D, Damez-Werno D, Cohen BM, Carlezon Jr WA. Exposure to the selective kappa-opioid receptor agonist salvinorin A modulates the behavioral and molecular effects of cocaine in rats. *Neuropsychopharmacology.* 2008;33(11): 2676–2687.
93. Chefer VI, Kieffer BL, Shippenberg TS. Contrasting effects of mu opioid receptor and delta opioid receptor deletion upon the behavioral and neurochemical effects of cocaine. *Neuroscience.* 2004;127(2):497–503.
94. Di Chiara G, Imperato A. Opposite effects of mu and kappa opiate agonists on dopamine release in the nucleus accumbens and in the dorsal caudate of freely moving rats. *J Pharmacol Exp Ther.* 1988;244(3):1067–1080.
95. Doyon WM, Howard EC, Shippenberg TS, Gonzales RA. Kappa-opioid receptor modulation of accumbal dopamine concentration during operant ethanol self-administration. *Neuropharmacology.* 2006;51(3):487–496.
96. Gerrits MA, Patkina N, Zvartau EE, van Ree JM. Opioid blockade attenuates acquisition and expression of cocaine-induced place preference conditioning in rats. *Psychopharmacology (Berl).* 1995;119(1):92–98.
97. Goeders NE, Lane JD, Smith JE. Self-administration of methionine enkephalin into the nucleus accumbens. *Pharmacol Biochem Behav.* 1984;20(3):451–455.
98. Heidbreder C, Shoaib M, Shippenberg TS. Differential role of delta-opioid receptors in the development and expression of behavioral sensitization to cocaine. *Eur J Pharmacol.* 1996;298(3):207–216.
99. Heidbreder CA, Goldberg SR, Shippenberg TS. The kappa-opioid receptor agonist U-69593 attenuates cocaine-induced behavioral sensitization in the rat. *Brain Res.* 1993;616(1–2):335–338.
100. Herz A. Opioid reward mechanisms: a key role in drug abuse? *Can J Physiol Pharmacol.* 1998;76(3):252–258.
101. Houdi AA, Bardo MT, Van Loon GR. Opioid mediation of cocaine-induced hyperactivity and reinforcement. *Brain Res.* 1989;497(1):195–198.
102. Hummel M, Ansonoff MA, Pintar JE, Unterwald EM. Genetic and pharmacological manipulation of mu opioid receptors in mice reveals a differential effect on behavioral sensitization to cocaine. *Neuroscience.* 2004;125(1):211–220.
103. Hummel M, Schroeder J, Liu-Chen LY, Cowan A, Unterwald EM. An antisense oligodeoxynucleotide to the mu opioid receptor attenuates cocaine-induced behavioral sensitization and reward in mice. *Neuroscience.* 2006;142(2):481–491.
104. Kuzmin AV, Gerrits MA, van Ree JM, Zvartau EE. Naloxone inhibits the reinforcing and motivational aspects of cocaine addiction in mice. *Life Sci.* 1997;60(18). PL-257-264.
105. Menkens K, Bilsky EJ, Wild KD, Portoghese PS, Reid LD, Porreca F. Cocaine place preference is blocked by the delta-opioid receptor antagonist, naltrindole. *Eur J Pharmacol.* 1992;219(2):345–346.
106. Shippenberg TS, Heidbreder C. The delta-opioid receptor antagonist naltrindole prevents sensitization to the conditioned rewarding effects of cocaine. *Eur J Pharmacol.* 1995;280(1):55–61.
107. Shippenberg TS, Heidbreder C. Sensitization to the conditioned rewarding effects of cocaine: pharmacological and temporal characteristics. *J Pharmacol Exp Ther.* 1995;273(2):808–815.
108. Shippenberg TS, LeFevour A, Heidbreder C. kappa-opioid receptor agonists prevent sensitization to the conditioned rewarding effects of cocaine. *J Pharmacol Exp Ther.* 1996;276(2):545–554.
109. Shippenberg TS, Zapata A, Chefer VI. Dynorphin and the pathophysiology of drug addiction. *Pharmacol Ther.* 2007;116(2):306–321.

110. Simmons D, Self DW. Role of mu- and delta-opioid receptors in the nucleus accumbens in cocaine-seeking behavior. *Neuropsychopharmacology.* 2009;34(8):1946–1957.
111. Carey AN, Borozny K, Aldrich JV, McLaughlin JP. Reinstatement of cocaine place-conditioning prevented by the peptide kappa-opioid receptor antagonist arodyn. *Eur J Pharmacol.* 2007;569(1–2):84–89.
112. Chartoff E, Sawyer A, Rachlin A, Potter D, Pliakas A, Carlezon WA. Blockade of kappa opioid receptors attenuates the development of depressive-like behaviors induced by cocaine withdrawal in rats. *Neuropharmacology.* 2012;62(1):167–176.
113. Chefer VI, Shippenberg TS. Paradoxical effects of prodynorphin gene deletion on basal and cocaine-evoked dopaminergic neurotransmission in the nucleus accumbens. *Eur J Neurosci.* 2006;23(1):229–238.
114. Koob G, Kreek MJ. Stress, dysregulation of drug reward pathways, and the transition to drug dependence. *Am J Psychiatry.* 2007;164(8):1149–1159.
115. Land BB, Bruchas MR, Lemos JC, Xu M, Melief EJ, Chavkin C. The dysphoric component of stress is encoded by activation of the dynorphin kappa-opioid system. *J Neurosci.* 2008;28(2):407–414.
116. McLaughlin JP, Land BB, Li S, Pintar JE, Chavkin C. Prior activation of kappa opioid receptors by U50,488 mimics repeated forced swim stress to potentiate cocaine place preference conditioning. *Neuropsychopharmacology.* 2006;31(4):787–794.
117. McLaughlin JP, Marton-Popovici M, Chavkin C. Kappa opioid receptor antagonism and prodynorphin gene disruption block stress-induced behavioral responses. *J Neurosci.* 2003;23(13):5674–5683.
118. Bruchas MR, Land BB, Lemos JC, Chavkin C. CRF1-R activation of the dynorphin/kappa opioid system in the mouse basolateral amygdala mediates anxiety-like behavior. *PloS One.* 2009;4(12):e8528.
119. Cador M, Ahmed SH, Koob GF, Le Moal M, Stinus L. Corticotropin-releasing factor induces a place aversion independent of its neuroendocrine role. *Brain Res.* 1992;597(2):304–309.
120. Goeders NE. The HPA axis and cocaine reinforcement. *Psychoneuroendocrinology.* 2002;27(1–2):13–33.
121. Goeders NE. Stress and cocaine addiction. *J Pharmacol Exp Ther.* 2002;301(3):785–789.
122. Goeders NE, Clampitt DM. Potential role for the hypothalamo–pituitary–adrenal axis in the conditioned reinforcer-induced reinstatement of extinguished cocaine seeking in rats. *Psychopharmacology (Berl).* 2002;161(3):222–232.
123. Haney M, Maccari S, Le Moal M, Simon H, Piazza PV. Social stress increases the acquisition of cocaine self-administration in male and female rats. *Brain Res.* 1995;698(1–2):46–52.
124. Koob GF, Le Moal M. Addiction and the brain antireward system. *Annu Rev Psychol.* 2008;59:29–53.
125. Shaham Y, Erb S, Stewart J. Stress-induced relapse to heroin and cocaine seeking in rats: a review. *Brain Res Brain Res Rev.* 2000;33(1):13–33.
126. Smith JS, Schindler AG, Martinelli E, Gustin RM, Bruchas MR, Chavkin C. Stress-induced activation of the dynorphin/kappa-opioid receptor system in the amygdala potentiates nicotine conditioned place preference. *J Neurosci.* 2012;32(4):1488–1495.
127. Sorg BA, Kalivas PW. Effects of cocaine and footshock stress on extracellular dopamine levels in the ventral striatum. *Brain Res.* 1991;559(1):29–36.
128. Stohr T, Almeida OF, Landgraf R, Shippenberg TS, Holsboer F, Spanagel R. Stress- and corticosteroid-induced modulation of the locomotor response to morphine in rats. *Behav Brain Res.* 1999;103(1):85–93.
129. Kuhar MJ, Ritz MC, Sharkey J. Cocaine receptors on dopamine transporters mediate cocaine-reinforced behavior. *NIDA Res Monogr.* 1988;88:14–22.

130. Ritz MC, Lamb RJ, Goldberg SR, Kuhar MJ. Cocaine receptors on dopamine transporters are related to self-administration of cocaine. *Science.* 1987;237 (4819):1219–1223.
131. Ritz MC, Lamb RJ, Goldberg SR, Kuhar MJ. Cocaine self-administration appears to be mediated by dopamine uptake inhibition. *Prog Neuropsychopharmacol Biol Psychiatry.* 1988;12(2–3):233–239.
132. Di Chiara G, Imperato A. Drugs abused by humans preferentially increase synaptic dopamine concentrations in the mesolimbic system of freely moving rats. *Proc Natl Acad Sci USA.* 1988;85(14):5274–5278.
133. Amalric M, Koob GF. Functionally selective neurochemical afferents and efferents of the mesocorticolimbic and nigrostriatal dopamine system. *Prog Brain Res.* 1993;99:209–226.
134. Delfs JM, Schreiber L, Kelley AE. Microinjection of cocaine into the nucleus accumbens elicits locomotor activation in the rat. *J Neurosci.* 1990;10(1):303–310.
135. Pijnenburg AJ, Honig WM, Van der Heyden JA, Van Rossum JM. Effects of chemical stimulation of the mesolimbic dopamine system upon locomotor activity. *Eur J Pharmacol.* 1976;35(1):45–58.
136. Kelley AE, Berridge KC. The neuroscience of natural rewards: relevance to addictive drugs. *J Neurosci.* 2002;22(9):3306–3311.
137. Kelly PH, Seviour PW, Iversen SD. Amphetamine and apomorphine responses in the rat following 6-OHDA lesions of the nucleus accumbens septi and corpus striatum. *Brain Res.* 1975;94(3):507–522.
138. Kelsey JE, Carlezon Jr WA, Falls WA. Lesions of the nucleus accumbens in rats reduce opiate reward but do not alter context-specific opiate tolerance. *Behav Neurosci.* 1989;103(6):1327–1334.
139. Berridge KC. The debate over dopamine's role in reward: the case for incentive salience. *Psychopharmacology (Berl).* 2007;191(3):391–431.
140. Lemon N, Manahan-Vaughan D. Dopamine D1/D5 receptors gate the acquisition of novel information through hippocampal long-term potentiation and long-term depression. *J Neurosci.* 2006;26(29):7723–7729.
141. Swant J, Wagner JJ. Dopamine transporter blockade increases LTP in the CA1 region of the rat hippocampus via activation of the D3 dopamine receptor. *Learn Mem.* 2006;13 (2):161–167.
142. Thompson AM, Swant J, Wagner JJ. Cocaine-induced modulation of long-term potentiation in the CA1 region of rat hippocampus. *Neuropharmacology.* 2005;49(2):185–194.
143. Berridge KC, Kringelbach ML. Pleasure systems in the brain. *Neuron.* 2015;86(3): 646–664.
144. Guarraci FA, Frohardt RJ, Young SL, Kapp BS. A functional role for dopamine transmission in the amygdala during conditioned fear. *Ann NY Acad Sci.* 1999;877: 732–736.
145. Guarraci FA, Kapp BS. An electrophysiological characterization of ventral tegmental area dopaminergic neurons during differential Pavlovian fear conditioning in the awake rabbit. *Behav Brain Res.* 1999;99(2):169–179.
146. Horvitz JC. Mesolimbocortical and nigrostriatal dopamine responses to salient non-reward events. *Neuroscience.* 2000;96(4):651–656.
147. Joshua M, Adler A, Mitelman R, Vaadia E, Bergman H. Midbrain dopaminergic neurons and striatal cholinergic interneurons encode the difference between reward and aversive events at different epochs of probabilistic classical conditioning trials. *J Neurosci.* 2008;28(45):11673–11684.
148. Land BB, Bruchas MR, Schattauer S, et al. Activation of the kappa opioid receptor in the dorsal raphe nucleus mediates the aversive effects of stress and reinstates drug seeking. *Proc Natl Acad Sci USA.* 2009;106(45):19168–19173.

149. Sora I, Wichems C, Takahashi N, et al. Cocaine reward models: conditioned place preference can be established in dopamine- and in serotonin-transporter knockout mice. *Proc Natl Acad Sci USA*. 1998;95(13):7699–7704.
150. Hnasko TS, Sotak BN, Palmiter RD. Cocaine-conditioned place preference by dopamine-deficient mice is mediated by serotonin. *J Neurosci*. 2007;27(46):12484–12488.
151. Marquez P, Baliram R, Dabaja I, Gajawada N, Lutfy K. The role of beta-endorphin in the acute motor stimulatory and rewarding actions of cocaine in mice. *Psychopharmacology (Berl)*. 2008;197(3):443–448.
152. Marquez P, Bebawy D, Lelievre V, et al. The role of endogenous PACAP in motor stimulation and conditioned place preference induced by morphine in mice. *Psychopharmacology (Berl)*. 2009;204(3):457–463.
153. Sora I, Hall FS, Andrews AM, et al. Molecular mechanisms of cocaine reward: combined dopamine and serotonin transporter knockouts eliminate cocaine place preference. *Proc Natl Acad Sci USA*. 2001;98(9):5300–5305.
154. Khroyan TV, Barrett-Larimore RL, Rowlett JK, Spealman RD. Dopamine D1- and D2-like receptor mechanisms in relapse to cocaine-seeking behavior: effects of selective antagonists and agonists. *J Pharmacol Exp Ther*. 2000;294(2):680–687.
155. Koob GF, Le HT, Creese I. The D1 dopamine receptor antagonist SCH 23390 increases cocaine self-administration in the rat. *Neurosci Lett*. 1987;79(3):315–320.
156. Self DW, Barnhart WJ, Lehman DA, Nestler EJ. Opposite modulation of cocaine-seeking behavior by D1- and D2-like dopamine receptor agonists. *Science*. 1996;271 (5255):1586–1589.
157. Di Giannuario A, Pieretti S, Catalani A, Loizzo A. Orphanin FQ reduces morphine-induced dopamine release in the nucleus accumbens: a microdialysis study in rats. *Neurosci Lett*. 1999;272(3):183–186.
158. Gavioli EC, Calo G. Nociceptin/orphanin FQ receptor antagonists as innovative antidepressant drugs. *Pharmacol Ther*. 2013;140(1):10–25.
159. Le Maitre E, Dourmap N, Vilpoux C, et al. Acute and subchronic treatments with selective serotonin reuptake inhibitors increase Nociceptin/Orphanin FQ (NOP) receptor density in the rat dorsal raphe nucleus; interactions between nociceptin/NOP system and serotonin. *Brain Res*. 2013;1520:51–60.
160. Lu N, Han M, Yang ZL, Wang YQ, Wu GC, Zhang YQ. Nociceptin/Orphanin FQ in PAG modulates the release of amino acids, serotonin and norepinephrine in the rostral ventromedial medulla and spinal cord in rats. *Pain*. 2010;148(3):414–425.
161. Mela F, Marti M, Ulazzi L, et al. Pharmacological profile of nociceptin/orphanin FQ receptors regulating 5-hydroxytryptamine release in the mouse neocortex. *Eur J Neurosci*. 2004;19(5):1317–1324.
162. Nazzaro C, Marino S, Barbieri M, Siniscalchi A. Inhibition of serotonin outflow by nociceptin/orphanin FQ in dorsal raphe nucleus slices from normal and stressed rats: role of corticotropin releasing factor. *Neurochem Int*. 2009;54(5–6):378–384.
163. Tao R, Ma Z, Thakkar MM, McCarley RW, Auerbach SB. Nociceptin/orphanin FQ decreases serotonin efflux in the rat brain but in contrast to a kappa-opioid has no antagonistic effect on mu-opioid-induced increases in serotonin efflux. *Neuroscience*. 2007;147(1):106–116.
164. Ciccocioppo R, Angeletti S, Sanna PP, Weiss F, Massi M. Effect of nociceptin/orphanin FQ on the rewarding properties of morphine. *Eur J Pharmacol*. 2000;404(1–2):153–159.
165. Kuzmin A, Kreek MJ, Bakalkin G, Liljequist S. The nociceptin/orphanin FQ receptor agonist Ro 64-6198 reduces alcohol self-administration and prevents relapse-like alcohol drinking. *Neuropsychopharmacology*. 2007;32(4):902–910.
166. Kuzmin A, Sandin J, Terenius L, Ogren SO. Acquisition, expression, and reinstatement of ethanol-induced conditioned place preference in mice: effects of opioid receptor-like 1 receptor agonists and naloxone. *J Pharmacol Exp Ther*. 2003;304(1):310–318.

167. Murphy NP, Lee Y, Maidment NT. Orphanin FQ/nociceptin blocks acquisition of morphine place preference. *Brain Res.* 1999;832(1–2):168–170.
168. Rademacher DJ, Steinpreis RE. Effects of the selective mu(1)-opioid receptor antagonist, naloxonazine, on cocaine-induced conditioned place preference and locomotor behavior in rats. *Neurosci Lett.* 2002;332(3):159–162.
169. Olive MF, Koenig HN, Nannini MA, Hodge CW. Stimulation of endorphin neurotransmission in the nucleus accumbens by ethanol, cocaine, and amphetamine. *J Neurosci.* 2001;21(23):RC184.
170. Roth-Deri I, Mayan R, Yadid G. A hypothalamic endorphinic lesion attenuates acquisition of cocaine self-administration in the rat. *Eur Neuropsychopharmacol.* 2006;16(1): 25–32.
171. Roth-Deri I, Schindler CJ, Yadid G. A critical role for beta-endorphin in cocaine-seeking behavior. *Neuroreport.* 2004;15(3):519–521.
172. Roth-Deri I, Zangen A, Aleli M, et al. Effect of experimenter-delivered and self-administered cocaine on extracellular beta-endorphin levels in the nucleus accumbens. *J Neurochem.* 2003;84(5):930–938.
173. Dikshtein Y, Barnea R, Kronfeld N, et al. beta-endorphin via the delta opioid receptor is a major factor in the incubation of cocaine craving. *Neuropsychopharmacology.* 2013;38 (12):2508–2514.
174. Nguyen AT, Marquez P, Hamid A, Kieffer B, Friedman TC, Lutfy K. The rewarding action of acute cocaine is reduced in beta-endorphin deficient but not in mu opioid receptor knockout mice. *Eur J Pharmacol.* 2012;686(1–3):50–54.
175. Bailey A, Yuferov V, Bendor J, et al. Immediate withdrawal from chronic "binge" cocaine administration increases mu-opioid receptor mRNA levels in rat frontal cortex. *Brain Res Mol Brain Res.* 2005;137(1–2):258–262.
176. Fagergren P, Smith HR, Daunais JB, Nader MA, Porrino LJ, Hurd YL. Temporal upregulation of prodynorphin mRNA in the primate striatum after cocaine self-administration. *Eur J Neurosci.* 2003;17(10):2212–2218.
177. Hurd YL, Svensson P, Ponten M. The role of dopamine, dynorphin, and CART systems in the ventral striatum and amygdala in cocaine abuse. *Ann NY Acad Sci.* 1999;877: 499–506.
178. Schlussman SD, Zhang Y, Yuferov V, LaForge KS, Ho A, Kreek MJ. Acute 'binge' cocaine administration elevates dynorphin mRNA in the caudate putamen of C57BL/ 6J but not 129/J mice. *Brain Res.* 2003;974(1–2):249–253.
179. Turchan J, Maj M, Przewlocka B, Przewlocki R. Effect of cocaine and amphetamine on biosynthesis of proenkephalin and prodynorphin in some regions of the rat limbic system. *Pol J Pharmacol.* 2002;54(4):367–372.
180. Werme M, Thoren P, Olson L, Brene S. Running and cocaine both upregulate dynorphin mRNA in medial caudate putamen. *Eur J Neurosci.* 2000;12(8):2967–2974.
181. Zhang Y, Schlussman SD, Butelman ER, Ho A, Kreek MJ. Effects of withdrawal from chronic escalating-dose binge cocaine on conditioned place preference to cocaine and striatal preproenkephalin mRNA in C57BL/6J mice. *Neuropharmacology.* 2012;63(2): 322–329.
182. Mogil JS, Grisel JE, Reinscheid RK, Civelli O, Belknap JK, Grandy DK. Orphanin FQ is a functional anti-opioid peptide. *Neuroscience.* 1996;75(2):333–337.
183. Zakarian S, Smyth D. Distribution of active and inactive forms of endorphins in rat pituitary and brain. *Proc Natl Acad Sci USA.* 1979;76(11):5972–5976.
184. Zakarian S, Smyth DG. Distribution of beta-endorphin-related peptides in rat pituitary and brain. *Biochem J.* 1982;202(3):561–571.
185. Spanagel R, Herz A, Shippenberg TS. Identification of the opioid receptor types mediating beta-endorphin-induced alterations in dopamine release in the nucleus accumbens. *Eur J Pharmacol.* 1990;190(1–2):177–184.

186. Koob GF, Le Moal M. Drug abuse: hedonic homeostatic dysregulation. *Science.* 1997;278(5335):52–58.
187. Dembo R, Dertke M, Borders S, Washburn M, Schmeidler J. The relationship between physical and sexual abuse and tobacco, alcohol, and illicit drug use among youths in a juvenile detention center. *Int J Addict.* 1988;23(4):351–378.
188. Dembo R, Williams L, Berry E, et al. The relationship between physical and sexual abuse and illicit drug use: a replication among a new sample of youths entering a juvenile detention center. *Int J Addict.* 1988;23(11):1101–1123.
189. Harrison PA, Fulkerson JA, Beebe TJ. Multiple substance use among adolescent physical and sexual abuse victims. *Child Abuse Negl.* 1997;21(6):529–539.
190. Widom CS. Posttraumatic stress disorder in abused and neglected children grown up. *Am J Psychiatry.* 1999;156(8):1223–1229.
191. Kreek MJ, Koob GF. Drug dependence: stress and dysregulation of brain reward pathways. *Drug Alcohol Depend.* 1998;51(1–2):23–47.
192. Kosten TA, Miserendino MJ, Kehoe P. Enhanced acquisition of cocaine self-administration in adult rats with neonatal isolation stress experience. *Brain Res.* 2000;875 (1–2):44–50.
193. Sinha R. How does stress increase risk of drug abuse and relapse? *Psychopharmacology (Berl).* 2001;158(4):343–359.
194. Zorrilla EP, Logrip ML, Koob GF. Corticotropin releasing factor: a key role in the neurobiology of addiction. *Front Neuroendocrinol.* 2014;35(2):234–244.
195. Ciccocioppo R, Martin-Fardon R, Weiss F, Massi M. Nociceptin/orphanin FQ inhibits stress- and CRF-induced anorexia in rats. *Neuroreport.* 2001;12(6):1145–1149.
196. Fernandez F, Misilmeri MA, Felger JC, Devine DP. Nociceptin/orphanin FQ increases anxiety-related behavior and circulating levels of corticosterone during neophobic tests of anxiety. *Neuropsychopharmacology.* 2004;29(1):59–71.
197. Green MK, Barbieri EV, Brown BD, Chen KW, Devine DP. Roles of the bed nucleus of stria terminalis and of the amygdala in N/OFQ-mediated anxiety and HPA axis activation. *Neuropeptides.* 2007;41(6):399–410.
198. Jenck F, Wichmann J, Dautzenberg FM, et al. A synthetic agonist at the orphanin FQ/nociceptin receptor ORL1: anxiolytic profile in the rat. *Proc Natl Acad Sci USA.* 2000;97 (9):4938–4943.
199. Wichmann J, Adam G, Rover S, et al. Synthesis of (1S,3aS)-8-(2,3,3a, 4,5, 6-hexahydro-1H-phenalen-1-yl)-1-phenyl-1,3,8-triaza-spiro[4. 5]decan-4-one, a potent and selective orphanin FQ (OFQ) receptor agonist with anxiolytic-like properties. *Eur J Med Chem.* 2000;35(9):839–851.
200. Gavioli EC, Rizzi A, Marzola G, Zucchini S, Regoli D, Calo G. Altered anxiety-related behavior in nociceptin/orphanin FQ receptor gene knockout mice. *Peptides.* 2007;28 (6):1229–1239.
201. Aujla H, Nedjadrasul D. Low-dose Nociceptin/Orphanin FQ reduces anxiety-like performance in alcohol-withdrawn, but not alcohol-naive, male Wistar rats. *Neuropharmacology.* 2015;93:1–6.
202. Karler R, Calder LD, Chaudhry IA, Turkanis SA. Blockade of "reverse tolerance" to cocaine and amphetamine by MK-801. *Life Sci.* 1989;45(7):599–606.
203. Hemby SE, Horman B, Tang W. Differential regulation of ionotropic glutamate receptor subunits following cocaine self-administration. *Brain Res.* 2005;1064(1–2):75–82.
204. Koob GF, Nestler EJ. The neurobiology of drug addiction. *J Neuropsychiatry Clin Neurosci.* 1997;9(3):482–497.
205. Purgianto A, Scheyer AF, Loweth JA, Ford KA, Tseng KY, Wolf ME. Different adaptations in AMPA receptor transmission in the nucleus accumbens after short vs long access cocaine self-administration regimens. *Neuropsychopharmacology.* 2013;38(9): 1789–1797.

206. Bouton ME. Context, ambiguity, and unlearning: sources of relapse after behavioral extinction. *Biol Psychiatry.* 2002;52(10):976–986.
207. Bouton ME. The other learning process in substance abuse: comment on Alessi, Roll, Reilly, and Johanson (2002). *Exp Clin Psychopharmacol.* 2002;10(2):84–86. discussion 101–103..
208. Childress AR, Mozley PD, McElgin W, Fitzgerald J, Reivich M, O'Brien CP. Limbic activation during cue-induced cocaine craving. *Am J Psychiatry.* 1999;156(1):11–18.
209. Fuchs RA, Branham RK, See RE. Different neural substrates mediate cocaine seeking after abstinence versus extinction training: a critical role for the dorsolateral caudate-putamen. *J Neurosci.* 2006;26(13):3584–3588.
210. McLaughlin J, See RE. Selective inactivation of the dorsomedial prefrontal cortex and the basolateral amygdala attenuates conditioned-cued reinstatement of extinguished cocaine-seeking behavior in rats. *Psychopharmacology (Berl).* 2003;168(1–2):57–65.
211. O'Brien CP, Childress AR, McLellan T, Ehrman R. Integrating systemic cue exposure with standard treatment in recovering drug dependent patients. *Addict Behav.* 1990;15 (4):355–365.
212. Hunt WA, Barnett LW, Branch LG. Relapse rates in addiction programs. *J Clin Psychol.* 1971;27(4):455–456.
213. Oslin D, Liberto JG, O'Brien J, Krois S, Norbeck J. Naltrexone as an adjunctive treatment for older patients with alcohol dependence. *Am J Geriatr Psychiatry.* 1997;5 (4):324–332.
214. Wydra K, Golembiowska K, Zaniewska M, et al. Accumbal and pallidal dopamine, glutamate and GABA overflow during cocaine self-administration and its extinction in rats. *Addict Biol.* 2013;18(2):307–324.
215. Fischer KD, Houston AC, Rebec GV. Role of the major glutamate transporter GLT1 in nucleus accumbens core versus shell in cue-induced cocaine-seeking behavior. *J Neurosci.* 2013;33(22):9319–9327.
216. Li X, Markou A. Metabotropic Glutamate Receptor 7 (mGluR7) as a target for the treatment of psychostimulant dependence. *CNS Neurol Disord Drug Targets.* 2015;14 (6):738–744.
217. Loweth JA, Scheyer AF, Milovanovic M, et al. Synaptic depression via mGluR1 positive allosteric modulation suppresses cue-induced cocaine craving. *Nat Neurosci.* 2014;17 (1):73–80.
218. Loweth JA, Tseng KY, Wolf ME. Adaptations in AMPA receptor transmission in the nucleus accumbens contributing to incubation of cocaine craving. *Neuropharmacology.* 2014;76 Pt B:287–300.
219. Goeldner C, Reiss D, Wichmann J, Kieffer BL, Ouagazzal AM. Activation of nociceptin opioid peptide (NOP) receptor impairs contextual fear learning in mice through glutamatergic mechanisms. *Neurobiol Learn Mem.* 2009;91(4):393–401.
220. Goeldner C, Reiss D, Wichmann J, Meziane H, Kieffer BL, Ouagazzal AM. Nociceptin receptor impairs recognition memory via interaction with NMDA receptor-dependent mitogen-activated protein kinase/extracellular signal-regulated kinase signaling in the hippocampus. *J Neurosci.* 2008;28(9):2190–2198.
221. Mamiya T, Yamada K, Miyamoto Y, et al. Neuronal mechanism of nociceptin-induced modulation of learning and memory: involvement of *N*-methyl-D-aspartate receptors. *Mol Psychiatry.* 2003;8(8):752–765.
222. Wei WZ, Xie CW. Orphanin FQ suppresses NMDA receptor-dependent long-term depression and depotentiation in hippocampal dentate gyrus. *Learn Mem.* 1999;6(5): 467–477.
223. Yu TP, Xie CW. Orphanin FQ/nociceptin inhibits synaptic transmission and long-term potentiation in rat dentate gyrus through postsynaptic mechanisms. *J Neurophysiol.* 1998;80(3):1277–1284.

224. Hyman SE, Malenka RC, Nestler EJ. Neural mechanisms of addiction: the role of reward-related learning and memory. *Annu Rev Neurosci.* 2006;29:565–598.
225. Nestler EJ, Landsman D. Learning about addiction from the genome. *Nature.* 2001;409 (6822):834–835.
226. Hyman SE, Malenka RC. Addiction and the brain: the neurobiology of compulsion and its persistence. *Nat Rev Neurosci.* 2001;2(10):695–703.
227. Foltin RW, Haney M. Conditioned effects of environmental stimuli paired with smoked cocaine in humans. *Psychopharmacology (Berl).* 2000;149(1):24–33.
228. Grant S, London ED, Newlin DB, et al. Activation of memory circuits during cue-elicited cocaine craving. *Proc Natl Acad Sci USA.* 1996;93(21):12040–12045.
229. Yu TP, Fein J, Phan T, Evans CJ, Xie CW. Orphanin FQ inhibits synaptic transmission and long-term potentiation in rat hippocampus. *Hippocampus.* 1997;7(1):88–94.
230. Redrobe JP, Calo G, Guerrini R, Regoli D, Quirion R. [Nphe(1)]-Nociceptin (1-13)-NH(2), a nociceptin receptor antagonist, reverses nociceptin-induced spatial memory impairments in the Morris water maze task in rats. *Br J Pharmacol.* 2000;131(7): 1379–1384.
231. Sandin J, Georgieva J, Schott PA, Ogren SO, Terenius L. Nociceptin/orphanin FQ microinjected into hippocampus impairs spatial learning in rats. *Eur J Neurosci.* 1997;9 (1):194–197.
232. Fornari RV, Soares JC, Ferreira TL, Moreira KM, Oliveira MG. Effects of nociceptin/orphanin FQ in the acquisition of contextual and tone fear conditioning in rats. *Behav Neurosci.* 2008;122(1):98–106.
233. Higgins GA, Kew JN, Richards JG, et al. A combined pharmacological and genetic approach to investigate the role of orphanin FQ in learning and memory. *Eur J Neurosci.* 2002;15(5):911–922.
234. Roozendaal B, Lengvilas R, McGaugh JL, Civelli O, Reinscheid RK. Orphanin FQ/nociceptin interacts with the basolateral amygdala noradrenergic system in memory consolidation. *Learn Mem.* 2007;14(1–2):29–35.
235. Mamiya T, Noda Y, Nishi M, Takeshima H, Nabeshima T. Enhancement of spatial attention in nociceptin/orphanin FQ receptor-knockout mice. *Brain Res.* 1998;783 (2):236–240.
236. Manabe T, Noda Y, Mamiya T, et al. Facilitation of long-term potentiation and memory in mice lacking nociceptin receptors. *Nature.* 1998;394(6693):577–581.
237. Renthal W, Nestler EJ. Histone acetylation in drug addiction. *Semin Cell Dev Biol.* 2009;20(4):387–394.
238. Kurdistani SK, Tavazoie S, Grunstein M. Mapping global histone acetylation patterns to gene expression. *Cell.* 2004;117(6):721–733.
239. Strahl BD, Allis CD. The language of covalent histone modifications. *Nature.* 2000;403 (6765):41–45.
240. Schmidt HD, McGinty JF, West AE, Sadri-Vakili G. Epigenetics and psychostimulant addiction. *Cold Spring Harb Perspect Med.* 2013;3(3):a012047.
241. Kumar A, Choi KH, Renthal W, et al. Chromatin remodeling is a key mechanism underlying cocaine-induced plasticity in striatum. *Neuron.* 2005;48(2):303–314.
242. Renthal W, Maze I, Krishnan V, et al. Histone deacetylase 5 epigenetically controls behavioral adaptations to chronic emotional stimuli. *Neuron.* 2007;56(3):517–529.
243. Malvaez M, Sanchis-Segura C, Vo D, Lattal KM, Wood MA. Modulation of chromatin modification facilitates extinction of cocaine-induced conditioned place preference. *Biol Psychiatry.* 2010;67(1):36–43.
244. Levine AA, Guan Z, Barco A, Xu S, Kandel ER, Schwartz JH. CREB-binding protein controls response to cocaine by acetylating histones at the fosB promoter in the mouse striatum. *Proc Natl Acad Sci USA.* 2005;102(52):19186–19191.

CHAPTER SIX

Recent Advances in Nicotinic Receptor Signaling in Alcohol Abuse and Alcoholism

Shafiqur Rahman[*,1], **Eric A. Engleman**[†], **Richard L. Bell**[†]
[*] Department of Pharmaceutical Sciences, South Dakota State University, Brookings, South Dakota, USA
[†] Department of Psychiatry, Indiana University School of Medicine, Indianapolis, Indiana, USA
[1] Corresponding author: e-mail address: shafiqur.rahman@sdstate.edu

Contents

Abstract

Alcohol is the most commonly abused legal substance and alcoholism is a serious public health problem. It is a leading cause of preventable death in the world. The cellular and molecular mechanisms of alcohol reward and addiction are still not well understood. Emerging evidence indicates that unlike other drugs of abuse, such as nicotine, cocaine, or opioids, alcohol targets numerous channel proteins, receptor molecules, and signaling pathways in the brain. Previously, research has identified brain nicotinic acetylcholine receptors (nAChRs), a heterogeneous family of pentameric ligand-gated cation channels expressed in the mammalian brain, as critical molecular targets for alcohol abuse and dependence. Genetic variations encoding nAChR subunits have been shown to increase the vulnerability to develop alcohol dependence. Here, we review recent insights into the rewarding effects of alcohol, as they pertain to different nAChR subtypes, associated signaling molecules, and pathways that contribute to the molecular mechanisms of alcoholism and/or comorbid

Progress in Molecular Biology and Translational Science, Volume 137
ISSN 1877-1173
http://dx.doi.org/10.1016/bs.pmbts.2015.10.004

brain disorders. Understanding these cellular changes and molecular underpinnings may be useful for the advancement of brain nicotinic–cholinergic mechanisms, and will lead to a better translational and therapeutic outcome for alcoholism and/or comorbid conditions.

1. INTRODUCTION

Alcohol is the most commonly abused legal substance by humans. Alcoholism is a complex and chronic relapsing disorder that represents a serious global public health problem.[1] Alcoholism related deaths are estimated to account for 4% of all deaths worldwide and this number will be higher in the coming decades.[2] The prevalence of fetal alcohol syndrome, caused by mother's alcohol abuse and dependence, is also a significant problem in the United States and other industrialized nations.[3] The estimated healthcare and economic cost from alcohol abuse and alcoholism in the United States, is significantly higher than many other diseases, including cancer.[4] Thus, there is a need for better understanding of this complex brain disorder, for better therapeutic approaches to reduce alcohol abuse and relapse. Like nicotine addiction, emerging evidence suggests that neuronal nicotinic acetylcholine receptors (nAChRs), in the mesocorticolimbic-dopamine system, mediate the rewarding effects of alcohol and the development of alcohol dependence.[5–9] Therefore, nAChRs are potential molecular targets for alcohol abuse and alcoholism. The primary focus of this review is to provide new insights into the rewarding effects of alcohol, as they pertain to different nAChR subtypes and other cholinergic factors that contribute to the molecular mechanisms of alcoholism. Understanding these cellular changes and molecular underpinnings, will lead to a better translational and therapeutic outcome of alcoholism. Considering the diversity of nAChRs in the brain, this chapter will discuss the role of different nAChR subunits, involving alcohol abuse and alcoholism obtained from genetic, preclinical, and clinical studies. In addition, the role of nAChRs in alcohol, nicotine coabuse, and other comorbid brain disorders will be discussed. The molecular targets and neurobiological mechanisms within the nAChRs system may offer a better understanding and effective therapeutic alternative to combat this complex relapsing brain disorder.

2. MESOLIMBIC-DOPAMINE SYSTEM AND nAChRs

Genetic, preclinical, and clinical studies have identified the brain nAChRs and membrane bound ion channels in the mesolimbic-dopamine system, as being involved in alcohol dependence.[10–18] The nAChRs are pentameric ligand-gated cation channels with a diverse composition, which is widely expressed throughout the central nervous system (CNS).[19–22] The nAChR-mediated signaling plays a critical role in many drug addiction processes, including the development of alcoholism.[23–27] Twelve neuronal nAChR subunits, which are classified as either alpha subunits ($\alpha2$–$\alpha10$) or beta subunits ($\beta2$–$\beta4$), have been identified.[19,28–31] These subunits assemble to form diverse functional nAChRs, which can be further subdivided into two major groups of brain nAChR subtypes.[23,24,27] For example, heteromeric receptors are assembled from both alpha ($\alpha2-\alpha6$) and beta subunits. The functional properties of such heteromeric receptors, depend on both the specific α and β subunits within the receptor complex and the α:β ratio of the subunits.[28–31] Previous research indicates that the most abundant nAChRs in the CNS are $\alpha4\beta2^\star$, receptors containing both $\alpha4$ and $\beta2$ subunits, and sometimes additional subunits (denoted by the asterisk). These receptors account for >90% of the high affinity nicotine binding sites in the brain.[30] The $\alpha4$ and $\beta2$ subunits are colocalized in many parts of the brain, including thalamic nuclei, cortex, and ventral striatum.[19,30]

On the other hand, homomeric receptors, such as, $\alpha7$ nAChR subtypes, are predominantly located in the hippocampal regions and cortical or subcortical regions, including the ventral striatum.[31] In addition, neuronal nAChRs containing $\alpha6$ subunits are expressed in dopamine-rich areas (e. g., the mesocorticolimbic reward neurocircuit). Thus, $\alpha6^\star$ receptors may be a new/novel drug target to treat many forms of drug addiction, including alcohol dependence.[25] A number of *invitro* or *invivo* studies have confirmed that alcohol activates the mesolimbic-dopamine system and elevates the synaptic release of dopamine in the ventral striatum, which partially mediates the rewarding effects of alcohol and other drugs of abuse.[32–35] Therefore, mesocorticolimbic nAChRs are considered to be a molecular switch, activated by addictive behaviors.[9,15,36–38] Neuronal nAChRs are widely expressed at the synapse, cell body, and axons in the CNS.[21] Presynaptic nAChRs are involved in regulating the release of ACh, monoamines, and amino acids.[39–43] In particular, dopamine release is regulated by $\alpha4\beta2^\star$, α3β2*, and α6* nAChRs (* indicates a possible involvement of other receptor subunits) in nigrostriatal/mesocorticolimbic terminals.[44–47]

3. KNOCKOUT AND TRANSGENIC MODELS AND nAChRs

Recent reports with nAChR subtype knockout (KO) mice have provided important information about both brain nAChR function and the mediation of addiction-related behavior.[23,48,49] For example, early research showed that mice lacking the β2 subunit do not display too many nicotine-associated responses, including nicotine-induced DA release in the dorsal and ventral striatum, as well as nicotine-elicited increases in the firing rate of associated DA neurons.[50,51] The lack of nicotine effect on the mesolimbic DA systems in β2 subtype nAChR KO mice is consistent with the absence of nicotine self-administration by these animals.[51] The α4 subunit requires a β2 subunit for assembly in a majority of heteromeric nAChRs in the brain. These and other studies using genetically modified mice suggest that α4/β2* nAChRs are critical for nicotine-related reward behaviors.[52,53] Despite the distribution of the α7 subunit in the brain, in particular its presence in the mesocorticolimbic system, studies in α7 KO mice are not definitive about a role for the α7 subunit in nicotine reward and conditioning.[48] However, α7* nAChRs are important for long-term potentiation, neuroplasticity associated with learning, and memory in the mesolimbic reward pathway.[48] KO mouse studies targeting the α6 subunit indicate that α6 partners with β2 nAChRs and may play an important role in nicotine addiction related behavior.[54] Like nicotine addiction, genetic studies have revealed that nAChRs are involved in ethanol self-administration, and reward behavior as well.[11] For example, a number of genetic studies have been conducted to identify the role of nAChR subtypes in ethanol-drinking behavior. Acute ethanol drinking behavior is reduced in α4 KO mice, compared to wild type (WT), indicating a role for the nAChR α4*subunit in ethanol abuse.[11,26] Similarly, ethanol-related behavior and ethanol-induced midbrain dopaminergic function get decreased in α4 KO mice.[55] On the other hand, β2 KO mice behave similar to WT type mice in ethanol drinking behaviors.[56] In addition, α6 KO and β3 KO mice also display ethanol drinking behavior that is similar to WT mice in a two-bottle ethanol drinking paradigm.[56] Moreover, α7 KO and WT mice consume similar amounts of ethanol, although there was a potential gender effect regarding α7 nAChRs effects on ethanol consumption.[56] Again, α5 KO mice do not differ in acute ethanol consumption, compared to WT mice.[57] Like nicotine-related behavior (as mentioned earlier), studies with transgenic over expression of the α5, α3, and β4 receptor subunit genes indicate these subunits have a complex role in the modulation of ethanol-related behaviors.[58] Together, these data indicate that nAChRs containing α5, α6, β2, or β3 subunits may not be critical in ethanol drinking behaviors. Overall, the evidence suggests that α4 receptors in the midbrain may be essential for ethanol-related behavior. These studies represent genetic mechanisms in ethanol dependence, involving brain nAChRs and associated neurobiological mechanisms.

It is widely recognized that brain nAChR subtypes are important mediators of the rewarding effects of ethanol and drugs of abuse.[55,59–66] For example, systemic or local administration of nAChR ligands reduce ethanol drinking in a number of animal models.[32,60,61,65] Furthermore, nAChRs in the ventral tegmental area (VTA) regulate ethanol consumption and associated mesolimbic neurochemical effects (e.g., dopamine release in the nucleus accumbens (Acb)), as shown in various preclinical studies.[60,62] However, some of these ligands produce mixed effects in alcohol/ethanol drinking behavior in humans,[59,63] indicating mixed efficacy for treating ethanol dependence through nAChRs in the mesolimbic-dopamine system. Similarly, specific nAChR ligands may be resistant to pharmacologically efficacious reductions in ethanol drinking behavior, thus suggesting a role for other nAChR targets, such as α6β2*.[33,36] Additional studies with the α7 nAChR ligand were found to be ineffective in reducing ethanol-taking behavior of animal models.[56] In other works, a partial α4β2* nAChR agonist was shown to reduce alcohol drinking in both animal models and humans.[16,26,56,65,67–75] Given the mixed results with pharmacological efficacy on ethanol drinking behavior, the role of specific nAChR subtypes needs further investigation. Additional studies with cytisine, a partial agonist at α4β2* and lobeline, which is a nonselective antagonist, were found to reduce ethanol-taking behavior in a number of preclinical models.[17,66,73–77] Interestingly, these nAChR ligands also altered alcohol-induced increases in mesolimbic tissue DA levels in mice,[77] supporting the important role of mesolimbic nAChRs in alcohol dependence. Confirming this, the nAChR ligands were found to reduce ethanol-taking behavior in a genetic animal model for alcohol abuse and dependence[78] or inbred mice,[74] indicating nAChRs are potential molecular targets for individuals with a genetic predisposition to develop alcohol dependence.

Thus, evidence suggests that selective desensitization of nAChRs with partial agonistic activity[79,80] reduces ethanol-taking behavior in rats selectively bred for an alcohol preference.[80,81] In addition with the involvement of nAChRs in ethanol-taking behavior, some of these ligands[26,75,81,82] also decrease the alcohol deprivation effect, which is an animal model of relapse behavior.[78,83–87] Emerging preclinical studies suggest that nicotine exposure reinstates alcohol seeking behaviors in rodents following extinction of alcohol reinforcement.[88,89] Therefore, examination of cholinergic mechanisms associated with relapse is also important for new drug development to treat ethanol abuse and dependence.[5,90] Overall, the existing animal and human studies suggest that ethanol-induced activation of the mesolimbic DA system involves brain diverse nAChRs stimulation, including $\alpha4\beta2^{\star}$, $\alpha6\beta2^{\star}$, as well as $\alpha3\beta4^{\star}$ subtypes.[91] Thus, central nAChRs continue to be critical targets for the reinforcing and DA-activating effects of ethanol and underscore the need to conduct more subunit-specific nAChR research regarding ethanol abuse and dependence.

4. ALCOHOL, NICOTINE COABUSE, AND nAChRs

Due to the prevalence of ethanol and nicotine couse,[92] and the extremely high rates of smoking in individuals diagnosed with alcohol dependence,[93,94,95] investigators have been working to develop models of coabuse in animals. However, till date very few such models have been validated. In one such model, within-session intravenous nicotine and oral ethanol drinking was demonstrated in Wistar rats.[96] More recently, ethanol-preferring (P) rats show oral operant intake of combined ethanol and nicotine solutions, that results in blood ethanol concentrations of approximately 100 mg% and blood nicotine levels of approximately 25 ng/mL. These levels are similar to levels obtained in human binge drinking and smoking.[97] Additionally, in adolescent C57BL/6 mice, exposure to cigarette smoke for 6 h/day for 16 days increased intake of 10% ethanol three to five fold in a scheduled access paradigm.[98] Also, chronic nicotine exposure in C3H mice during adolescence enhances ethanol withdrawal effects in adulthood.[99]

In a study using rats selectively bred for high versus low locomotor activity induced by a novel field, it was found that adolescent exposure to nicotine facilitated an ethanol-induced conditioned place preference in early adulthood.[100] Regarding $\alpha4\beta2^{\star}$ receptors, a recent study revealed that chronic ethanol exposure/intake, including *in utero* exposure, by rhesus monkeys significantly decreased $\alpha4\beta2^{\star}$ levels in the frontal and insular cortex.[101] In a study involving *in utero* exposure of rats, it was found that gestational exposure to ethanol and nicotine significantly increased nicotine self-administration during adolescence, and this effect seemed to be due to glutamatergic modulation of the mesolimbic-dopamine system.[102] These data suggest a possible interaction between stages of development and cigarette smoke/nicotine exposure to increase ethanol and/or nicotine intake through different mechanisms that may extend into adulthood.

Animals selected for ethanol preference also show an increased sensitivity to the reinforcing effects of various drugs, including nicotine.[97] For instance, the P rat shows elevated sensitivity to the self-administration of intravenous nicotine, compared to the nonpreferring (NP) rat.[103] Consistent with this finding, P-rats are more sensitive than Wistar rats to the self-administration of nicotine directly into the posterior ventral tegmental area (pVTA) using the intracranial self-administration technique,[104] and ethanol and nicotine are co-self-administered into the pVTA at low concentrations that do not support self-administration individually.[105] However, in a recent study employing DBA/2J mice, nicotine was found to enhance the locomotor stimulating, but not the conditioned rewarding effects of ethanol, suggesting that in this model, combined stimulant effects of nicotine and ethanol do not predict enhanced reward.[106,107]

Overall, current data suggest that ethanol and nicotine coabuse is rooted in the basic biological underpinnings of both drugs, supposedly by targeting $\alpha4\beta2^*$ nAChR subtype in the mesolimbic-dopamine system. The drugs may function synergistically on some measures and each drug affects the other in terms of abuse liability. Emerging findings would suggest that the selection of ethanol-preference may increase the potential intake of nicotine coabuse, and adolescent coabuse may be lead to elevated levels of coabuse in adulthood. Additional studies are clearly needed to provide a better understanding of the mechanisms that are involved in ethanol and nicotine coabuse.

5. GENETIC POLYMORPHISMS, ALCOHOL DEPENDENCE AND nAChRs

Emerging evidence indicates that cholinergic genes may play a significant role in ethanol-dependent behavior.[108] For example, significant associations between CHRNA6 polymorphisms (rs1072003, rs2304297, and rs892413) as well as CHRNB3 polymorphism (rs13280604) and excessive ethanol-drinking behavior have been reported.[109] In a recent study, using Hispanic and non-Hispanic white subjects from the Social and Emotional Contexts of Adolescent Smoking Patterns, revealed multiple polymorphisms of CHRNA4 were associated with a significantly elevated risk for adolescent binge drinking.[110] These authors reported that polymorphisms for other nAChR genes were not associated with this risk, which may be related to the ethnicity of the sample. Additional studies from the Nicotine Addiction Genetics consortium in Finland, reported a significant association between the CHRNB4 polymorphism rs11636753 and regular ethanol drinking with comorbidity for depression, which may have been sex dependent.[111] Regarding addiction in general, although polysubstance abuse may be a factor, the CHRNA5 risk polymorphism (rs16969968) is not only associated with nicotine dependence,[112] but also associated with other drugs of abuse, such as cocaine.[113] Similarly, other CHRNA5 polymorphisms (rs615470 and rs684513) have significant associations with ethanol and cocaine dependence, respectively.[113] Given an early onset of ethanol, drug use and abuse increases the probability of developing dependence later in life, it is important to examine whether these effects are under genetic control. One study examining polymorphisms within the CHRNA5-A3-B4 gene cluster found a significant association with the age of initiating the use of multiple abused substances.[114] Similarly, recent findings from the San Diego Sibling Pair study indicate that, variants for CHRNA5 within this cluster are significantly associated with subjective level of response (i.e., intoxication) to ethanol.[115] Interestingly, a recent study indicated missense variants in CHRNA3 may confer resistance to cocaine dependence in African Americans.[116] However, it was also reported that a missense variant in CHRNB3 called rs149775276, was significantly associated with ethanol and cocaine dependence in European Americans.[116] Combined findings indicate that multiple polymorphisms associated with nAChR genes and addictions, have been identified. Moreover, these associations predict dependence on a number of abused substances and/or associated behaviors, across national, ethnic, and psychiatric groups. Taken together, these studies suggest that nAChRs could be important targets for the development of therapeutics targeting multiple addictions, including alcoholism.

6. ALCOHOL DEPENDENCE, COMORBID PSYCHIATRIC CONDITION AND nAChRs

Recent evidence indicates that nAChRs are involved in drug addiction and comorbid psychiatric disorders, such as anxiety or depression.[117] Just as in nicotine addiction,[118] it has been proposed that there is a relationship between ethanol dependence and depression.[119,120] For example, patients with depression have higher rates of ethanol-related problems, than the general population.[121,122] Ethanol abstinence-related depression increases the chance of relapse because people may use ethanol for self-medication.[123] A number of studies have determined how genetic predisposition to high ethanol intake affects depression-like behavior, and how genetic predisposition to depression-like behavior affects ethanol intake in rats, but the results are not well established,[124] probably due to the challenges of defining criteria for valid animal models of these disorders. Previous studies suggest that chronic ethanol consumption in mice increases depression-like behavior during abstinence from ethanol.[125,126] Therefore, alcoholism and depression supposedly share common molecular targets and associated neurobiological mechanisms. For example, abstinence from ethanol increases ACh release in the Acb,[127] and microdialysis studies indicate that ethanol abstinence induces rapid and sustained increases in extracellular ACh levels in the hippocampus.[128] A reduction in serotonergic neurotransmission has also been implicated in subpopulations of alcoholic patients.[129,130] And, the efficacy of selective serotonin reuptake inhibitors (SSRIs) provides strong support for the role of serotonin in depression, although SSRI treatment efficacy is highly variable for maintaining long-term abstinence from ethanol.[131] Alcoholism,[132] nicotine addiction,[133] and depression[134,135] are disorders mediated by neuroplasticity including transcriptional control, especially with brain-derived neurotrophic factor (BDNF) and activity-regulated cytoskeleton-associated protein (Arc)[136] pathways. Regarding this, prolonged exposure to and abstinence from ethanol causes long-lasting neuroadaptations, which may underlie the development of depression-like behavior.[126] Moreover, chronic ethanol consumption reduces the expression of BDNF in the rat hippocampus, which is an effect seen in depression-like behavior in animal models.[137] Given the common neurobiological mechanisms in ethanol dependence and comorbid depression, it is possible that β_2*-nAChRs modulation and associated molecular mechanisms will be critical in reducing depression, in the presence of ethanol abuse and dependence. Therefore, the contribution of specific nAChR subtypes, and their molecular signaling in comorbid conditions, needs considerable additional investigation.

7. ALCOHOL, TRANSCRIPTION FACTORS, AND nAChRs

Evidence indicates that nAChRs modulate acute ethanol-induced increased expression of Fos family in the immediate early genes, such as c-Fos in midbrain dopaminergic neurons.[17,26] In addition, both partial agonists and antagonists targeting nAChRs, reduce acute ethanol-induced c-Fos expression in the VTA and Acb.[17,26] Similarly deltaFosB, which is a truncated splice variant of FosB gene, activity is associated with chronic ethanol exposure and neuroplasticity in the mesolimbic reward neurocircuit.[138,139,140,141] Also, recent research suggests that nAChRs modulate deltaFosB upregulation in both ventral and dorsal striatum.[76] For example, partial agonists at nAChRs reduce chronic ethanol drinking behavior and its associated striatal deltaFosB upregulation.[76] Overall, these studies support the possible role of nAChRs in modulating long-lasting behavioral and molecular neuroadaptations that may be related to alcoholism. However, the role of nAChRs on other important transcription factors, such as cAMP-responsive element binding (CREB) expression requires further investigation.[142,143]

8. CONCLUSIONS

In this review, evidence has been presented for the fact that AChRs in the mesolimbic-dopamine system are important molecular targets for ethanol abuse and alcoholism. It is evident from *in vitro* and *in vivo* studies that nAChRs are critically involved with synaptic activity of mesolimbic dopamine as well as associated cellular and molecular mechanisms, which underlie the addictive properties produced by ethanol. Given the variety of nAChR subtypes, localization and functions of these nAChRs are potential mediators of the complex neurobiological effects of ethanol. While the nAChR subtypes share a common structure, their pharmacological properties on drug addiction, including models of alcoholism, may depend on nAChR subunit composition. We believe that multiple nAChR subtypes could be targets for neurobiological effects of ethanol and their contribution to alcoholism. Therefore, understanding subtype-selective mechanisms will be critical for future translational research. Thus, brain nAChRs represent a potential molecular target for treating ethanol abuse and dependence. The fact that ethanol and nicotine addiction often cooccur in humans, is widely recognized in this literature. Recent evidence indicates a genetic corelation between these two addictive disorders. Preclinical and clinical studies suggest that both nicotine and ethanol can, either directly or indirectly, activate the mesolimbic-dopamine system, which putatively mediates the rewarding effects and ethanol-induced addictive behavior, associated with nAChRs.

Emerging data also indicate that neuronal nAChRs are involved in addiction and comorbid psychiatric disorders, such as anxiety or depression. Finally, it is important to note that nAChRs modulate important transcription factors, such as c-Fos or deltaFosB, in the mesolimbic-dopamine system that may be associated with chronic effects of ethanol. However, the role of nAChRs on CREB or BDNF expression associated with alcoholism remains to be determined. Overall, the evidence supporting a role for nAChRs in the neurobiological effects of ethanol is significant from a translational perspective, as are the implications of nAChRs in addictive behaviors, and a variety of comorbid psychiatric and cognitive conditions. Additional research and a refined understanding of the specific contribution of nAChR subtypes, and associated neurobiological mechanisms involving alcoholism and comorbid neuropsychiatric conditions, may identify new molecular targets and signaling pathways for the development of better treatment and prevention strategies.

ACKNOWLEDGMENTS

The authors acknowledge support from the grants in part from Juhnke Endowment Fund and South Dakota State University Foundation (SR); AA13522 (RLB) and AA020396 (EAE, RLB) from the National Institutes of Health (NIH)/National Institute on Alcohol Abuse and Alcoholism (NIAAA). The views expressed herein are solely those of the authors and do not necessarily reflect the views of the funding organizations, NIH or NIAAA.

REFERENCES

1. Dwyer-Lindgren L, Flaxman AD, Ng M, Hansen GM, Murray CJ, Mokdad AH. Drinking patterns in US counties from 2002 to 2012. *Am J Public Health.* 2015;105:1120–1127.
2. *Report on Alcohol.* Geneva: World Health Organization, Department of Mental Health and Substance Abuse; 2011
3. Leibson T, Neuman G, Chudley AE, Koren G. The differential diagnosis of fetal alcohol spectrum disorder. *J Popul Ther Clin Pharmacol.* 2014;21:e1–e30.
4. CDC, Center for Disease Control. *The Power of Prevention Chronic Disease. The Public Health Challenge of the 21st Century.* Washington, DC: CDC; 2009
5. Koob GF, Volkow ND. Neurocircuitry of addiction. *Neuropsychopharmacology.* 2010;35:217–238.
6. Volkow ND, Baler RD. Addiction science: uncovering neurobiological complexity. *Neuropharmacology.* 2014;76(Pt B):235–249.
7. Ericson M, Löf E, Stomberg R, Söderpalm B. The smoking cessation medication varenicline attenuates alcohol and nicotine interactions in the rat mesolimbic dopamine system. *J Pharmacol Exp Ther.* 2009;329:225–230.
8. Wise RA, Koob GF. The development and maintenance of drug addiction. *Neuropsychopharmacology.* 2014;39:254–262.

9. Rahman S, EnglemanEA. Bell RL. Nicotinic receptor modulation to treat alcohol and drug dependence. *Front Neurosci.* 2015;8:1–11.
10. Feduccia AA. Chatterjee S, Bartlett SE. Neuronal nicotinic acetylcholine receptor: neuroplastic changes underlying alcohol and nicotine addictions. *Front Mol Neurosci.* 2012;5:1–18.
11. Hendrickson LM, Guildford MJ, Tapper AR. Neuronal nicotinic acetylcholine receptors: common molecular substrates of nicotine and alcohol dependence. *FrontPsychiatry.* 2013;4:29.
12. Rahman S, Prendergast MA. Brain cholinergic receptor system as a target for treating alcohol dependence. *Recent Patents CNS Drug Discov.* 2012;7:145–150.
13. Rahman S. Nicotinic receptors as therapeutic targets for drug addictive disorders. *CNS Neurol Dis DrugTargets.* 2013;12:633–640.
14. Wu J, Gao M, Taylor DH. Neuronal nicotinic acetylcholine receptors are important targets for alcohol reward and dependence. *Acta Pharmacol Sinica.* 2014;35: 311–315.
15. Chatterjee S, Bartlett SE. Neuronal nicotinic acetylcholine receptors as pharmacotherapeutic targets for the treatment of alcohol use disorders. *CNS Neurol Dis DrugTargets.* 2010;9:60–76.
16. Chatterjee S, Steensland P, Rollema H, Bartlett SE, et al. Partial agonists of the alpha3 beta4 neuronal nicotinic acetylcholine receptor reduce ethanol consumption and seeking in rats. *Neuropsychopharmacology.* 2011;36:603–615.
17. Hendrickson LM, Zhao-Shea R, Tapper AR. Modulation of ethanol drinking-in-the-dark by mecamylamine and nicotinic acetylcholine receptor agonists in C57BL/6J mice. *Psychopharmacology.* 2009;204:563–572.
18. Tizabi Y, Copeland Jr RL, Louis VA, Taylor RE. Effects of combined systemic alcohol and central nicotine administration into ventral tegmental area on dopamine release in the nucleus accumbens. *Alcohol Clin Exp Res.* 2002;26:394–399.
19. Zoli M, Pistillo F, Gotti C. Diversity of native nicotinic receptor subtypes in mammalian brain. *Neuropharmacology.* 2015;96:302–311.
20. Millar NS, Gotti C. Diversity of vertebrate nicotinic acytylcholine receptors. *Neuropharmacology.* 2009;56:237–246.
21. Livingstone PD, Wonnacott S. Nicotinic acetylcholine receptors and the ascending dopamine pathways. *Biochem Pharmacol.* 2009;78:744–755.
22. Picciotto MR, Addy NA, Mineur YS, Brunzell DH. It is not "either/or": activation and desensitization of nicotinic acetylcholine receptors both contribute to behaviors related tonicotine addiction and mood. *Prog Neurobiol.* 2008;84:329–342.
23. Fowler CD, Arends MA, Kenny PJ. Subtypes of nicotinic acetylcholine receptors in nicotine reward, dependence, and withdrawal: evidence from genetically modified mice. *Behav Pharmacol.* 2008;19:461–484.
24. Rahman S, Lopez-Hernandez G, Corrigall WA, Papke RL. Neuronal nicotinic receptors as brain targets for the pharmacotherapy of drug addiction. *CNS Neurol Dis Drug Targets.* 2008;7:422–441.
25. Kuzmin A, Jerlhag E, Liljequist S, Engel JA. Effects of subunit selective nAChR receptors on operant ethanol self-administration and relapse-like ethanol-drinking behavior. *Psychopharmacology.* 2009;203:99–108.
26. Hendrickson LM, Zhao-Shea R, Pang X, Gardner PD, Tapper AR. Activation of alpha4* nAChRs is necessary and sufficient for varenicline-induced reduction of alcohol consumption. *J Neurosci.* 2010;30:10169–10176.
27. Tuesta LM, Fowler CD, Kenny PJ. Recent advances in understanding nicotinic receptor signaling mechanisms that regulate drug-self administration behavior. *Biochem Pharmacol.* 2011;82:984–995.

28. Le Novere N, Zoli M, Changeux JP. Neuronal nicotinic receptor alpha 6 subunit mRNA is selectively concentrated in catecholaminergic nuclei of the rat brain. *Eur J Neurosci.* 1996;8:2428–2439.
29. Buisson B, Bertrand D. Nicotine addiction: the possible role of functional upregulation. *Trends Pharmacol Sci.* 2002;23:130–136.
30. Lindstrom JM. Nicotinic acetylcholine receptors of muscles and nerves: comparison of their structures, functional roles, and vulnerability to pathology. *Ann NY Acad Sci.* 2003;998:41–52.
31. Gotti C, Clementi F. Neuronal nicotinic receptors: from structure to pathology. *Prog Neurobiol.* 2004;74:363–396.
32. Soderpalm B, Ericson M, Olausson P, Blomqvist O, Engel JA. Nicotinic mechanisms involved in the dopamine activating and reinforcing properties of ethanol. *Behav Brain Res.* 2000;113:85–96.
33. Larsson A, Svensson L, Söderpalm B, Engel JA. Role of different nicotinic acetylcholine receptors in mediating behavioral and neurochemical effects of ethanol in mice. *Alcohol.* 2002;28:157–167.
34. McBride WJ, Li TK. Animal models of alcoholism: neurobiology of high alcohol-drinking behavior in rodents. *Crit Rev Neurobiol.* 1998;12:339–369.
35. Tizabi Y, Bai L, Copeland Jr RL, Taylor RE. Combined effects of systemic alcohol and nicotine on dopamine release in the nucleus accumbens shell. *Alcohol.* 2007;42: 413–416.
36. Larsson A, Engel JA. Neurochemical and behavioral studies on ethanol and nicotine interactions. *Neurosci Biobehav Rev.* 2004;27:713–720.
37. Schlaepfer IR, Hoft NR, Ehringer MA. The genetic components of alcohol and nicotine co-addiction: from genes to behavior. *Curr Drug Abuse Rev.* 2008;1: 124–134.
38. Crunelle CL, Miller ML, Booji J, Van den Brink W. The nicotinic acetylcholine receptor partial agonist varencline and the treatment of drug dependence: a review. *Eur Neuropsychopharmacol.* 2010;20:69–79.
39. Wilkie GI, Hutson PH, Stephens MW, Whiting P, Wonnacott S. Hippocampal nicotinic autoreceptors modulate acetylcholine release. *Biochem Soc Trans.* 1993;21: 429–431.
40. Alkondon M, Pereira EF, Barbosa CT, Albuquerque EX. Neuronal nicotinic acetylcholine receptor activation modulates γ-aminobutyric acid release from CA1 neurons of rat hippocampal slices. *J Pharmacol Exp Ther.* 1997;283:1396–1411.
41. Clarke P, Reuben M. Release of [^{3}H]-noradrenaline from rat hippocampal synaptosomes by nicotine: mediation by different nicotinic receptor subtypes from striatal [^{3}H]-dopamine release. *Br J Pharmacol.* 1996;117:595–606.
42. Grady S, Marks MJ, Wonnacott S, Collins AC. Characterization of nicotinic receptor-mediated [^{3}H] dopamine release from synaptosomes prepared from mouse striatum. *J Neurochem.* 1992;59:848–856.
43. Yang X, Criswell HE, Breese GR. Nicotine-induced inhibition in medial septum involves activation of presynaptic nicotinic cholinergic receptors on gamma-aminobutyric acid containing neurons. *J Pharmacol Exp Ther.* 1996;276:482–489.
44. Le Novere N, Zoli M, Changeux JP. Neuronal nicotinic receptor alpha 6 subunit mRNA is selectively concentrated in catecholaminergic nuclei of the rat brain. *Eur J Neurosci.* 1996;8:2428–2439.
45. Luo S, Kulak JM, Cartier GE, Jacobsen RB, Yoshikami D, Olivera BM, McIntosh JM. α-ConotoxinAuIB selectively blocks α3β4 nicotinic acetylcholine receptors and nicotine-evoked norepinephrine release. *J Neurosci.* 1998;18: 8571–8579.

46. Salminen O, Murphy KL, McIntosh JM, Drago J, Marks MJ, Collins AC, Grady SR. Subunit composition and pharmacology of two classes of striatal presynaptic nicotinic acetylcholine receptors mediating dopamine release in mice. *Mol Pharmacol.* 2004;65:1526–1535.
47. Wonnacott S, Kaiser S, Mogg A, Soliakov L, Jones IW. Presynaptic nicotinic receptors modulating dopamine release in the rat striatum. *Eur J Pharmacol.* 2000;393:51–58.
48. Mineur YS, Picciotto MR. Genetics of nicotinic acetylcholine receptors: relevance to nicotine addiction. *Biochem Pharmacol.* 2008;75:323–333.
49. Changeux JP. Nicotine addiction and nicotinic receptors: lessons from genetically modified mice. *Nat Rev Neurosci.* 2010;11:389–401.
50. Picciotto MR, Zoli M, Lena C, Bessis A, Lallemand Y, Le Novere N, et al. Abnormal avoidance learning in mice lacking functional high-affinity nicotine receptor in the brain. *Nature.* 1995;374:65–67.
51. Picciotto MR, Zoli M, Rimondini R, Lena C, Marubio LM, Pich EM, et al. Acetylcholine receptors containing the beta2 subunit are involved in the reinforcing properties of nicotine. *Nature.* 1998;391:173–177.
52. Ross SA, Wong JY, Clifford JJ, Kinsella A, Massalas JS, Horne MK, et al. Phenotypic characterization of an alpha 4 neuronal nicotinic acetylcholine receptor subunit knock-out mouse. *J Neurosci.* 2000;20:6431–6441.
53. Tapper AR, McKinney SL, Nashmi R, Schwarz J, Deshpande P, Labarca C, et al. Nicotine activation of alpha4* receptors: sufficient for reward, tolerance, and sensitization. *Science.* 2004;306:1029–1032.
54. Champtiaux N, Gotti C, Cordero-Erausquin M, David DJ, Przybylski C, Léna C, Clementi F, Moretti M, Rossi FM, Le Novère N, McIntosh JM, Gardier AM, Changeux JP. Subunit composition of functional nicotinic receptors in dopaminergic neurons investigated with knock-out mice. *J Neurosci.* 2003;23:7820–7829.
55. Liu L, Hendrickson LM, Guildford MJ, Zhao-Shea R, Gardner PD, Tapper AR. Nicotinic acetylcholine receptors containing the α4 subunit modulate alcohol reward. *Biol Psychiatry.* 2013;73:738–746.
56. Kamens HM, Anderson J, Picciotto MR. Modulation of ethanol consumption by genetic and pharmacological manipulation of nicotinic acetylcholine receptors in mice. *Psychopharmacology.* 2010;208:613–626.
57. Santos N, Chatterjee S, Henry A, Holgate J, Bartlett SE. The alpha5 neuronal nicotinic acetylcholine receptor subunit plays an important role in the sedative effects of ethanol but does not modulate consumption in mice. *Alcohol Clin Exp Res.* 2012;37:655–662.
58. Gallego X, Ruiz-Medina J, Valverde O, Molas S, Robles N, Sabrià J, Crabbe JC, Dierssen M. Transgenic over expression of nicotinic receptor alpha 5, alpha 3, and beta 4 subunit genes reduces ethanol intake in mice. *Alcohol.* 2012;46:205–215.
59. Blomqvist O, Ericson M, Johnson DH, Engel JA, Söderpalm B. Voluntary ethanol intake in the rat: effects of nicotinic acetylcholine receptor blockade or subchronic nicotine treatment. *Eur J Pharmacol.* 1996;314:257–267.
60. Ericson M, Blomqvist O, Engel JA, Söderpalm B. Voluntary ethanol intake in the rat and the associated accumbal dopamine overflow are blocked by ventral tegmental mecamylamine. *Eur J Pharmacol.* 1998;358:189–196.
61. Lê AD, Corrigall WA, Harding JW, Juzytsch W, Li TK. Involvement of nicotinic receptors in alcohol self-administration. *Alcohol Clin Exp Res.* 2000;24:155–163.
62. Chi H, de Wit H. Mecamylamine attenuates the subjective stimulant-like effects of alcohol in social drinkers. *Alcohol Clin Exp Res.* 2003;27:780–786.
63. Young EM, Mahler S, Chi H, de Wit H. Mecamylamine and ethanol preference in healthy volunteers. *Alcohol Clin Exp Res.* 2005;29:58–65.
64. Reus VI, Obach RS, Coe JW, Faessel H, Rollema H, Watsky E, Reeves K. Varenicline: new treatment with efficacy in smoking cessation. *Drugs Today.* 2007;43:65–75.

65. Steensland P, Simms JA, Holgate J, Richards JK, Bartlett SE. Varenicline, an alpha4beta2 nicotinic acetylcholine receptor partial agonist, selectively decreases ethanol consumption and seeking. *Proc Natl Acad Sci USA*. 2007;104:12518–12523.
66. Bell RL, Eiler BJ, Cook JB, Rahman S. Nicotinic receptor ligands reduce ethanol intake by high alcohol-drinking HAD-2 rats. *Alcohol*. 2009;43:581–592.
67. McKee SA, Harrison EL, O'Malley SS, Krishnan-Sarin S, Shi J, Tetrault JM, et al. Varenicline reduces alcohol self-administration in heavy-drinking smokers. *Biol Psychiatry*. 2009;66:185–190.
68. Bito-Onon JJ, Simms JA, Chatterjee S, Holgate J, Bartlett SE. Varenicline, a partial agonist at neuronal nicotinic acetylcholine receptors, reduces nicotine-induced increases in 20% ethanol operant self-administration in Sprague-Dawley rats. *Addict Biol*. 2011;16:440–449.
69. Mitchell JM, Teague CH, Kayser AS, Bartlett SE, Fields HL. Varenicline decreases alcohol consumption in heavy-drinking smokers. *Psychopharmacology*. 2012;223: 299–306.
70. Kaminski BJ, Weerts EM. The effects of varenicline on alcohol seeking and self-administration in baboons. *Alcohol Clin Exp Res*. 2014;38:376–383.
71. Litten RZ, Egli M, Heilig M, Cui C, Fertig JB, Ryan ML, Falk DE, Moss H, Huebner R, Noronha A. Medications development to treat alcohol dependence: a vision for the next decade. *Addict Biol*. 2012;17:513–527.
72. Sotomayor-Zarate R, Gysling K, Busto UE, Cassels BK, Tampier L, Quintanilla ME. Varenicline and cytisine: two nicotinic acetylcholine receptor ligands reduce ethanol intake in University of Chile bibulous rats. *Psychopharmacology*. 2013;227:287–298.
73. Sajja RK, Rahman S. Lobeline and cytisine reduce voluntary ethanol drinking behavior in male C57BL/6J mice. *Prog Neuropsychopharmacol Biol Psychiatry*. 2011;35:257–264.
74. Sajja RK, Rahman S. Neuronal nicotinic receptor ligands modulate chronic nicotine-induced ethanol consumption in C57BL/6J mice. *Pharmacol Biochem Behav*. 2012;102: 36–43.
75. Sajja RK, Rahman S. Nicotinic receptor partial agonists modulate alcohol deprivation effect in C57BL/6J in mice. *Pharmacol Biochem Behav*. 2013;110:161–167.
76. Sajja RK, Rahman S. Cytisine modulates chronic voluntary ethanol consumption and ethanol-induced striatal up-regulation of Delta FosB in mice. *Alcohol*. 2013;47: 299–307.
77. Sajja RK, Dwivedi C, Rahman S. Nicotinic ligands modulate ethanol-induced dopamine functions in mice. *Pharmacology*. 2010;86:168–173.
78. Bell RL, Sable HJK, Colombo G, Hyytia P, Rodd ZA, Lumeng L. Animal models for medications development targeting alcohol abuse using selectively bred rat lines: neurobiological and pharmacological validity. *Pharmacol Biochem Behav*. 2012;103: 119–155.
79. Rezvani AH, Cauley M, Xiao Y, Kellar KJ, Levin ED. Effects of chronic sazetidine-A, a selective $\alpha4\beta2$ neuronal nicotinic acetylcholine receptors desensitizing agent on pharmacologically-induced impaired attention in rats. *Psychopharmacology*. 2013;226:35–43.
80. Xiao Y, Fan H, Musachio JL, Wei ZL, Chellappan SK, Kozikowski AP, et al. Sazetidine-A, a novel ligand that desensitizes$\alpha4\beta2$ nicotinic acetylcholine receptors without activating them. *Mol Pharmacol*. 2006;70:1454–1460.
81. Rezvani AH, Slade S, Wells C, Petro A, Lumeng L, Li T-K, et al. Effects of sazetidine-A a selective alpha4beta2 nicotinic acetylcholine receptor desensitizing agent on alcohol and nicotine self-administration in selectively bred alcohol-preferring (P) rats. *Psychopharmacology*. 2010;211:161–174.
82. Wouda JA, Riga D, De Vries W, Stegeman M, van Mourik Y, Schetters D, et al. Varenicline attenuates cue-induced relapse to alcohol but not nicotine seeking while reducing inhibitory response control. *Psychopharmacology*. 2011;216:267–277.

83. McKinzie DL, Nowak KL, Yorger L, McBride WJ, Murphy JM, Lumeng L, et al. The alcohol deprivation effect in the alcohol preferring P rat under free-drinking and operant conditions. *Alcohol Clin Exp Res.* 1998;22:1170–1176.
84. Rodd ZA, Bell RL, Sable HJK, Murphy JM, McBride WJ. Recent advances in animal models of alcohol craving and relapse. *Pharmacol Biochem Behav.* 2004;79:439–450.
85. Spanagel R, Hölter SM. Long-term alcohol self-administration with repeated alcohol deprivation phases: an animal model of alcoholism? *Alcohol.* 1999;34:231–243.
86. Sparta DR, Ferraro III FM, Fee JR, Knapp DJ, Breese GR, Thiele TE. The alcohol deprivation effect in C57BL/6J mice is observed using operant self-administration procedures and is modulated by CRF-1 receptor signaling. *Alcohol Clin Exp Res.* 2009;233:31–42.
87. Melendez RI, Middaugh LD, Kalivas PW. Development of an alcohol deprivation and escalation effect in C57BL/6J mice. *Alcohol Clin Exp Res.* 2006;30:2017–2025.
88. Lê AD, Li Z, Funk D, Shram M, Li T-K, Shaham Y. Nicotine increases alcohol self-administration and reinstates alcohol seeking in rats. *Psychopharmacology.* 2003;68: 216–221.
89. Hauser SR, Getachew B, Oster SM, Dhaher R, Ding Z-M, Bell RL, et al. Nicotine modulates alcohol-seeking and relapse by alcohol-preferring (P) rats in a time dependent manner. *Alcohol Clin Exp Res.* 2012;36:43–54.
90. Weiss F, Porrino LJ. Behavioral neurobiology of alcohol addiction: recent advances and challenges. *J Neurosci.* 2002;222:3332–3337.
91. Rollema H, Coe JW, Chambers LK, Hurst RS, Stahl SM, Williams KE. Rationale, pharmacology and clinical efficacy of partial agonists of alpha4beta2 nACh receptors for smoking cessation. *Trends Pharmacol Sci.* 2007;28:316–325.
92. Kandel D, Chen K, Warner LA, Kessler RC, Grant B. Prevalence and demographic correlates of symptoms of last year dependence on alcohol, nicotine, marijuana and cocaine in the U.S. population. *Drug Alcohol Depen.* 1997;44:11–29.
93. Gulliver SB, Rohsenow DJ, Colby SM, Dey AN, Abrams DB, Niaura RS, et al. Interrelationship of smoking and alcohol dependence, use and urges to use. *J Stud Alcohol.* 1995;56:202–206.
94. Hughes JR. Treating smokers with current or past alcohol dependence. *Am J Health Behav.* 1996;20:286–290.
95. John U, Meyer C, Rumpf H, Schumann JA, Thyrian JR, Hapke U. Strength of the relationship between tobacco smoking, nicotine dependence and the severity of alcohol dependence syndrome criteria in a population-based sample. *Alcohol Alcohol.* 2003;38: 606–612.
96. Lê AD, Lo S, Harding S, Juzytsch W, Marinelli PW, Funk D. Co-administration of intravenous nicotine and oral alcohol in rats. *Psychopharmacology.* 2010;208:475–486.
97. McBride WJ, Rodd ZA, Bell RL, Lumeng L, Li TK. The alcohol-preferring (P) and high-alcohol-drinking (HAD) rats—animal models of alcoholism. *Alcohol.* 2014;48: 209–215.
98. Burns BE, Proctor WR. Cigarette smoke exposure greatly increases alcohol consumption in adolescent C57BL/6 mice. *Alcohol Clin Exp Res.* 2013;37:E364–E372.
99. Riley HH, Zalud AW, Diaz-Granados JL. The influence of a chronic adolescent nicotine exposure on ethanol withdrawal severity during adulthood in C3H mice. *Alcohol.* 2010;44:81–87.
100. Philpot RM, Engberg ME, Wecker L. Ethanol conditioned place preference and alterations in deltaFosB following adolescent nicotine administration differ in rats exhibiting high or low behavioral reactivity to a novel environment. *Behav Brain Res.* 2014;262:101–108.
101. Hillmer AT, Tudorascu DL, Wooten DW, Lao PJ, Barnhart TE, Ahlers EO, Resch LM, Larson JA, Converse AK, Moore CF, Schneider ML, Christian BT. Changes in the

alpha4beta2* nicotinic acetylcholine system during chronic controlled alcohol exposure in nonhuman primates. *Drug Alcohol Depen.* 2014;138:216–219.

102. Roguski EE, Sharp BM, Chen H, Matta SG. Full-gestational exposure to nicotine and ethanol augments nicotine self-administration by altering ventral tegmental dopaminergic function due to NMDA receptors in adolescent rats. *J Neurochem.* 2014;128: 701–712.
103. Lê AD, Li Z, Funk D, Shram M, Li TK, Shaham Y. Increased vulnerability to nicotine self-administration and relapse in alcohol-naive offspring of rats selectively bred for high alcohol intake. *The J Neurosci.* 2006;26:1872–1879.
104. Hauser SR, Bracken AL, Deehan Jr GA, Toalston JE, Ding ZM, Truitt WA, Bell RL, McBride WJ, Rodd ZA. Selective breeding for high alcohol preference increases the sensitivity of the posterior VTA to the reinforcing effects of nicotine. *Addict Biol.* 2014;19:800–811.
105. Truitt WA, Hauser SR, Deehan Jr GA, Toalston JE, Wilden JA, Bell RL, McBride WJ, Rodd ZA. Ethanol and nicotine interaction within the posterior ventral tegmental area in male and female alcohol-preferring rats: evidence of synergy and differential gene activation in the nucleus accumbens shell. *Psychopharmacology.* 2015;232:639–649.
106. Gubner NR, McKinnon CS, Reed C, Phillips TJ. Accentuating effects of nicotine on ethanol response in mice with high genetic predisposition to ethanol-induced locomotor stimulation. *Drug Alcohol Depend.* 2013;127:108–114.
107. Gubner NR, Cunningham CL, Phillips TJ. Nicotine enhances the locomotor stimulating but not the conditioned rewarding effect of ethanol in DBA/2J mice. *Alcohol Clin Exp Res.* 2015;39:64–72.
108. Madden PA, Heath AC. Shared genetic vulnerability in alcohol and cigarette use and dependence. *Alcohol Clin Exp Res.* 2002;26:1919–1921.
109. Hoft NR, Corley RP, McQueen MB, Huizinga D, Menard S, Ehringer MA. SNPs in CHRNA6 and CHRNB3 are associated with alcohol consumption in a nationally representative sample. *Genes Brain Behav.* 2009;8:631–637.
110. Coon H, Piasecki TM, Cook EH, Dunn D, Mermelstein RJ, Weiss RB, Cannon DS. Association of the CHRNA4 neuronal nicotinic receptor subunit gene with frequency of binge drinking in young adults. *Alcohol Clin Exp Res.* 2014;38:930–937.
111. Broms U, Wedenoja J, Largeau MR, Korhonen T, Pitkaniemi J, Keskitalo-Vuokko K, Happola A, Heikkila KH, Heikkila K, Ripatti S, Sarin AP, et al. Analysis of detailed phenotype profiles reveals CHRNA5-CHRNA3-CHRNB4 gene cluster association with several nicotine dependence traits. *Nic Tob Res.* 2012;4:720–733.
112. Anatharaman D, Chabrier A, Gaborieau V, Franceschi S, Herrero R, Rajkumar T, Samant T, Mahimkar MB, Brennan P, McKay JD. Genetic variants in nicotine addiction and alcohol metabolism genes, oral cancer risk and the propensity to smoke and drink alcohol: a replication study in India. *PLoS ONE.* 2014;9:e88240.
113. Sherva R, Kranzler HR, Yu Y, Logue MW, Poling J, Arias AJ, Anton RF, Oslin D, Farrer LA, Gelernter J. Variation in nicotinic acetylcholine receptor genes is associated with multiple substance dependence phenotypes. *Neuropsychopharmacology.* 2010;35: 1921–1931.
114. Lubke GH, Stephens SH, Lessem JM, Hewitt JK, Ehringer MA. The CHRNA5/A3/B4 gene cluster and tobacco, alcohol, cannabis, inhalants and other substance initiation: replication and new findings using mixture analyses. *Behav Genet.* 2012;42:636–646.
115. Choquet H, Joslyn G, Lee A, Kasberger J, Robertson M, Brush G, Schuckit MA, White R, Jorgenson E. Examination of rare missense variants in the CHRNA5-A3-B4 gene cluster to level of response to alcohol in the San Diego Sibling Pair study. *Alcohol Clin Exp Res.* 2013;37:1311–1316.
116. Haller G, Kapoor M, Budde J, Xuei X, Edenberg H, Nurnberger J, Kramer J, Brokks A, Tischfield J, Almasy L, Agrawal A, Bucholz K, Rice J, Saccone N, Bierut L, Goate A.

Rare missense variants in CHRNB3 and CHRNA3 are associated with risk of alcohol and cocaine dependence. *Hum Mol Genet.* 2014;23:810–819.

117. Rahman S. Targeting brain nicotinic acetylcholine receptors to treat major depression and co-morbid alcohol or nicotine addiction. *CNS Neurol Disord Drug Targets.* 2015;14: 647–653.
118. Roni MA, Rahman S. The effects of lobeline on nicotine withdrawal-induced depression-like behavior in mice. *Psychopharmacology.* 2014;231:2989–2998.
119. Sullivan LE, Fiellin DA, O'Connor PG. The prevalence and impact of alcohol problems in major depression: a systematic review. *Am J Med.* 2005;118:330–341.
120. Boschloo L, Vogelzangs N, Smit JH, van den Brink W, Veltman DJ, Beekman ATF, Penninx BWJH. Comorbidity and risk indicators for alcohol use disorders among persons with anxiety and/or depressive disorders: findings from the Netherlands Study of Depression and Anxiety (NESDA). *J Affect Disord.* 2011;131:233–242.
121. Kessler RC, Nelson CB, McGonagle KA, Edlund MJ, Frank RG, Leaf PJ. The epidemiology of co-occurring addictive and mental disorders: implications for prevention and service utilization. *Am J Orthopsychiatry.* 1996;66:17–31.
122. Garbutt JC, West SL, Carey TS, Lohr KN, Crews FT. Pharmacological treatment of alcohol dependence. *JAMA.* 1999;281:1318–1325.
123. Hasin DS, Grant BF. Major depression in 6050 former drinkers: association with past alcohol dependence. *Arch Gen Psychiatry.* 2002;59:794–800.
124. Godfrey CD, Froehlich JC, Stewart RB, Li TK, Murphy JM. Comparison of rats selectively bred for high and low ethanol intake in a forced-swim-test model of depression: effects of desipramine. *Physiol Behav.* 1997;62:729–733.
125. Pang TY, Renoir T, Du X, Lawrence AJ, Hannan AJ. Depression-related behaviours displayed by female C57BL/6J mice during abstinence from chronic ethanol consumption are rescued by wheel running. *Eur J Neurosci.* 2013;37:1803–1810.
126. Stevenson JR, Schroeder JP, Nixon K, Besheer J, Crews FT, Hodge CW. Abstinence following alcohol drinking produces depression-like behavior and reduced hippocampal neurogenesis in mice. *Neuropsychopharmacology.* 2009;34: 1209–1222.
127. Rada P, Johnson D, Lewis M, Hoebel B. In alcohol-treated rats, naloxone decreases extracellular dopamine and increases acetylcholine in the nucleus accumbens: evidence of opioid withdrawal. *Pharmacol Biochem Behav.* 2004;79:599–605.
128. Imperato A, Dazzi L, Carta G, Colombo G, Biggio G. Rapid increase in basal acetylcholine release in the hippocampus of freely moving rats induced by withdrawal from long-term ethanol intoxication. *Brain Res.* 1998;784:347–350.
129. Patkar AA, Gopalakrishnan R, Naik PC, Murray HW, Vergare MJ, Marsden CA. Changes in plasma noradrenaline and serotonin levels and craving during alcohol withdrawal. *Alcohol.* 2003;38:224–231.
130. Cadete-Leite A, Brandao F, Andrade J, Ribeiro-da-Silva A, Paula-Barbosa M. The GABAergic system of the dentate gyrus after withdrawal from chronic alcohol consumption: effects of intracerebral grafting and putative neuroprotective agents. *Alcohol.* 1997;32:471–484.
131. Narnajo CA, Knoke DM. The role of selective serotonin reuptake inhibitors in reducing alcohol consumption. *J Clin Psychiatry.* 2001;62:18–25.
132. Yakovleva T, Bazov I, Watanabe H, Hauser KF, Bakalkin G. Transcriptional control of maladaptive and protective responses in alcoholics: a role of the NF-kB system. *Brain Behav Immun.* 2011;25:S29–S38.
133. Gao Y, Peng S, Wen Q, Zheng C, Lin J, Tan Y, Ma Y, Luo Y, Xue Y, Wu P, Ding Z, Lu L, Li Y. The mammalian target of rapamycin pathway in the basolateral amygdala is critical for nicotine-induced behavioural sensitization. *Intl J Neuropsychopharmacol.* 2014;17: 1881–1894.

134. Carlberg L, Scheibelreiter J, Hassler MR, Schloegelhofer M, Schmoeger M, Ludwig Kasper S, Aschauer H, Egger G, Schosser A. Brain-derived neurotrophic factor (BDNF)—epigenetic regulation in unipolar and bipolar affective disorder. *J Affect Dis.* 2014;168:399–406.
135. Mitchelmore C, Gede L. Brain derived neurotrophic factor: epigenetic regulation in psychiatric disorders. *Brain Res.* 2014;1586:162–172.
136. Rosi S. Neuroinflammation and the plasticity-related immediate-early gene. *Arc Brain Behav Immun.* 2011;25:S39–S49.
137. Hauser SR, Getachew B, Taylor RE, Tizabi Y. Alcohol induced depressive-like behavior is associated with a reduction in hippocampal BDNF. *Pharmacol Biochem Behav.* 2011;100:253–258.
138. McClung CA, Nestler EJ. Neuroplasticity mediated by altered gene expression. *Neuropsychopharmacol Rev.* 2008;33:3–17.
139. Li J, Cheng Y, Bian W, Liu X, Zhang C, Ye JH. Region-specific induction of FosB/ΔFosB by voluntary alcohol intake: effects of naltrexone. *Alcohol Clin Exp Res.* 2010;34:1742–1750.
140. Robison AJ, Nestler EJ. Transcriptional and epigenetic mechanisms of addiction. *Nat Rev Neurosci.* 2011;12:623–637.
141. Perrotti LI, Hadeishi Y, Ulery PG, Barrot M, Monteggia L, Duman RS, Nestler EJ. Induction of ΔFosB in reward-related brain structures after chronic stress. *J Neurosci.* 2004;24:10594–10602.
142. Moonat S, Starkman BG, Sakharkar A, Pandey SC. Neuroscience of alcoholism: molecular and cellular mechanisms. *Cell Mol Life Sci.* 2010;67:73–88.
143. Krishnan HR, Sakharkar AJ, Teppen TL, Berkel TD, Pandey SC. The epigenetic landscape of alcoholism. *Int Rev Neurobiol.* 2014;115:75–116.

CHAPTER SEVEN

Molecular Changes in Opioid Addiction: The Role of Adenylyl Cyclase and cAMP/PKA System

Patrick Chan[*,1], **Kabirullah Lutfy**[†]
[*] Department of Pharmacy and Pharmacy Administration, Western University of Health Sciences, College of Pharmacy, Pomona, California, USA
[†] Department of Pharmaceutical Sciences, College of Pharmacy, Western University of Health Sciences, Pomona, California, USA
[1] Corresponding author: e-mail address: chanp@westernu.edu.

Contents

Abstract

For centuries, opiate analgesics have had a considerable presence in the treatment of moderate to severe pain. While effective in providing analgesia, opiates are notorious in exerting many undesirable adverse reactions. The receptor targets and the intracellular effectors of opioids have largely been identified. Furthermore, much of the mechanisms underlying the development of tolerance, dependence, and withdrawal have been delineated. Thus, there is a focus on developing novel compounds or strategies in mitigating or avoiding the development of tolerance, dependence, and withdrawal. This review focuses on the adenylyl cyclase and cyclic adenosine 3,5-monophosphate (cAMP)/protein kinase A (AC/cAMP/PKA) system as the central player in mediating the acute and chronic effects of opioids. This chapter also reviews

Progress in Molecular Biology and Translational Science, Volume 137
ISSN 1877-1173
http://dx.doi.org/10.1016/bs.pmbts.2015.10.005

the neuronal adaptive changes in the locus coeruleus, amygdala, periaqueductal gray, and ventral tegmental area induced by acute and chronic actions of opioid because these neuronal adaptive changes in these regions may underlie the behavioral changes observed in opiate users and abusers.

1. INTRODUCTION

Opioid use and abuse has increased significantly over the past several decades. Using the Drug Abuse Warning Network (DAWN) and Automation of Reports and Consolidated Orders System (ARCOS), a study published in 2014 showed higher rates of use for seven of the eight most commonly used opioid analgesics (buprenorphine, fentanyl, hydrocodone, hydromorphone, methadone, morphine, and oxycodone) between 1996 and 2011.[1] The use of these drugs had increased over 40-fold during the same time period. As a result of the increased use and misuse of opioid analgesics, substance abuse rehabilitation increased nearly 200%, with the largest increases observed in buprenorphine, hydromorphone, and oxycodone. In 2012, it was estimated that 2.1 million people were dependent on or abused opioid analgesics, representing a 50% increase from 2004 to 2012. It is second only to marijuana use.[2] Among illicit users of opioids, addiction rates may be as high as 30%. Notably, the clinical use of opioids has been on the rise. In 2012, 259 million prescriptions for painkillers were written for patients in the United States.[3]

Opioid analgesics have tremendous utility in the relief of moderate to severe pain. However, chronic opioid treatment has a high propensity to induce tolerance and dependence, and perhaps addiction, which is a serious and prevalent medical condition in the United States, resulting in a tremendous economic burden. The phenomenon of tolerance is defined as a rightward shift in the analgesic and other actions of morphine and related analgesics. Therefore, it is not unusual to see patients taking doses substantially larger than their original dose over a period of time.[4] For patients suffering from opioid dependency, there is both a physiological and psychological need to continue taking the drug. Opioid withdrawal occurs with abrupt cessation of chronic use. In the absence of the drug, withdrawal syndrome emerges that produces negative-affective states and physical symptoms, which can be exaggerated by stress. This often leads patients to relapse to seeking opioid consumption. Not only stress, but environmental cues associated with drug administration can elicit cravings and reinstate drug-seeking and drug-taking behaviors in abstinence individuals. In animal models of withdrawal, symptoms include hyperalgesia, ptosis, wet-dog shakes, escape attempts, tachypnea, and diarrhea. The various symptoms can be correlated with changes in neuronal functions.[5] Furthermore, an "incubation" phenomenon in which cravings for drug use is increased and may linger for weeks to months following exposure to cues associated with drug administration in subjects undergoing withdrawal from opiates.[6] Overall, all of these neuronal adaptive changes contribute to the development and maintenance of drug addiction. Much research has focused on the neurobiological processes involved in opioid tolerance, dependence, withdrawal, and addiction. The adenylyl cyclase (AC) pathway has been implicated in these processes and is the focus of this review.

2. THE ADENYLYL CYCLASE PATHWAY

There are at least three types of opioid receptors—mu (μ; MOP), kappa (κ; KOP), and delta (δ; DOP), which are also referred to as traditional or classical opioid receptors. In addition, there is a related receptor, nociceptin receptor (NOP), also known as the opioid receptor-like (ORL1) receptor. Sigma (σ) receptors were previously categorized as opioid receptors but they are no longer classified as opioid receptors. Opioid receptors are G-protein-coupled receptors containing seven transmembrane domains with an extracellular N-terminus with various glycosylation sites and an intracellular C-terminus.[7] The MOP, KOP, and DOP receptors are coupled with heterotrimeric, pertussis toxin (PTX)-sensitive, inhibitory G (G_i/G_o) proteins, with all three receptors having the ability to activate both G_i and G_o with similar potencies and subsequently, regulate AC, Ca^{2+} channels, phospholipase C, and mitogen-activated protein kinases.[8–10] The involvement of each component of AC/cAMP/PKA pathway in analgesia, tolerance, and dependence induced by opioids.

2.1 Adenylyl Cyclase

AC is an enzyme that is activated following stimulation of receptors that are coupled to the AC via Gs, such as beta-adrenergic receptors. On the other hand, the activity of the enzyme is inhibited by activation of receptors that are coupled to $G_{i/o}$, such as opioid receptors. Following the stimulation of opioid receptors, the G_α subunit inhibits AC.[11,12] Inhibition of AC results in reduced production of intracellular cAMP.[13] The decrease in intracellular cAMP levels leads to a reduction in protein kinase A (PKA) activity. The activity of AC is altered differentially with acute and chronic exposure to opioids.

Multiple studies have demonstrated the involvement of AC and cAMP in pain and opioid-induced analgesia. Increased nociception is correlated with elevated levels of cAMP.[14] Administration of inhibitors (rolipram and RP-73401) of phosphodiesterase (PDE), which break down cAMP, and activators of AC (forskolin) caused hyperalgesia in rat models of pain.[15] Mammalian genes encode for ten AC isoforms denoted 1–10. AC isoforms I, III, V, VI, and IX are involved in spinal pain transmission. AC V is involved in the mediation of morphine analgesia.[16] High densities of AC V are expressed in the striatum along with MOP, KOP, and DOP receptors.[17] It was further established that AC V mediates the effects of MOP and DOP but not KOP agonists.[16] In wild-type mice, the administration of MOP agonist, DAMGO (D-Ala^2, NME-Phe^4, Gly-ol^5]-enkephalin) or DOP agonist, DPDPE ([D-Pen^{2,5}]Enkephalin), both reduced forskolin-stimulated AC activity in striatal tissues. Following acute administration of opioids, the activity of AC isoforms I, V, VI, and VIII were reduced due to MOP activation. DOP activation by endogenous opioid peptide enkephalin reduced cAMP levels via inhibition of ACII.[18] In AC knockout (isoforms 1, 8, and 1 and 8) mice, forskolin-induced nociception was reduced.[19,20] Morphine-induced antinociception was not different between wild-type and AC knockout (isoforms 1, 8, and 1 and 8) mice. However, as expected, tolerance to morphine was blunted in the knockout compared to wild-type mice.[21]

While the acute effects of opioid administration involve inhibitory effects on AC mediated by $G\alpha_i$ and $G\alpha_o$ subunits, chronic administration reveals superactivation of AC (also termed supersensitization or cAMP overshoot) and a compensatory increase in cAMP production. Sharma and coworkers first demonstrated that chronic morphine exposure increases AC activity, as measured by increased cAMP levels, following the initial decrease in AC activity.[22] At the time, the group proposed that a yet unidentified mechanism underlying the adaptive changes to positively regulate AC activity is related to the development of tolerance and withdrawal observed clinically in patients. Similarly, Collier and Francis hypothesized that elevated cAMP level following chronic morphine exposure is related to opioid dependence.[23] The superactivation of AC was observed in a recombinant CHO (Chinese hamster ovary) cell line expressing DOP.[24] Treatment with a DOP agonist reduced DOP density, and moreover, inhibited forskolin-stimulated cAMP accumulation. The inhibition and subsequent superactivation of AC was also isoform specific. The various isoforms of AC have different intracellular signaling mechanisms in response to chronic opioid activation.[25]

With chronic morphine administration, several steps of the AC/cAMP/PKA pathway are amplified, notably the activity of AC.[26,27] Withdrawal symptoms associated with morphine dependence are reduced in AC V knockout mice.[16] The locus coeruleus (LC) plays a central role in opiate dependence and withdrawal and its activity is regulated by opioids.[28] While neuronal firing in the LC is reduced with acute opioid administration, AC activity and cAMP levels return toward basal levels during chronic opiate exposure.[29] Overshoots of cAMP occur following naloxone-precipitated withdrawal along with increased firing rates of LC neurons.[28] Furthermore, chronic morphine stimulates G_i and G_o subunits in the LC.[30] Chronic administration of opioids causes an uncoupling of the G_α and $G_{\beta\gamma}$ proteins.[31]

PTX-induced inhibition of $G\alpha_i$ and $G\alpha_o$ proteins prevented sensitization of AC as demonstrated with the lack of cAMP accumulation, suggesting that $G\alpha_i$ and $G\alpha_o$ proteins are required for this process.[32] Additional insight into the response of $G\alpha_i$ and $G\alpha_o$ subunits under constant stimulation were studied in NS20Y-2_{2L} cells expressing dopamine (D_{2L}) receptors.[33] $G\alpha_i$ and $G\alpha_o$ subunits were genetically altered to be PTX-insensitive, allowing for the examination of which G subunit was involved in the sensitization process. $G\alpha_o$ but not $G\alpha_i$ proteins produced sensitization of forskolin-induced accumulation of cAMP. Under these conditions, $G\alpha_s$ may be stimulated, leading to $G_{\beta\gamma}$-subunit-enhanced AC activity.[34] Furthermore, chronic morphine has been shown to produce a superactive state of AC[34] via $G_{\beta\gamma}$ stimulation of plasma-membrane-bound AC 1 and 8.[35] Under continuous agonist exposure, receptors appear to be desensitized or down-regulated. However, removal of the agonist unveils the overshoot of cAMP due to $G\alpha_s$ activation. A reduction in PDE activity has also been observed with chronic exposure.[36] This process may contribute to the withdrawal symptoms observed upon removal of the agonist.

The sensitization process of AC also involves $G\alpha_s$.[37] Following chronic morphine administration, $G\alpha_s$ subunits coupling to PGE_1 (prostaglandin) receptors are increased.[37] This resulted in PGE-stimulated AC activity and increased cAMP levels. While most work with MOP signaling involves $G\alpha_i$ and $G\alpha_o$, other studies have shown that the stimulation of MOP with chronic morphine reduces $G\alpha_s$ phosphorylation.[38]

2.2 Protein Kinase A

PKA is a cAMP-dependent kinase that is a major downstream effector of opioid receptor activation. The phosphorylation of PKA leads to the activation of the enzyme and has numerous intracellular roles, including the upregulation of gene expression through cAMP response element-binding protein (CREB), upregulation of tyrosine kinase, and increasing Na^+-dependent inward current.

Acute exposure to opioids inhibits AC activity and reduces the production of cAMP, and therefore, decreases the activity of PKA. With chronic exposure of opioids, AC activity returns toward basal levels, resulting in increased cAMP levels and PKA activity. Chronic morphine exposure decreases phosphorylation of G_s, and enhances its coupling with MOP, increasing AC activity.[39] Several studies indicate that PKA phosphorylates and interferes with MOP coupling to G_i subunits.[40,41] Other studies have shown a reduction in PKA-mediated phosphorylation of MOP.[42] Phosphorylation of GM_1 ganglioside by PKA converts MOP coupling to G_s from $G_{i/o}$. This conversion from inhibitory $G_{i/o}$ to stimulatory G_s via GM_1 ganglioside phosphorylation may be one of the underlying mechanisms for the development of opioid dependence.[43] Furthermore, we have previously shown that this shift may also enhance the toxic effects of morphine. Morphine-induced antinociception was reduced but seizure-like activity increased in mice treated intracerebroventricularly with PTX (ADP-ribosylates the inhibitory G proteins).[44] The role of GM_1 was further substantiated by the work of Crain and Shen who showed that low-dose naloxone produced hyperalgesia in the tail-flick assay in naive mice treated with GM_1 ganglioside, a response that is seen in subjects chronically treated with morphine and exposed to the same dose of naloxone.[45] This response was shown to be mediated via the KOP, as administration of high doses of naltrexone or KOP antagonist nor BNI blocked hyperalgesia induced by naloxone in these mice.

The role of PKA in the development of tolerance and withdrawal is not completely clear. PKA activators directly infused into the NAc increased drug-seeking behaviors observed in animals while PKA inhibitors mitigated these behaviors.[46] Some studies demonstrate that inhibition of PKA has minimal or no effect on the development of acute tolerance. Injections of MOP agonist DAMGO in mice produced acute tolerance.[47] Administration of PKC inhibitors prevented the development of acute tolerance produced by intrathecal DAMGO. Injection of PKA inhibitors, however, did not attenuate the development of acute tolerance to DAMGO. Similar findings regarding the role of PKA and PKC in acute tolerance to the antinociceptive effect of peripherally administered morphine were reported in the bradykinin-induced flinching test.[48] Calphostin C, an inhibitor of PKC, prevented morphine-induced tolerance, whereas KT-5720, an inhibitor of PKA, did not. Bilsky and coworkers treated mice with morphine (100 mg/kg) to induce acute tolerance to its analgesic effect and 4 h post morphine treatment injected them with naloxone to precipitate withdrawal. They determined that the effect of a neutral antagonist, such as, D-Phe-Cys-Tyr-D-Trp-Arg-Thr-Pen-Thr-NH2 (CTAP), and naloxone, a nonselective opioid antagonist, which acts as an inverse agonist in opioid-dependent subject, in the presence and absence of the protein kinase C inhibitors, H7 and its related analog, H8. Central administration of naloxone elicited withdrawal jumping and this response was blocked by H7 but not H8. While CTAP induced limited jumping, it suggested that mu-constitutive activity required PKC in this response.[49]

The role of PKA in opioid withdrawal was assessed with the application of Rp-cAMPS, a PKA inhibitor.[50] Infusion of Rp-cAMPS directly into the LC and PAG attenuated naloxone-induced withdrawal symptoms whereas infusion into the amygdala did not. Furthermore, the infusion of PKA activator Sp-cAMPS into the LC or PAG in opioid-naive animals produced withdrawal-like symptoms. Rp-cAMPS significantly increased phosphorylation of tyrosine hydroxylase in the LC.

3. OPIOID EFFECT ON cAMP-RESPONSIVE ELEMENT-BINDING PROTEIN

A variety of upstream signals regulate gene expression through phosphorylation of CREB. Phosphorylation of CREB by PKA promotes a high-affinity binding site with transcriptional coactivators CREB-binding protein and p300.[51,52] The secondary structure of CREB is unaltered. The serine residue of the kinase-inducible domain is the site for phosphorylation by Ca^{2+} and PKA.[53] CREB plays a functional role in the regulation of gene expression.[54] Acute morphine administration reduces AC activity, which ultimately leads to reduced PKA-induced phosphorylation of CREB. However, the level of phosphorylated CREB is altered differentially by acute and chronic morphine and this alteration is brain region-dependent.

Chronic morphine administration reduced CREB immunoreactivity in the NAc[55] but increased it in the LC.[56] Acute opioid administration inhibits AC activity, therefore reducing cAMP-dependent phosphorylation of CREB.[57] Maldonado and coworkers created CREB-mutant mice to further examine the role of CREB in opioid dependence.[58] Wild-type animals treated with chronic morphine displayed expected signs of withdrawal upon precipitation by naloxone. However, in CREB-mutant mice, all signs of withdrawal were reduced. Seven-day treatment with morphine via pellets implantation showed elevated levels of phosphorylated CREB in the lateral hypothalamus, ventral tegmental area (VTA), and the hypothalamus.[59] Alterations in CREB levels were not observed in the NAc and amygdala.

Mapping of CREB response to opioid administration was obtained using transgenic mice encoding β-galactosidase (β-gal).[60] β-Gal levels were increased in the LC and amygdala but decreased in the VTA or dorsal raphe nucleus (DRN). The reduction in CREB activity in the VTA during withdrawal may be due to inhibition of dopaminergic activity by γ-aminobutyric acid (GABA) mechanisms. The reduction in serotonergic firing in the DRN is also mediated by GABAergic interneurons.

Prohormone convertases, PC1/3 and PC2, are enzymes that cleave prohormones to hormones. The expression of PCs is dependent upon CREB activation.[61,62] These enzymes are found in high concentrations in the pituitary gland and some brain regions.[63] Most of the endogenous peptides are secreted as inactive precursors and converted into an active form by PC1/3 and PC2. PC1/3 and PC2 response to opioid exposure appears to be biphasic. We have previously shown that short-term morphine exposure downregulated the enzymes while chronic (7 days) morphine administration upregulated PC1/3 and PC2 levels in the midbrain.[64,65] The level of phosphorylated CREB (P-CREB) is increased with chronic morphine exposure in the hypothalamus.[64] Short-term exposure of morphine (6 h) reduced P-CREB levels, indicating that P-CREB is involved in the downregulation of PC1/3 and PC2 in the pituitary.[64] Lastly, pituitary proopiomelanocortin (POMC) and β-endorphin levels were reduced by morphine administration. This may be due to feedback mechanisms responding to exogenous opioid exposure by reducing endogenous opioid peptide levels via the CREB system. Naltrexone, a nonselective opioid antagonist, increased POMC, PC1/3, PC2, and β-endorphin levels. Thus, CREB may be a target for the action of exogenous opioid to regulate the level of endogenous opioids although alterations in the POMC levels could lead to changes in the expression of nonopioid peptides, such as adrenocorticotropic hormones and α-melanocyte-stimulating hormone. During extinction of morphine-induced conditioned place preference (CPP), CREB levels in the hippocampus, NAc, and PFC return toward baseline levels.[66] However, extracellular signal-regulated protein kinase (ERK)–CREB activity is elevated during extinction of conditioned place aversion following morphine withdrawal.[67]

Downstream effectors of $G\alpha_i$ and $G\alpha_o$ proteins regulate AC through phosphorylation, thereby, providing additional mechanisms for sensitization. Following activation of $G\alpha_i$ and $G\alpha_o$, the $G_{\beta\gamma}$ subunit interacts with various protein kinases. Protein kinase Raf was shown to phosphorylate AC VI and sensitize the enzyme.[68] Varga *et al.* proposed that Raf-1 is the site of convergence for the effects of Ras and PKC.[69] $G_{\beta\gamma}$ activates receptor tyrosine kinase (RTK), β-arrestin, and PKC. RTK and β-arrestin both increase the activity of nonreceptor tyrosine kinase, leading to RAS activation. RAS and PKC increase the activity of Raf-1 during the chronic opioid exposure. Given that ERK is downstream of Raf-1 and RAS, the alteration in RAS and Raf-1 may play a role in the incubation of opiate craving. Indeed, a low dose of morphine (3 mg/kg) induced a long-lasting CPP in rats. Along with this enduring CPP response, there was a significant increase in the level of phosphorylated ERK as well as CREB in the central amygdala. Interestingly, injection of ERK and CREB inhibitors in this brain region prevented the incubation of opiate craving.[70] Thus, ERK could be a potential target for the development of medications to treat opiate addiction and particularly opiate craving. However, ERK is also implicated in neuronal plasticity and thus the task to develop small-molecule ERK inhibitors that selectively reduce opiate craving needs further investigations.

4. MOLECULAR CHANGES IN BRAIN REGIONS THAT MAY UNDERLIE OPIATE DEPENDENCE

4.1 Molecular Changes in the Locus Coeruleus

The LC has been studied extensively as a site of mediating opiate withdrawal.[71–74] The LC is densely populated with adrenergic neurons [75,76] and has a substantial role in manifestation of behavioral signs of withdrawal.[77] The LC primarily receives neuronal input from the hippocampus, prefrontal cortex (PFC), and the amygdala. The LC projects to the limbic system, cerebral and cerebellar cortices as well as various nuclei throughout the CNS such as the hypothalamus, spinal cord, and brainstem (paragigantocellularis) nuclei.[78] All three opioid receptors are expressed in this brain area. However, MOP is the predominate opioid receptor in the region.[79] Chronic administration of opioids induces substantial changes in intracellular second messenger systems in the LC. Behavioral studies in morphine-dependent rats showed that the LC is the most sensitive structure in the CNS during the withdrawal phase induced by opioid antagonist methylnaloxonium.[80]

Acute administration of levorphanol, an opioid receptor agonist, produced rapid suppression of LC firing.[81] Naloxone prevented and reversed the inhibitory effects of levorphanol, showing that opioid receptors are involved in this response. Morphine and clonidine, an α_2-adrenoreceptor receptor agonist, each decreases LC firing in brain slices.[82] Chronic morphine (but not clonidine) treatment induces tolerance, that is, LC neurons show reduced sensitivity to morphine.

Increases in LC neuronal firing have been demonstrated following opioid withdrawal.[71] Direct administration of kynurenate, an antagonist of excitatory amino acids, 6-cyano-7-dinitroquinoxaline-2,3-dione (CNQX), a non-NMDA receptor antagonist, or AP5, a selective NMDA-receptor antagonist, into the LC attenuated withdrawal induced neuronal firing.[83] Microdialysis studies following administration of naltrexone showed increases in excitatory amino acids (EAA) glutamate and aspartate release, which correlated with increased neuronal firing in the LC.[84] The changes in LC firing during withdrawal are accounted primarily by EAA mechanisms within the LC.[85]

The α_2-adrenoreceptors are present in the LC as both presynaptic and postsynaptic receptors and are negatively coupled to AC.[86] The effects of chronic opioids on α_2-adrenoreceptor density and binding have produced mixed results. Chronic morphine showed a reduction in receptor binding[87] and density.[88] In contrast, no change in α_2-adrenoreceptor density was observed in a neurotumor cell line[89] while another study showed an increase in density.[90] In hippocampal slices, chronic administration of morphine reduced α_2-adrenoreceptor binding.[88]

Clonidine has been used successfully in the treatment of acute opiate withdrawal.[91] Both objective signs and subjective symptoms were reduced in a human trial using clonidine, which suggests that α_2-adrenoreceptors were involved in mediating opioid dependence and withdrawal.[92] However, the LC was not investigated as the CNS site for mediating the signs of physical dependence on opiates as the site of action for clonidine until a decade later. Naloxone-induced withdrawal in subjects exposed to chronic morphine was reversed by local infusions of clonidine into the LC.[93] As stated earlier, clonidine is an agonist at presynaptic α_2-adrenergic receptors in the LC, decreasing adrenergic outflow responsible for the observed withdrawal symptoms (wet-dog shakes, weight loss, diarrhea, and ptosis). Naloxone, a nonselective opioid receptor antagonist, directly injected into the LC induced withdrawal symptoms in rats chronically administered with morphine for 11 days.[94] The application of methylnaloxonium intracerebroventricularly precipitated withdrawal symptoms in subjects with chronic opioid exposure.[95] Withdrawal symptoms were attenuated following intraperitoneal injection of clonidine, indicating that the symptoms were mediated through the adrenergic neurons of the LC.[96]

Rasmussen and Aghajanian provided insight to the role of LC by examining behavioral, biochemical, and electrophysiological changes.[97] A behavioral assessment, composed of 14 behaviors characteristic of opioid withdrawal, showed a rapid reduction in withdrawal symptoms during the first 4 h after naloxone administration in opioid-dependent rats. However, withdrawal symptoms gradually declined and continued to last for 72 h. Within several minutes of naltrexone exposure, neuronal firing in the LC increased to a sixfold peak within 15–30 min in comparison to prenaltrexone exposure in opioid-dependent rats. In congruence with the withdrawal symptoms, the elevation in LC firing rats remained elevated for 72 h compared to controls. Biochemical studies of AC and PKA showed elevated activity 20 min after naltrexone administration, but no differences in activity were observed at this time point compared to morphine-dependent subjects exposed to abstinence withdrawal. The behavioral and electrophysiological parameters showed changes after the induction of opioid withdrawal. In contrast, the biochemical results were already elevated prior to the induction of withdrawal. The rendering of a nonfunctional LC by electrolytic lesion mitigated the withdrawal symptoms.[95] Lesions of the nucleus paragigantocellularis (PGi) reduced withdrawal induced by LC activation.[97]

Neuronal firing in the LC of anesthetized rats was gradually suppressed with implantation of morphine pellets, with complete suppression of firing achieved 1–2 h.[98] Within 48–72 h, firing rates returned toward basal level. The administration of MOP antagonist induced neuronal firing substantially above baseline levels. Nestler and Tallman also investigated the effects of chronic opioid administration on LC neuronal firing.[30] Rats treated with chronic but not acute morphine showed elevated PKA activity. This paralleled the time course of the development of tolerance and dependence on morphine as demonstrated by electrophysiologic studies. Chronic morphine exposure increases $G\alpha_{i/o}$, AC, and tyrosine hydroxylase in the LC, all of which may mediate the development of tolerance and withdrawal.[99] Protooncogene c-FOS levels are increased within several hours during withdrawal from opioid use.[100]

PKA plays a pivotal role in mediating opiate withdrawal in the LC. Administration of a PKA antagonist directly into the LC reduced withdrawal behaviors induced by naloxone in morphine-treated animals.[101] Injection into the PAG produced a lesser effect to attenuate withdrawal behaviors compared to injections into the LC. The use of a tyrosine kinase inhibitor did not attenuate the withdrawal symptoms.

Changes in CREB response in LC have been observed.[102] Acute morphine exposure reduced CREB phosphorylation. However, CREB phosphorylation returned toward basal levels with chronic morphine administration. Furthermore, increased phosphorylation of CREB was demonstrated following induction of withdrawal with a MOP antagonist. Changes in CREB phosphorylation in the LC may be linked to gene expression involved in addiction. AC levels and gene expression did not increase during withdrawal in mice with CREB gene disruption.[58] Thus, to examine the role of CREB, antisense oligonucleotides were infused directly into the LC of rats under anesthesia.[103] The antisense sequence was directed at the start site of the CREB mRNA translation. Morphine-induced upregulation of AC 8 and tyrosine kinase, but not PKA or G_i, were attenuated in animals receiving CREB antisense oligonucleotides. Electrophysiologic studies showed a reduction in LC firing in CREB antisense-treated animals. Results from behavioral studies showed decreased withdrawal symptoms induced by naltrexone in CREB antisense-treated animals.

Long-term glutamate receptor desensitization was observed during opioid withdrawal due to elevated glutamate levels.[84,104] Desensitization of glutamate receptors occurred within 20 min of continuous glutamate administration, with resensitization occurring within several hours. Phosphorylation of glutamate receptors by PKA[105] or PKC[106] may regulate the desensitization process.

While the LC has been shown to play an important role in dependence and withdrawal, it does not appear to mediate the development of opioid tolerance.[107] Pretreatment with *N*-(2-chloroethyl)-*N*-ethyl-2-bromobenzylamine (DSP-4), a noradrengeric neurotoxin, produced lesions specific to LC projections. Behavioral studies using the tail-flick test showed no changes in the development of tolerance to chronic morphine administration in DSP-4-lesioned animals. However, DSP-4-induced lesions of the LC attenuated opioid withdrawal.

4.2 Molecular Changes in the Amygdala

The amygdala plays a functional role in integrating stress and arousal and producing negative reinforcement related to addictive disorders.[108] Furthermore, the amygdala mediates signs of withdrawal in subjects with either acute or chronic opioid exposure.[109] Acute administration of morphine stimulates MOP-activated K^+ channels in the amygdala.[110] The resultant hyperpolarization inhibits GABAergic neurotransmission. DAMGO application decreased inhibitory postsynaptic currents (IPSCs) in neurons projecting to the PAG, an effect that was blocked with the administration of bicuculline, a $GABA_A$ receptor antagonist.[111] The modulation of GABAergic neurotransmission by acute morphine appears to be mediated by a PTX-insensitive process.[112] The administration of RP-cAMPS triethylammonium, a cAMP inhibitor, reduced IPSC and attenuated morphine-induced GABAergic neurotransmission. The authors proposed that a switch to $G_{\alpha s}$ along with $G_{\beta\gamma}$-stimulated AC mediated the superactivation of cAMP observed in the amygdala.

There is evidence of a differential response or switch by dopamine (DA) in the pyramidal neurons of the amygdala under acute versus chronic opioid exposure.[113,114] In particular, the pyramidal neurons of the basolateral amygdala receive input from the VTA. In saline-treated amygdala slices, DA reduced the amplitude of excitatory postsynaptic currents (EPSCs) by inhibiting α-amino-3-hydroxy-5-mehtyl-4-isoxazoleproprionic acid (AMPA) receptors.[113] However, in slices exposed to chronic morphine, DA increased the amplitude of EPSCs by inducing presynaptic glutamate release. This effect was abolished by the application of SCH23390, a D1 (dopamine) receptor antagonist. The enhanced glutamate release is mediated via increased PKA activity in the presynaptic neuron. The switch from an inhibitory to excitatory modulation of DA receptors is dependent on cAMP. Behavioral studies showed that administration of SCH23990, directly into the amygdala prevented naloxone-induced aversion.

CREB levels following chronic morphine exposure remain unchanged or elevated. Phosphorylated CREB levels were not increased in the amygdala with chronic morphine exposure.[59] The role of the central amygdala during the "incubation" phenomenon was examined after chronic morphine administration. Interestingly, animals treated with low but not high doses of morphine exhibited increases in CPP in a three-chambered apparatus after 14 days of withdrawal.[70] Prior studies using the two-chambered apparatus showed decreased CPP after withdrawal.[115,116] ERK expression is increased by glutamate release during withdrawal, enhancing ERK phosphorylation.[70] The authors postulated that ERK-inhibited K^+ channels cause depolarization, which perhaps exposed the neurons to be more sensitive to drug cues as observed with increased CPP. UO126, a MEK inhibitor, reduced ERK and CREB activity and reversed CPP, thus substantiating a role for ERK and

CREB in the mediation of drug-seeking behaviors following withdrawal. As stated earlier, ERK then could be a potential target for the development of novel pharmacotherapy of opiate addiction.

4.3 Molecular Changes in the Periaqueductal Gray

The periaqueductal gray (PAG) plays a vital role in mediating pain transmission as injections of morphine or partial MOP-agonist enkephalin produced analgesia.[117,118] The analgesic effects are mediated by serotonergic fibers arising from the PAG. Serotonergic neurotransmission in the PAG is inhibited by local GABAergic neurons.[119,120] Exogenous opioids and endogenous peptide endomorphin-activating MOP in the PAG, which inhibits GABAergic neurons, resulting in disinhibition of serotonergic neurotransmission.[117,121]

The electrical stimulation[122] of neurons or direct injection of morphine[123] into the PAG produces analgesia. The PAG receives input from the amygdala and PFC. DAMGO suppressed AC activity, an effect that was reversed by the application of naloxone. Direct injection of PKA inhibitors, H7 or Rp-cAMPS, into the PAG reduced precipitated withdrawal symptoms.[50,101] Electrophysiologic data support the development of tolerance by PAG neurons to chronic morphine exposure.[124] PKC but not PKA modulates Ca^{2+} channels in the PAG following MOP activation.[125] Administration of a protein kinase C activator, phorbol-12-myristate-13-acetate (PMA), prevented DAMGO-induced inhibition of Ca^{2+} currents.

Acute treatment with morphine increased MOP coupling to $G_{i/o}$ in the PAG.[126] However, chronic exposure reduced MOP coupling to $G_{i/o}$ and increased coupling to G_s. $G_{\beta\gamma}$ interaction with AC II and IV also occurred during chronic treatment. Concurrent administration of low-dose naloxone (10 ng/kg, subcutaneously or 0.05 ng, intrathecally) inhibited morphine-induced G_s coupling and $G_{\beta\gamma}$ activation of AC. Low-dose naloxone coadministered with morphine enhanced acute analgesic effect of morphine and attenuated its dependence and tolerance liability.

GABA transporter-1 (GAT) activity in the PAG is increased during withdrawal from chronic opioid use. $GABA_B$ receptors are expressed extensively in the PAG[124] and are coupled to $G_{i/o}$ subunits,[127] AC is a downstream effector of $GABA_B$ receptors.[128] Elevated PKA activity increased GABA currents, causing depolarization and stimulation of GABAergic neurons in the PAG.[129] Thus, PAG neuronal activity in the hypothalamus and medulla is reduced by GABAergic activation in the PAG.[130] However, it does not appear that the activation of $GABA_B$ receptors affect GAT activity.[131]

Baclofen, a $GABA_B$ receptor agonist, mitigates some of the behavioral symptoms of withdrawal. However, Bagley *et al.* suggested that the $GABA_B$ receptors display differential coupling to AC during chronic opioid treatment.[131]

5. MOLECULAR CHANGES IN THE VENTRAL TEGMENTAL AREA

The VTA plays a prominent role in the reward-related behaviors. Therefore, the effects of drugs of abuse have been studied extensively in the VTA. The VTA contains a high density of dopaminergic neurons projecting to the NAc, which is likewise involved in drug reward. The VTA also projects to the PFC and amygdala, which are thought to mediate thought process, decision-making, and the emotional aspects of reward. The VTA primarily receives excitatory neuronal input from the PFC while receiving inhibitory input from the amygdala.[132]

Acute administration of morphine directly into the VTA induces release of DA in the NAc as measured by microdialysis in rats.[133] This coincides with increased firing of dopaminergic neurons in the VTA with acute IV morphine.[134] Naloxone reversed this effect. The putative mechanism by which morphine increases dopaminergic firing is via coupling of MOP to $G\alpha_i$ and $G\alpha_o$ proteins and causing hyperpolarization of local GABAergic interneurons in the VTA. Morphine administration disinhibits GABAergic influence on dopaminergic neurons, allowing DA release. Furthermore, intracellular recordings demonstrate that morphine does not directly affect dopaminergic neurons in the VTA, but primarily through disinhibition of GABAergic neurons. KOP, on the other hand, produces inhibition of dopaminergic neurons in the VTA.[135] Intracellular recordings of the VTA neurons show that KOP agonists decrease neuronal firing of dopaminergic neurons mediated by the opening of inward rectifying K^+ channels. Thus, acute administration of opioids can stimulate and inhibit DA activity in the VTA.

Chronic administration of morphine produces sustained dopaminergic activity, with a return toward baseline levels during withdrawal. Intracellular recordings from midbrain slices containing the VTA showed increased IPSCs in GABA neurons following withdrawal from morphine.[136] The increase in GABA activity may be due to the sensitization of AC and the buildup of cAMP levels. The application of forskolin increases GABA IPSCs in morphine-treated brain slices while dideoxyforskolin, which does not have an effect on AC, did not increase GABA IPSC.[136,137]

6. MOLECULAR CHANGES IN OTHER CNS REGIONS

Within the amygdala, the bed nucleus of the stria terminalis (BNST) serves as the output pathway. The BNST receives considerable noradrenergic input and has been shown to mediate the neurocircuitry and behaviors of reward and stress.[138] Immunoreactivity of tyrosine hydroxylase increased considerably in morphine-dependent animals in this brain region following the induction of withdrawal by naltrexone.[139] Withdrawal-induced aversion assessed by CPP was also significantly higher in morphine-treated animals. Lesion of axons from the LC to BNST did not affect place aversion during withdrawal. However, lesion of axons from A1 and A2 noradrenergic neurons reduced withdrawal-induced aversion, implicating the influence of the ventral noradrenergic bundle in stimulating behavioral symptoms mediated by the BNST. Local application of ST-91 into BNST, a clonidine analog, reduced aversion. Taken together, the BNST may be a site of action for the therapeutic effects of clonidine. The BNST projects inhibitory fibers to the VTA, GABAergic, and GABAergic/enkephalinergic, to modulate response to MOP activation.[140] While behavioral changes are associated with BNST modulation upon chronic morphine administration, AC activity is not affected.[141]

While opioid antagonists have been employed experimentally to induce withdrawal in animals treated chronically with opioids, several studies have examined the chronic use of low-dose opioid antagonists to alleviate withdrawal symptoms.[142–144] Withdrawal jumping was reduced significantly in both acutely and chronically treated animals with morphine and low-dose naltrexone.[144] Moreover, tail-flick tests showed that low-dose naltrexone appeared to potentiate the analgesic effects of morphine compared to animals treated with morphine alone. The hypothesis was that blocking the Gs-coupled MOP facilitates morphine analgesia and dependence liability, which is in line with the theory of opiate tolerance and involves superactivation of AC. Low-dose naltrexone (5 mg/L) was dissolved in drinking water for rats treated for eight days with morphine.[142] On the eighth day, animals were injected with naltrexone to induce withdrawal. In the naltrexone pretreated animals, withdrawal symptoms were significantly reduced. Furthermore, c-FOS labeling, PKA protein levels, and phosphorylated CREB levels were lower in the LC and the nucleus of the solitary tract (NTS) compared to nontreated animals. The reduction in c-FOS, PKA, and P-CREB is correlated with the reduction in withdrawal symptoms. Further examination under similar conditions showed increased MOP expression in the NTS but not in the LC in animals pretreated with naltrexone.[143] However, DOP expression remained unchanged in both the NTS and LC. No changes in PKA and P-CREB were observed in the frontal cortex, striatum, amygdala, and VTA.

7. CONCLUSIONS

The need for the use of opioids in the treatment of pain associated with various disease-states will continue in healthcare. This increases the propensity for the development of tolerance, dependence, withdrawal, and addiction in patients who are prescribed to use opioid chronically. The molecular targets and changes due to acute and chronic opioid administration have been delineated. Earlier studies showed that chronic opioid administration induces desensitization and internalization. However, changes in the number of opioid receptors following chronic opioid administration have yielded mixed results. Targeting second messenger systems may be proven useful to treat opioid withdrawal. In particular, the AC/cAMP/PKA pathway may be a potential target to reduce pain and prevent symptoms of opioid withdrawal. Indeed, the use of α_2-adrenergic receptor agonists, such as clonidine, have been proven useful in this regard. Thus, small-molecule compounds that target the AC/cAMP/PKA pathway can be useful as adjunct therapeutic agents to reduce pain and the undesirable side effect of the opioid that may develop following their chronic use.

REFERENCES

1. Atluri S, Sudarshan G, Manchikanti L. Assessment of the trends in medical use and misuse of opioid analgesics from 2004 to 2011. *Pain Physician*. 2014;17(2): E119–E128.
2. Substance Abuse and Mental Health Services Administration, Results from the 2012 National Survey on Drug Use and Health: Summary of National Findings, NSDUH Series H-46, HHS Publication No. (SMA) 13-4795. Rockville, MD: Substance Abuse and Mental Health Services Administration; 2013
3. Paulozzi LJ, Mack KA, Hockenberry JM. Vital signs: variation among States in prescribing of opioid pain relievers and benzodiazepines—United States, 2012. *MMWR Morb Mortal Wkly Rep*. 2014;63(26):563–568.
4. Goodman LS, Brunton LL, Chabner B, Knollmann BrCC. *Goodman & Gilmans Pharmacological Basis of Therapeutics*. 12th ed New York: McGraw-Hill; 2011.
5. Williams JT, Christie MJ, Manzoni O. Cellular and synaptic adaptations mediating opioid dependence. *Physiol Rev*. 2001;81(1):299–343.
6. Lu L, Grimm JW, Hope BT, Shaham Y. Incubation of cocaine craving after withdrawal: a review of preclinical data. *Neuropharmacology*. 2004;47(suppl 1):214–226.

7. Childers SR, Snyder SH. Guanine nucleotides differentiate agonist and antagonist interactions with opiate receptors. *Life Sci.* 1978;23(7):759–761.
8. Roerig SC, Loh HH, Law PY. Identification of three separate guanine nucleotide-binding proteins that interact with the delta-opioid receptor in NG108-15 neuroblastoma glioma hybrid cells. *Mol Pharmacol.* 1992;41(5):822–831.
9. Prather PL, McGinn TM, Erickson LJ, Evans CJ, Loh HH, Law PY. Ability of delta-opioid receptors to interact with multiple G-proteins is independent of receptor density. *J Biol Chem.* 1994;269(33):21293–21302.
10. Chakrabarti S, Prather PL, Yu L, Law PY, Loh HH. Expression of the mu-opioid receptor in CHO cells: ability of mu-opioid ligands to promote alpha-azidoanilido [32P]GTP labeling of multiple G protein alpha subunits. *J Neurochem.* 1995;64(6): 2534–2543.
11. Hsia JA, Moss J, Hewlett EL, Vaughan M. ADP-ribosylation of adenylate cyclase by pertussis toxin. Effects on inhibitory agonist binding. *J Biol Chem.* 1984;259(2): 1086–1090.
12. Wong YH, Federman A, Pace AM, et al. Mutant alpha subunits of Gi2 inhibit cyclic AMP accumulation. *Nature.* 1991;351(6321):63–65.
13. Minneman KP, Iversen IL. Enkephalin and opiate narcotics increase cyclic GMP accumulation in slices of rat neostriatum. *Nature.* 1976;262(5566):313–314.
14. Hucho T, Levine JD. Signaling pathways in sensitization: toward a nociceptor cell biology. *Neuron.* 2007;55(3):365–376.
15. Cunha FQ, Teixeira MM, Ferreira SH. Pharmacological modulation of secondary mediator systems—cyclic AMP and cyclic GMP—on inflammatory hyperalgesia. *Br J Pharmacol.* 1999;127(3):671–678.
16. Kim KS, Lee KW, Lee KW, et al. Adenylyl cyclase type 5 (AC5) is an essential mediator of morphine action. *Proc Natl Acad Sci USA.* 2006;103(10):3908–3913.
17. Glatt CE, Snyder SH. Cloning and expression of an adenylyl cyclase localized to the corpus striatum. *Nature.* 1993;361(6412):536–538.
18. Tsu RC, Chan JS, Wong YH. Regulation of multiple effectors by the cloned delta-opioid receptor: stimulation of phospholipase C and type II adenylyl cyclase. *J Neurochem.* 1995;64(6):2700–2707.
19. Wei F, Qiu CS, Kim SJ, et al. Genetic elimination of behavioral sensitization in mice lacking calmodulin-stimulated adenylyl cyclases. *Neuron.* 2002;36(4):713–726.
20. Wei F, Vadakkan KI, Toyoda H, et al. Calcium calmodulin-stimulated adenylyl cyclases contribute to activation of extracellular signal-regulated kinase in spinal dorsal horn neurons in adult rats and mice. *J Neurosci.* 2006;26(3):851–861.
21. Li S, Lee ML, Bruchas MR, Chan GC, Storm DR, Chavkin C. Calmodulin-stimulated adenylyl cyclase gene deletion affects morphine responses. *Mol Pharmacol.* 2006;70 (5):1742–1749.
22. Sharma SK, Klee WA, Nirenberg M. Dual regulation of adenylate cyclase accounts for narcotic dependence and tolerance. *Proc Natl Acad Sci USA.* 1975;72(8):3092–3096.
23. Collier HO, Francis DL. Morphine abstinence is associated with increased brain cyclic AMP. *Nature.* 1975;255(5504):159–162.
24. Malatynska E, Wang Y, Knapp RJ, et al. Human delta opioid receptor: functional studies on stably transfected Chinese hamster ovary cells after acute and chronic treatment with the selective nonpeptidic agonist SNC-80. *J Pharmacol Exp Ther.* 1996;278(3): 1083–1089.
25. Watts VJ, Neve KA. Sensitization of adenylate cyclase by Galpha i/o-coupled receptors. *Pharmacol Ther.* 2005;106(3):405–421.
26. Terwilliger RZ, Beitner-Johnson D, Sevarino KA, Crain SM, Nestler EJ. A general role for adaptations in G-proteins and the cyclic AMP system in mediating the chronic actions of morphine and cocaine on neuronal function. *Brain Res.* 1991;548(1–2):100–110.

27. Nestler EJ, Aghajanian GK. Molecular and cellular basis of addiction. *Science*. 1997;278 (5335):58–63.
28. Nestler EJ. Molecular mechanisms of drug addiction. *J Neurosci*. 1992;12(7): 2439–2450.
29. Matsuoka I, Maldonado R, Defer N, Noel F, Hanoune J, Roques BP. Chronic morphine administration causes region-specific increase of brain type VIII adenylyl cyclase mRNA. *Eur J Pharmacol*. 1994;268(2):215–221.
30. Nestler EJ, Tallman JF. Chronic morphine treatment increases cyclic AMP-dependent protein kinase activity in the rat locus coeruleus. *Mol Pharmacol*. 1988;33(2):127–132.
31. Liu JG, Anand KJ. Protein kinases modulate the cellular adaptations associated with opioid tolerance and dependence. *Brain Res Brain Res Rev*. 2001;38(1–2):1–19.
32. Rubenzik M, Varga E, Stropova D, Roeske WR, Yamamura HI. Expression of alpha-transducin in Chinese hamster ovary cells stably transfected with the human delta-opioid receptor attenuates chronic opioid agonist-induced adenylyl cyclase superactivation. *Mol Pharmacol*. 2001;60(5):1076–1082.
33. Watts VJ, Wiens BL, Cumbay MG, Vu MN, Neve RL, Neve KA. Selective activation of Galphao by D2L dopamine receptors in NS20Y neuroblastoma cells. *J Neurosci*. 1998;18 (21):8692–8699.
34. Avidor-Reiss T, Nevo I, Levy R, Pfeuffer T, Vogel Z. Chronic opioid treatment induces adenylyl cyclase V superactivation. Involvement of Gbetagamma. *J Biol Chem*. 1996;271 (35):21309–21315.
35. Avidor-Reiss T, Nevo I, Saya D, Bayewitch M, Vogel Z. Opiate-induced adenylyl cyclase superactivation is isozyme-specific. *J Biol Chem*. 1997;272(8):5040–5047.
36. Law PY, Loh HH. delta-Opioid receptor activates cAMP phosphodiesterase activities in neuroblastoma glioma NG108-15 hybrid cells. *Mol Pharmacol*. 1993;43(5):684–693.
37. Ammer H, Schulz R. Morphine dependence in human neuroblastoma SH-SY5Y cells is associated with adaptive changes in both the quantity and functional interaction of PGE1 receptors and stimulatory G proteins. *Brain Res*. 1996;707(2):235–244.
38. Chakrabarti S, Gintzler AR. Phosphorylation of Galphas influences its association with the micro-opioid receptor and is modulated by long-term morphine exposure. *Mol Pharmacol*. 2007;72(3):753–760.
39. Chakrabarti S, Wang L, Tang WJ, Gintzler AR. Chronic morphine augments adenylyl cyclase phosphorylation: relevance to altered signaling during tolerance/dependence. *Mol Pharmacol*. 1998;54(6):949–953.
40. Harada H, Ueda H, Wada Y, Katada T, Ui M, Satoh M. Phosphorylation of mu-opioid receptors—a putative mechanism of selective uncoupling of receptor–Gi interaction, measured with low-Km GTPase and nucleotide-sensitive agonist binding. *Neurosci Lett*. 1989;100(1–3):221–226.
41. Harada H, Ueda H, Katada T, Ui M, Satoh M. Phosphorylated mu-opioid receptor purified from rat brains lacks functional coupling with Gi1, a GTP-binding protein in reconstituted lipid vesicles. *Neurosci Lett*. 1990;113(1):47–49.
42. Bernstein MA, Welch SP. mu-Opioid receptor down-regulation and cAMP-dependent protein kinase phosphorylation in a mouse model of chronic morphine tolerance. *Brain Res Mol Brain Res*. 1998;55(2):237–242.
43. Crain SM, Shen KF. Antagonists of excitatory opioid receptor functions enhance morphine's analgesic potency and attenuate opioid tolerance/dependence liability. *Pain*. 2000;84(2–3):121–131.
44. Lutfy K, Chang SC, Candido J, Jang Y, Sierra V, Yoburn BC. Modification of morphine-induced analgesia and toxicity by pertussis toxin. *Brain Res*. 1991;544(2):191–195.
45. Crain SM, Shen KF. Naloxone rapidly evokes endogenous kappa opioid receptor-mediated hyperalgesia in naive mice pretreated briefly with GM1 ganglioside or in chronic morphine-dependent mice. *Brain Res*. 2007;1167:31–41.

46. Self DW, Genova LM, Hope BT, Barnhart WJ, Spencer JJ, Nestler EJ. Involvement of cAMP-dependent protein kinase in the nucleus accumbens in cocaine self-administration and relapse of cocaine-seeking behavior. *J Neurosci.* 1998;18(5): 1848–1859.
47. Narita M, Narita M, Mizoguchi H, Tseng LF. Inhibition of protein kinase C, but not of protein kinase A, blocks the development of acute antinociceptive tolerance to an intrathecally administered mu-opioid receptor agonist in the mouse. *Eur J Pharmacol.* 1995;280(2):R1–R3.
48. Inoue M, Ueda H. Protein kinase C-mediated acute tolerance to peripheral mu-opioid analgesia in the bradykinin-nociception test in mice. *J Pharmacol Exp Ther.* 2000;293(2): 662–669.
49. Bilsky EJ, Bernstein RN, Wang Z, Sadee W, Porreca F. Effects of naloxone and D-Phe-Cys-Tyr-D-Trp-Arg-Thr-Pen-Thr-NH2 and the protein kinase inhibitors H7 and H8 on acute morphine dependence and antinociceptive tolerance in mice. *J Pharmacol Exp Ther.* 1996;277(1):484–490.
50. Punch LJ, Self DW, Nestler EJ, Taylor JR. Opposite modulation of opiate withdrawal behaviors on microinfusion of a protein kinase A inhibitor versus activator into the locus coeruleus or periaqueductal gray. *J Neurosci.* 1997;17(21):8520–8527.
51. Richards JP, Bachinger HP, Goodman RH, Brennan RG. Analysis of the structural properties of cAMP-responsive element-binding protein (CREB) and phosphorylated CREB. *J Biol Chem.* 1996;271(23):13716–13723.
52. Chrivia JC, Kwok RP, Lamb N, Hagiwara M, Montminy MR, Goodman RH. Phosphorylated CREB binds specifically to the nuclear protein CBP. *Nature.* 1993;365 (6449):855–859.
53. Gonzalez GA, Montminy MR. Cyclic AMP stimulates somatostatin gene transcription by phosphorylation of CREB at serine 133. *Cell.* 1989;59(4):675–680.
54. Bilecki W, Przewlocki R. Effect of opioids on Ca^{2+}/cAMP responsive element binding protein. *Acta Neurobiol Exp.* 2000;60(4):557–567.
55. Widnell KL, Self DW, Lane SB, et al. Regulation of CREB expression: *in vivo* evidence for a functional role in morphine action in the nucleus accumbens. *J Pharmacol Exp Ther.* 1996;276(1):306–315.
56. Widnell KL, Russell DS, Nestler EJ. Regulation of expression of cAMP response element-binding protein in the locus coeruleus *in vivo* and in a locus coeruleus-like cell line *in vitro*. *Proc Natl Acad Sci USA.* 1994;91(23):10947–10951.
57. Duman RS, Tallman JF, Nestler EJ. Acute and chronic opiate-regulation of adenylate cyclase in brain: specific effects in locus coeruleus. *J Pharmacol Exp Ther.* 1988;246(3): 1033–1039.
58. Maldonado R, Blendy JA, Tzavara E, et al. Reduction of morphine abstinence in mice with a mutation in the gene encoding CREB. *Science.* 1996;273(5275):657–659.
59. Ren X, Lutfy K, Mangubat M, et al. Alterations in phosphorylated CREB expression in different brain regions following short- and long-term morphine exposure: relationship to food intake. *J Obes.* 2013;2013:764742.
60. Shaw-Lutchman TZ, Barrot M, Wallace T, et al. Regional and cellular mapping of cAMP response element-mediated transcription during naltrexone-precipitated morphine withdrawal. *J Neurosci.* 2002;22(9):3663–3672.
61. Lamas M, Molina C, Foulkes NS, Jansen E, Sassone-Corsi P. Ectopic ICER expression in pituitary corticotroph AtT20 cells: effects on morphology, cell cycle, and hormonal production. *Mol Endocrinol.* 1997;11(10):1425–1434.
62. Jansen E, Ayoubi TA, Meulemans SM, Van de Ven WJ. Neuroendocrine-specific expression of the human prohormone convertase 1 gene. Hormonal regulation of transcription through distinct cAMP response elements. *J Biol Chem.* 1995;270(25): 15391–15397.

63. Schafer MK, Day R, Cullinan WE, Chretien M, Seidah NG, Watson SJ. Gene expression of prohormone and proprotein convertases in the rat CNS: a comparative *in situ* hybridization analysis. *J Neurosci.* 1993;13(3):1258–1279.
64. Espinosa VP, Liu Y, Ferrini M, et al. Differential regulation of prohormone convertase 1/3, prohormone convertase 2 and phosphorylated cyclic-AMP-response element binding protein by short-term and long-term morphine treatment: implications for understanding the "switch" to opiate addiction. *Neuroscience.* 2008;156(3):788–799.
65. Nillni EA, Lee A, Legradi G, Lechan RM. Effect of precipitated morphine withdrawal on post-translational processing of prothyrotropin releasing hormone (proTRH) in the ventrolateral column of the midbrain periaqueductal gray. *J Neurochem.* 2002;80(5):874–884.
66. Zhou LF, Zhu YP. Changes of CREB in rat hippocampus, prefrontal cortex and nucleus accumbens during three phases of morphine induced conditioned place preference in rats. *J Zhejiang Univ Sci B.* 2006;7(2):107–113.
67. Wang WS, Kang S, Liu WT, et al. Extinction of aversive memories associated with morphine withdrawal requires ERK-mediated epigenetic regulation of brain-derived neurotrophic factor transcription in the rat ventromedial prefrontal cortex. *J Neurosci.* 2012;32(40):13763–13775.
68. Tan CM, Kelvin DJ, Litchfield DW, Ferguson SS, Feldman RD. Tyrosine kinase-mediated serine phosphorylation of adenylyl cyclase. *Biochemistry.* 2001;40(6):1702–1709.
69. Varga EV, Rubenzik MK, Stropova D, et al. Converging protein kinase pathways mediate adenylyl cyclase superactivation upon chronic delta-opioid agonist treatment. *J Pharmacol Exp Ther.* 2003;306(1):109–115.
70. Li YQ, Li FQ, Wang XY, et al. Central amygdala extracellular signal-regulated kinase signaling pathway is critical to incubation of opiate craving. *J Neurosci.* 2008;28(49):13248–13257.
71. Nestler EJ, Alreja M, Aghajanian GK. Molecular and cellular mechanisms of opiate action: studies in the rat locus coeruleus. *Brain Res Bull.* 1994;35(5–6):521–528.
72. Reyes BA, Glaser JD, Van Bockstaele EJ. Ultrastructural evidence for co-localization of corticotropin-releasing factor receptor and mu-opioid receptor in the rat nucleus locus coeruleus. *Neurosci Lett.* 2007;413(3):216–221.
73. Reyes BA, Drolet G, Van Bockstaele EJ. Dynorphin and stress-related peptides in rat locus coeruleus: contribution of amygdalar efferents. *J Comp Neurol.* 2008;508(4):663–675.
74. Kreibich A, Reyes BA, Curtis AL, et al. Presynaptic inhibition of diverse afferents to the locus ceruleus by kappa-opiate receptors: a novel mechanism for regulating the central norepinephrine system. *J Neurosci.* 2008;28(25):6516–6525.
75. Dahlstroem A, Fuxe K. Evidence for the existence of monoamine-containing neurons in the central nervous system. I. Demonstration of monoamines in the cell bodies of brain stem neurons. *Acta Physiol Scand Suppl.* 1964;suppl 232:231–255.
76. Foote SL, Bloom FE, Aston-Jones G. Nucleus locus ceruleus: new evidence of anatomical and physiological specificity. *Physiol Rev.* 1983;63(3):844–914.
77. Dahlstrom A, Fuxe K. Evidence for the existence of an outflow of noradrenaline nerve fibres in the ventral roots of the rat spinal cord. *Experientia.* 1965;21(7):409–410.
78. Prokopova I. Noradrenaline and behavior. *Cesk Fysiol.* 2010;59(2):51–58.
79. Quirion R, Zajac JM, Morgat JL, Roques BP. Autoradiographic distribution of mu and delta opiate receptors in rat brain using highly selective ligands. *Life Sci.* 1983;33(suppl 1):227–230.
80. Maldonado R, Stinus L, Gold LH, Koob GF. Role of different brain structures in the expression of the physical morphine withdrawal syndrome. *J Pharmacol Exp Ther.* 1992;261(2):669–677.

81. Bird SJ, Kuhar MJ. Iontophoretic application of opiates to the locus coeruleus. *Brain Res.* 1977;122(3):523–533.
82. Andrade R, Vandermaelen CP, Aghajanian GK. Morphine tolerance and dependence in the locus coeruleus: single cell studies in brain slices. *Eur J Pharmacol.* 1983;91 (2–3):161–169.
83. Akaoka H, Aston-Jones G. Opiate withdrawal-induced hyperactivity of locus coeruleus neurons is substantially mediated by augmented excitatory amino acid input. *J Neurosci.* 1991;11(12):3830–3839.
84. Aghajanian GK, Kogan JH, Moghaddam B. Opiate withdrawal increases glutamate and aspartate efflux in the locus coeruleus: an *in vivo* microdialysis study. *Brain Res.* 1994;636 (1):126–130.
85. Aston-Jones G, Hirata H, Akaoka H. Local opiate withdrawal in locus coeruleus *in vivo*. *Brain Res.* 1997;765(2):331–336.
86. Strosberg AD. Structure, function, and regulation of adrenergic receptors. *Protein Sci.* 1993;2(8):1198–1209.
87. Smith CB, Hollingsworth PJ, Geer JJ, Moises HC. Changes in alpha2 adrenoreceptors in various areas of the rat brain after long-term administration of "mu" and "kappa" opiate agonists. *Life Sci.* 1983;33(suppl 1):369–372.
88. Smith CB, Moises HC, Spengler RN, Hollingsworth PJ. Changes in alpha 2-adrenoceptor number and function in brains of morphine-dependent rats. *Eur J Pharmacol.* 1989;161(2–3):111–119.
89. Vicentini LM, Miller RJ, Robertson MJ. Chronic opiate treatment does not modify alpha 2-adrenergic receptors in rat cerebral cortex, kidney and in the neurotumor cell line NCB20. *Eur J Pharmacol.* 1983;95(3–4):265–270.
90. Hamburg M, Tallman JF. Chronic morphine administration increases the apparent number of alpha 2-adrenergic receptors in rat brain. *Nature.* 1981;291(5815):493–495.
91. Gowing L, Farrell MF, Ali R, White JM. Alpha2-adrenergic agonists for the management of opioid withdrawal. *Cochrane Database Syst Rev.* 2014;3:CD002024.
92. Gold MS, Redmond Jr DE, Kleber HD. Clonidine blocks acute opiate-withdrawal symptoms. *Lancet.* 1978;2(8090):599–602.
93. Taylor JR, Elsworth JD, Garcia EJ, Grant SJ, Roth RH, Redmond Jr DE. Clonidine infusions into the locus coeruleus attenuate behavioral and neurochemical changes associated with naloxone-precipitated withdrawal. *Psychopharmacology (Berl).* 1988;96 (1):121–134.
94. Esposito E, Kruszewska A, Ossowska G, Samanin R. Noradrenergic and behavioural effects of naloxone injected in the locus coeruleus of morphine-dependent rats and their control by clonidine. *Psychopharmacology (Berl).* 1987;93(3):393–396.
95. Maldonado R, Koob GF. Destruction of the locus coeruleus decreases physical signs of opiate withdrawal. *Brain Res.* 1993;605(1):128–138.
96. Tseng LF, Loh HH, Wei ET. Effects of clonidine on morphine withdrawal signs in the rat. *Eur J Pharmacol.* 1975;30(1):93–99.
97. Rasmussen K, Aghajanian GK. Withdrawal-induced activation of locus coeruleus neurons in opiate-dependent rats: attenuation by lesions of the nucleus paragigantocellularis. *Brain Res.* 1989;505(2):346–350.
98. Aghajanian GK. Tolerance of locus coeruleus neurones to morphine and suppression of withdrawal response by clonidine. *Nature.* 1978;276(5684):186–188.
99. Guitart X, Nestler EJ. Second messenger and protein phosphorylation mechanisms underlying opiate addiction: studies in the rat locus coeruleus. *Neurochem Res.* 1993;18(1):5–13.
100. Hayward MD, Duman RS, Nestler EJ. Induction of the c-fos proto-oncogene during opiate withdrawal in the locus coeruleus and other regions of rat brain. *Brain Res.* 1990;525(2):256–266.

101. Maldonado R, Valverde O, Garbay C, Roques BP. Protein kinases in the locus coeruleus and periaqueductal gray matter are involved in the expression of opiate withdrawal. *Naunyn Schmiedebergs Arch Pharmacol.* 1995;352(5):565–575.
102. Guitart X, Thompson MA, Mirante CK, Greenberg ME, Nestler EJ. Regulation of cyclic AMP response element-binding protein (CREB) phosphorylation by acute and chronic morphine in the rat locus coeruleus. *J Neurochem.* 1992;58(3):1168–1171.
103. Lane-Ladd SB, Pineda J, Boundy VA, et al. CREB (cAMP response element-binding protein) in the locus coeruleus: biochemical, physiological, and behavioral evidence for a role in opiate dependence. *J Neurosci.* 1997;17(20):7890–7901.
104. Kogan JH, Aghajanian GK. Long-term glutamate desensitization in locus coeruleus neurons and its role in opiate withdrawal. *Brain Res.* 1995;689(1):111–121.
105. Raymond LA, Blackstone CD, Huganir RL. Phosphorylation and modulation of recombinant GluR6 glutamate receptors by cAMP-dependent protein kinase. *Nature.* 1993;361(6413):637–641.
106. Herbert JM, Augereau JM, Gleye J, Maffrand JP. Chelerythrine is a potent and specific inhibitor of protein kinase C. *Biochem Biophys Res Commun.* 1990;172(3):993–999.
107. Dossin O, Hanoun N, Zajac JM. Involvement of locus coeruleus projections in opiate withdrawal but not in opiate tolerance in mice. *Eur J Pharmacol.* 1996;308(3):271–274.
108. Stamatakis AM, Sparta DR, Jennings JH, McElligott ZA, Decot H, Stuber GD. Amygdala and bed nucleus of the stria terminalis circuitry: implications for addiction-related behaviors. *Neuropharmacology.* 2014;76 Pt B:320–328.
109. Koob GF, Le Moal M. Review. Neurobiological mechanisms for opponent motivational processes in addiction. *Philos Trans R Soc Lond B Biol Sci.* 2008;363(1507):3113–3123.
110. Zhu W, Pan ZZ. Synaptic properties and postsynaptic opioid effects in rat central amygdala neurons. *Neuroscience.* 2004;127(4):871–879.
111. Finnegan TF, Chen SR, Pan HL. Effect of the {mu} opioid on excitatory and inhibitory synaptic inputs to periaqueductal gray-projecting neurons in the amygdala. *J Pharmacol Exp Ther.* 2005;312(2):441–448.
112. Bajo M, Madamba SG, Roberto M, Siggins GR. Acute morphine alters GABAergic transmission in the central amygdala during naloxone-precipitated morphine withdrawal: role of cyclic AMP. *Front Integr Neurosci.* 2014;8:45.
113. Li Z, Luan W, Chen Y, et al. Chronic morphine treatment switches the effect of dopamine on excitatory synaptic transmission from inhibition to excitation in pyramidal cells of the basolateral amygdala. *J Neurosci.* 2011;31(48):17527–17536.
114. Lintas A, Chi N, Lauzon NM, et al. Identification of a dopamine receptor-mediated opiate reward memory switch in the basolateral amygdala-nucleus accumbens circuit. *J Neurosci.* 2011;31(31):11172–11183.
115. Lu L, Ceng X, Huang M. Corticotropin-releasing factor receptor type I mediates stress-induced relapse to opiate dependence in rats. *Neuroreport.* 2000;11(11):2373–2378.
116. Wang B, Luo F, Zhang WT, Han JS. Stress or drug priming induces reinstatement of extinguished conditioned place preference. *Neuroreport.* 2000;11(12):2781–2784.
117. Depaulis A, Morgan MM, Liebeskind JC. GABAergic modulation of the analgesic effects of morphine microinjected in the ventral periaqueductal gray matter of the rat. *Brain Res.* 1987;436(2):223–228.
118. Park C, Kim JH, Yoon BE, Choi EJ, Lee CJ, Shin HS. T-type channels control the opioidergic descending analgesia at the low threshold-spiking GABAergic neurons in the periaqueductal gray. *Proc Natl Acad Sci USA.* 2010;107(33):14857–14862.
119. Lovick TA. Involvement of GABA in medullary raphe-evoked modulation of neuronal activity in the periaqueductal grey matter in the rat. *Exp Brain Res.* 2001;137(2):214–218.
120. Griffiths JL, Lovick TA. Co-localization of 5-HT 2A-receptor- and GABA-immunoreactivity in neurones in the periaqueductal grey matter of the rat. *Neurosci Lett.* 2002;326(3):151–154.

121. Moreau JL, Fields HL. Evidence for GABA involvement in midbrain control of medullary neurons that modulate nociceptive transmission. *Brain Res.* 1986;397(1):37–46.
122. Millan MJ, Czlonkowski A, Millan MH, Herz A. Activation of periaqueductal grey pools of beta-endorphin by analgetic electrical stimulation in freely moving rats. *Brain Res.* 1987;407(1):199–203.
123. Jensen TS, Yaksh TL. Comparison of the antinociceptive action of mu and delta opioid receptor ligands in the periaqueductal gray matter, medial and paramedial ventral medulla in the rat as studied by the microinjection technique. *Brain Res.* 1986;372(2):301–312.
124. Chieng B, Christie MD. Local opioid withdrawal in rat single periaqueductal gray neurons *in vitro*. *J Neurosci.* 1996;16(22):7128–7136.
125. Cho YW, Han SH, Min BI, Rhee JS, Akaike N. Antagonizing effect of protein kinase C activation on the mu-opioid agonist-induced inhibition of high voltage-activated calcium current in rat periaqueductal gray neuron. *Brain Res.* 2001;916(1–2):61–69.
126. Wang HY, Friedman E, Olmstead MC, Burns LH. Ultra-low-dose naloxone suppresses opioid tolerance, dependence and associated changes in mu opioid receptor-G protein coupling and Gbetagamma signaling. *Neuroscience.* 2005;135(1):247–261.
127. Bettler B, Kaupmann K, Mosbacher J, Gassmann M. Molecular structure and physiological functions of GABA(B) receptors. *Physiol Rev.* 2004;84(3):835–867.
128. Gerber U, Gahwiler BH. GABAB and adenosine receptors mediate enhancement of the K+ current, IAHP, by reducing adenylyl cyclase activity in rat CA3 hippocampal neurons. *J Neurophysiol.* 1994;72(5):2360–2367.
129. Bagley EE, Chieng BC, Christie MJ, Connor M. Opioid tolerance in periaqueductal gray neurons isolated from mice chronically treated with morphine. *Br J Pharmacol.* 2005;146(1):68–76.
130. Bagley EE, Hacker J, Chefer VI, et al. Drug-induced GABA transporter currents enhance GABA release to induce opioid withdrawal behaviors. *Nat Neurosci.* 2011;14(12):1548–1554.
131. Bagley EE. Opioid and GABAB receptors differentially couple to an adenylyl cyclase/protein kinase A downstream effector after chronic morphine treatment. *Front Pharmacol.* 2014;5:148.
132. Geisler S, Derst C, Veh RW, Zahm DS. Glutamatergic afferents of the ventral tegmental area in the rat. *J Neurosci.* 2007;27(21):5730–5743.
133. Leone P, Pocock D, Wise RA. Morphine-dopamine interaction: ventral tegmental morphine increases nucleus accumbens dopamine release. *Pharmacol Biochem Behav.* 1991;39(2):469–472.
134. Gysling K, Wang RY. Morphine-induced activation of A10 dopamine neurons in the rat. *Brain Res.* 1983;277(1):119–127.
135. Margolis EB, Hjelmstad GO, Bonci A, Fields HL. Kappa-opioid agonists directly inhibit midbrain dopaminergic neurons. *J Neurosci.* 2003;23(31):9981–9986.
136. Bonci A, Williams JT. Increased probability of GABA release during withdrawal from morphine. *J Neurosci.* 1997;17(2):796–803.
137. Madhavan A, He L, Stuber GD, Bonci A, Whistler JL. micro-Opioid receptor endocytosis prevents adaptations in ventral tegmental area GABA transmission induced during naloxone-precipitated morphine withdrawal. *J Neurosci.* 2010;30(9):3276–3286.
138. Herman JP, Cullinan WE. Neurocircuitry of stress: central control of the hypothalamo–pituitary–adrenocortical axis. *Trends Neurosci.* 1997;20(2):78–84.
139. Delfs JM, Zhu Y, Druhan JP, Aston-Jones G. Noradrenaline in the ventral forebrain is critical for opiate withdrawal-induced aversion. *Nature.* 2000;403(6768):430–434.
140. Kudo T, Konno K, Uchigashima M, et al. GABAergic neurons in the ventral tegmental area receive dual GABA/enkephalin-mediated inhibitory inputs from the bed nucleus of the stria terminalis. *Eur J Neurosci.* 2014;39(11):1796–1809.

141. Contet C, Filliol D, Matifas A, Kieffer BL. Morphine-induced analgesic tolerance, locomotor sensitization and physical dependence do not require modification of mu opioid receptor, cdk5 and adenylate cyclase activity. *Neuropharmacology.* 2008;54(3): 475–486.
142. Mannelli P, Gottheil E, Peoples JF, Oropeza VC, Van Bockstaele EJ. Chronic very low dose naltrexone administration attenuates opioid withdrawal expression. *Biol Psychiatry.* 2004;56(4):261–268.
143. Van Bockstaele EJ, Rudoy C, Mannelli P, Oropeza V, Qian Y. Elevated mu-opioid receptor expression in the nucleus of the solitary tract accompanies attenuated withdrawal signs after chronic low dose naltrexone in opiate-dependent rats. *J Neurosci Res.* 2006;83(3):508–514.
144. Crain SM, Shen KF. Ultra-low concentrations of naloxone selectively antagonize excitatory effects of morphine on sensory neurons, thereby increasing its antinociceptive potency and attenuating tolerance/dependence during chronic cotreatment. *Proc Natl Acad Sci USA.* 1995;92(23):10540–10544.

CHAPTER EIGHT

Caenorhabditis elegans as a Model to Study the Molecular and Genetic Mechanisms of Drug Addiction

Eric A. Engleman*[,1], Simon N. Katner*, Bethany S. Neal-Beliveau[†]
[*] Department of Psychiatry, Indiana University School of Medicine, Indianapolis, Indiana, USA
[†] Department of Psychology, Purdue School of Science, Indiana University–Purdue University Indianapolis, Indianapolis, Indiana, USA
[1] Corresponding author: e-mail address: eenglema@iupui.edu

Contents

Abstract

Drug addiction takes a massive toll on society. Novel animal models are needed to test new treatments and understand the basic mechanisms underlying addiction. Rodent models have identified the neurocircuitry involved in addictive behavior and indicate that rodents possess some of the same neurobiologic mechanisms that mediate addiction in humans. Recent studies indicate that addiction is mechanistically and phylogenetically ancient and many mechanisms that underlie human addiction are also present in invertebrates. The nematode *Caenorhabditis elegans* has conserved neurobiologic systems with powerful molecular and genetic tools and a rapid rate of

Progress in Molecular Biology and Translational Science, Volume 137
ISSN 1877-1173
http://dx.doi.org/10.1016/bs.pmbts.2015.10.019

development that enables cost-effective translational discovery. Emerging evidence suggests that *C. elegans* is an excellent model to identify molecular mechanisms that mediate drug-induced behavior and potential targets for medications development for various addictive compounds. *C. elegans* emit many behaviors that can be easily quantitated including some that involve interactions with the environment. Ethanol (EtOH) is the best-studied drug-of-abuse in *C. elegans* and at least 50 different genes/targets have been identified as mediating EtOH's effects and polymorphisms in some orthologs in humans are associated with alcohol use disorders. *C. elegans* has also been shown to display dopamine and cholinergic system–dependent attraction to nicotine and demonstrate preference for cues previously associated with nicotine. Cocaine and methamphetamine have been found to produce dopamine-dependent reward-like behaviors in *C. elegans*. These behavioral tests in combination with genetic/molecular manipulations have led to the identification of dozens of target genes/systems in *C. elegans* that mediate drug effects. The one target/gene identified as essential for drug-induced behavioral responses across all drugs of abuse was the cat-2 gene coding for tyrosine hydroxylase, which is consistent with the role of dopamine neurotransmission in human addiction. Overall, *C. elegans* can be used to model aspects of drug addiction and identify systems and molecular mechanisms that mediate drug effects. The findings are surprisingly consistent with analogous findings in higher-level organisms. Further, model refinement is warranted to improve model validity and increase utility for medications development.

1. INTRODUCTION

Addictions represent a major and growing challenge in our society. Drug addiction takes a massive toll on both direct and indirect human and financial costs, including hundreds of thousands of deaths annually from alcohol and drug-related auto accidents, cancer, and other conditions resulting from alcohol and nicotine addiction, drug-associated homicides, and overdoses. The resulting costs to society are estimated to be hundreds of billions of dollars annually worldwide in lost productivity, drug enforcement, hospitalizations, and treatments for healthcare resulting directly or indirectly, from drug use.[1] Clearly there is an urgent need for effective treatments and prevention strategies that are developed from an understanding of the basic mechanisms that underlie addictive behavior. Much of what we know about the neurobiology of addictions has been either discovered or enhanced through the use of animal models.[2] This includes the discovery and characterization of some of the basic reward circuitry and the development of behavioral measures to model and study human addiction in animals.[3]

Here we discuss a relatively new model system to study addiction based on behavioral measures in a decidedly ancient and simple animal *Caenorhabditis elegans*. The review will briefly introduce animal modeling of addiction in vertebrates and invertebrates to provide some context for the following discussion of *C. elegans* models of addiction. Discussion then turns to specific drugs of abuse including alcohol, nicotine, cocaine, and methamphetamine, their effects in *C. elegans*, and some of the biologic systems and molecular targets identified with these models. Some ideas are presented indicating how *C. elegans* models may be developed and refined in the future to enhance model validity, increase utility for medications development, and improve model value for translational applications.

2. MODELS OF ADDICTION

Through the years, model systems, using primates and other mammals closely associated with humans, have provided essential information—particularly associated with the behavioral effects of drugs.[4,5] However, a review of the literature on animal models of addiction shows that rodents (mainly rats and mice) are a very popular species for addictions research.[2,6] This is due, in large part, to the fact that rodents share basic neurobiologic systems, both structurally and functionally, with humans. In this way, they have enabled the identification of circuitry underlying addictive behavior. In addition, these animals self-administer drugs of abuse including ethanol (EtOH),[6] stimulants including cocaine, nicotine, amphetamine, methamphetamine,[7,8] opiates such as heroin and morphine,[9] and many other drugs including caffeine and THC.[10,11] Self-administration provides validity to these models, indicating that such models recapitulate some aspects of drug-taking behavior.[12] Similarly, rodent models of addiction also show evidence of construct validity, in which some of the mechanisms and neurobiology that appear to mediate addictive behavior in humans also are present, and function through analogous systems in mammalian models.[12] Finally, rodent models also demonstrate predictive validity, indicating that drugs/treatments that reduce addictive behavior in humans, also show some efficacy in rodents.[13] Overall, these models have, and continue to provide essential information about the neurobiology of addiction and have been instrumental in the development of the few available pharmacologic treatments.[14] However, relatively little is understood about the molecular foundations of addiction, and animal models that can quickly, efficiently, and systematically examine the underlying mechanisms of addictions have yet to be developed.

3. INVERTEBRATE MODELS

Recent work shows that addiction is a phylogenetically ancient process and indicates that many mechanisms that underlie addictions are present in invertebrates. Elegant behavioral models of addiction such as conditioned place preference (CPP), and other tools historically used to study aspects of addictive behavior in rodent models,[15] have demonstrated that crayfish show drug reward, seeking, and withdrawal to cocaine, amphetamines, and opiates.[16] Similarly, EtOH self-administration and conditioning paradigms have demonstrated that *Drosophila melanogaster* show preference responses to cues that had been previously paired with EtOH.[17] Although some might find it surprising that such a simple animal can be used to model complex behaviors, behavioral models using invertebrates have played a central role in the discovery of the molecular mechanisms that underlie learning and memory.[18] The nematode *C. elegans* is an excellent model organism with conserved neurobiologic systems that is used to model various disease states.[19] It provides the researcher with numerous molecular and genetic tools, including a tractable and fully sequenced genome, the availability of thousands of mutants, and the ability to manipulate genes and their expression through transgenic approaches and RNAi techniques. In addition, a relatively short life cycle and a 3-day generation time from egg to adult can lead to a dramatic increase in the pace of discovery at a fraction of the cost of using higher level organisms. However, to date, there are few established *C. elegans* behavioral models of addiction. We have discovered that, like mammals and other invertebrates, *C. elegans* also develops a conditioned preference for cues after previous parings with methamphetamine or cocaine that is dependent on dopamine neurotransmission.[20] Moreover, with drug pre-exposure they demonstrate sensitization, cross-sensitization, tolerance, and cross-tolerance, all of which are hallmarks of addiction in humans. Validated *C. elegans* behavioral models of addiction designed to enable fast and accurate generation of data would provide the field with valuable and powerful tools to study the molecular mechanisms that underlie addiction, and open new avenues to identify new targets for medications development.

4. *C. ELEGANS* AS A MODEL SYSTEM TO STUDY ADDICTION

C. elegans is an obvious choice as a model system, as the first animal to have its genome completely sequenced,[21] and with approximately 19,000 genes, more is known about the genetics and molecular make-up of *C. elegans* than any other ambulatory organism.[22] *C. elegans* has been used to model many types of disorders in humans,[19] including neurologic and psychiatric disorders ranging from Parkinson's disease to Autism.[23–25] Although rodents have been used to model addiction for many decades,[26–29] the use of *C. elegans* as a model system to study addiction is a relatively recent development.[30–33] Clearly, using *C. elegans* as a model for psychiatric disorders (such as addiction) has some limitations including a lack of some neurotransmitter systems such as norepinephrine. Also, till date, a limited number of valid behavioral models have been developed and characterized to study the reinforcing properties of drugs. However, accumulating evidence indicates that *C. elegans* is an excellent model to identify molecular mechanisms that mediate drug effects and potential targets for medications development for various addictive compounds. *C. elegans* emit many behaviors that can be easily quantitated such as egg laying and defecation, as well as a host of movement and postural measures, including speed of locomotion and counting the number of body bends per unit time.[34] Other measures document how *C. elegans* interacts with its environment, including chemotaxic behavior,[35] as well as associative and nonassociative learning.[36] Many of these behaviors have been studied for decades and the neurobiologic systems and circuits that mediate them are well described, making *C. elegans* an excellent candidate model system to study the effects of drugs on behavior.

5. DRUGS OF ABUSE

5.1 Ethanol

Investigators have discovered that *C. elegans* can be used to study the effects of EtOH.[37,38] Several studies have established that *C. elegans* display concentration-dependent depression of a variety of behaviors, including locomotion, body bend amplitude, and egg laying, after exposure to EtOH.[38] Importantly, the depressant effects on the locomotor activity of *C. elegans* occur when the internal tissue concentration of EtOH reaches levels that correspond to intoxicating blood alcohol levels in humans.[39] To date, at least 50 genes have been identified that influence EtOH-associated behaviors in *C. elegans*, and several orthologs of these genes have been implicated in alcohol use disorders in humans.[40] As with vertebrates, dopamine systems appear to play a role in EtOH-induced behavioral effects in *C. elegans*. EtOH induces state-dependent learning in *C. elegans* that is absent in animals with functional mutations in the vesicular monoamine transporter (cat-1) or tyrosine hydroxylase (cat-2). In certain experimental paradigms, *C. elegans* also show an EtOH preference response in choice tests that appears to be mediated through the dopamine and serotonin systems.[41] Some of these early observations led to the development of a simple behavioral model to study the effects of EtOH in *C. elegans* and leverage the fully tractable molecular genetics available to researchers using this species. Investigators use some of these behavioral paradigms to identify mutations in individual genes that affect behavioral responses to EtOH in *C. elegans*. Once identified, conserved homologs of these genes may be examined for effects on alcohol-related behaviors in other animal models, and/or polymorphisms of such genes may be assessed in humans for possible roles in alcohol use disorders.

After isolating mutants that showed resistance to the behavioral effects of EtOH, Davies *et al.*[42] found that mutations in the gene slo-1, a highly conserved gene which codes for a BK potassium channel that is homologous to one found in humans, produced resistance to the locomotor effects of EtOH.[33,43] These effects of EtOH were found to be mediated through a direct action at the channel to increase current and to be selective for EtOH.[38] The BK potassium channel appears to subserve behavioral responses across multiple species including humans.[32,33,42] Additional work has identified a specific residue (T381I) on the channel that confers dramatic and selective resistance to the behavioral effects of EtOH.[32] Thus, the BK potassium channel is a verified mediator of the effects of EtOH across phyla and may serve as a target for the identification and development of new treatments for alcohol use disorders.

An essential characteristic of EtOH effects in animal models is the development of tolerance, which is an adaptation that occurs when the same concentration of EtOH produces a reduced behavioral response after chronic or repeated exposure.[33,44] After continued exposure to EtOH, acute functional tolerance becomes apparent in the Bristol (N2) wild-type *C. elegans* strain. However, such tolerance occurs much more rapidly in the Hawaiian CB4856 wild-type strain.[45] This effect was found to be mediated by a variation in the npr-1 gene. This gene codes for a neuropeptide Y (NPY) receptor homolog in *C. elegans* that was previously shown to underlie differences in social behavior and responses to food.[46] Moreover, NPY is known to regulate EtOH and food intake in vertebrate models.[47,48] Since tolerance is a key feature in the progression to alcohol dependence in humans,[49] NPY appears to be an excellent molecular target for treatment development[50,51] and may be aided through the study of the npr-1 gene in *C. elegans*. More recently, EtOH-induced muscle hypercontraction (EHC) was found to be dependent on cholinergic signaling as the effect was significantly reduced in cholinergic signaling mutants (cha-1 and unc-17) and with exposure of the nonselective nicotinic cholinergic receptor antagonist mecamylamine.[52] Tolerance to EHC was evident in wild-type worms, but was absent in a Na^+/K^+ ATPase mutant eat-6(eg200). Interestingly, cholinergic functioning is also affected by long-chain polyunsaturated fatty acids and mutants deficient in this type of fatty acids show deficits in both the initial sensitivity to EtOH and in the development of acute functional tolerance.[53] These data implicate cholinergic systems, fatty acid metabolism, and Na^+/K^+ ATPase function in the acute activation and/or tolerance effects of EtOH in *C. elegans*, and similar systems may mediate EtOH effects in vertebrates through orthologous mechanisms.

Recent efforts in many research domains have focused on epigenetics. It has become increasingly clear that epigenetic factors such as histone modification play important roles in various aspects of addiction.[54,55] A recent study employing EtOH response behaviors in *C. elegans* demonstrated that genes coding for components of the conserved switching defective/sucrose nonfermenting (SWI/SNF) chromatin-remodeling system are required for the development of acute functional alcohol tolerance in *C. elegans* and/or affect the initial sensitivity to EtOH.[56] This study identified 12 different genes within this system that are involved in mediating these effects in worms. Moreover, allelic variations in SWI/SNF genes (especially in bromodomain containing 7 (BRD7)—homolog to swsn-9 in *C. elegans*) were associated with a diagnosis of alcohol dependence in a human genome-wide association study. Although it is likely that this study only begins to explore epigenetic mechanisms mediating the behavioral effects of EtOH, it does demonstrate the utility of using *C. elegans* models to identify possible epigenetic factors and to identify genes/proteins that may serve as future targets for medications development.

5.2 Nicotine

C. elegans are thought to express at least 27 different nicotinic acetylcholine receptor (nAChR) subunits,[57,58] and thus a rich cholinergic pharmacology.

As in higher-level organisms, acetylcholine in *C. elegans* is critical for many essential behaviors involving muscle contraction, including movement, feeding, and egg laying and several nAChR genes have been identified.[57] Nicotine application induces muscular hypercontraction and egg laying. Continuous exposure to nicotine affects control of egg laying[59,60] which is dependent on the UNC-29 gene.[59] Such exposure also results in tolerance, and nicotine-adapted animals display uncoordinated locomotor activity when removed from nicotine[61] and tolerance is thought to be protein kinase C (PKC) dependent.[59] Further work has demonstrated that nicotine-dependent behavior in *C. elegans* is controlled by transient receptor potential (TRP) proteins TRP-1 and TRP-2.[62] A TRP channel (TRPA1) works to regulate the aversive responses to nicotine and has been identified as a potential target for nicotine pharmacotherapy development in humans.[63] Overall, these studies demonstrate that nicotinic systems subserve many analogous functions in *C. elegans* and nicotine exposure produces behavioral effects that are also consistent with nicotine effects in humans. Furthermore, some specific mediators of nicotine responses appear to be highly conserved in *C. elegans*.

C. elegans have also been shown to display a concentration, time, and age-dependent attraction to nicotine that is reduced by exposure to the nonselective nicotinic receptor antagonists mecamylamine or varenicline.[31] Worms with mutations in genes coding for the dop-1 or dop-2 dopamine receptors, or the acr-5 or the acr-15 nicotinic receptor subunit genes, also showed reduced approach to nicotine. The approach deficit in the acr-15 mutant could be rescued by selective re-expression in neurons but not muscle. *C. elegans* also show "reward-like" cue-conditioned preference for cues previously associated with nicotine which is absent in the acr-5 mutant.[31] Together, these findings provide additional evidence that *C. elegans* may model not only the basic physiologic effects of nicotine, but also the motivational and rewarding properties of nicotine. Use of such models may help to identify the molecular underpinnings of nicotine dependence and identify new targets for the development of new smoking cessation pharmacotherapies.

5.3 Cocaine

C. elegans have conserved monoamine systems and a dopamine system that functions with remarkable similarity to vertebrates, including humans, in terms of signaling.[64] Dopamine is involved in a wide variety of behaviors in the worm, including movement, egg laying, defecation, habituation to touch, as well as sensing and responding to food sources, and copulation behavior in males.[64,65] The dopamine transporter (DAT-1) is sensitive to cocaine[66] and the dopamine neurotoxin 6-hydroxydopamine induces selective dopamine neuronal degeneration in *C. elegans* as it does in vertebrates.[67] Also, cocaine at relatively high concentrations (1.0+ mM) can moderately reduce locomotion velocity and this effect appears to be mediated mainly through the serotonin system.[68] To determine if cocaine could induce reward-like behaviors in *C. elegans*, our group employed a Pavlovian chemosensory cue-conditioning paradigm in which cocaine was paired with an environmental stimulus (a distinctive food or salt cue).[20] After multiple pairings, worms were tested in the absence of cocaine to determine if the history of cocaine coexposure affected the worms' affinity for the cue. Cocaine pairing (5–50 μM) significantly increased preference for either a salt or food cue. The effect was absent in dopamine-deficient mutants, including cat-1 (defective in vesicular packaging of monoamines including dopamine) and cat-2 (tyrosine hydroxylase deficient). The cue-conditioned response was rescued in the mutant strains by conducting the conditioning and testing procedures in the presence of exogenous dopamine.[20] In all, these studies suggest that *C. elegans* may be an excellent model system to study the behavioral responses (including the rewarding properties) of cocaine.

5.4 Methamphetamine

Like cocaine, the addictive properties of methamphetamine in mammals are thought to be mediated, in large part, through its effects on dopamine transport.[69] However, in addition to inhibiting uptake, methamphetamine also induces release, and prolonged exposure can induce biogenic amine neurotoxicity.[70] In *C. elegans*, relatively high concentrations of methamphetamine were shown to have effects on egg laying (8.0+ mM), feeding (2.0+ mM), locomotion (16.0 mM), and survival (8.0+ mM) after 1 h of exposure.[71] The lethal effects of methamphetamine (but not its effects on egg laying) were significantly reduced in a mutant strain (nsy-1 [eg691]) that had a single nucleotide mutation in the NSY-1 protein. The mutant also demonstrated resistance to the lethal effects of exposure to high concentrations of dopamine and 3,4-methylenedioxymethamphetamine (MDMA). The nsy-1 gene is associated with sensory neurons,[72,73] has functions in the innate immune response,[74] and appears to be an ortholog of apoptosis signal-regulating kinase-1.[71] Little is known about the relationship between the neurotoxic and appetitive/addictive properties of methamphetamine, and *C. elegans* may serve as an appropriate model system to study this relationship and identify targets for treatments to possibly reduce the neurotoxic effects of methamphetamine.

As with cocaine, our group also investigated reward-like behaviors of methamphetamine in *C. elegans* using the Pavlovian chemosensory cue-conditioning paradigm.[20] Previous methamphetamine pairing (50–500 μM) significantly increased preference for either a salt or food cue. Also similar to cocaine, the effect was absent in both dopamine-deficient mutants cat-1 and cat-2. The methamphetamine cue-conditioned response was also rescued in the mutant strains by conducting the conditioning and testing procedures in the presence of exogenous dopamine.[20] These studies provide additional evidence that, as in mammals methamphetamine is mediating its rewarding effects through the dopamine system, and both the rewarding and neurotoxic effects of methamphetamine can be modeled in *C. elegans*.

6. CONVERGENT MECHANISMS OF DRUGS OF ABUSE IN *C. ELEGANS*

Although the majority of work examining behavioral responses to drugs of abuse in *C. elegans* thus far has been conducted in EtOH research, and additional work is needed with other abused drugs to better characterize the mechanisms of action across drug classes, some consistencies are apparent across the drugs examined in the current manuscript. In reviewing the molecular targets identified in the behavioral paradigms as described in Table 1, monoamine neurotransmission-associated genes are involved in mediating at least some behaviors induced by each drug.[20,31,41,68,75–77] In particular, within this classification, mutation of the tyrosine hydroxylase gene cat-2 resulted in reductions in drug-induced behaviors for each drug of abuse.[20,76,77] Although this type of mutation is lethal in rodents,[78] drugs of abuse have long been known to function, at least in part, through the dopamine neurotransmitter system[3,27] and all of the addictive drugs examined here have direct or indirect effect on dopamine systems. Similarly, alterations that affect cholinergic neurotransmission also inhibit behavioral responses of *C. elegans* to nicotine.[31,62] In this respect, these findings are consistent with known mechanisms of action of these drugs in higher-level organisms, including humans. Also like vertebrate species, drug effects in *C. elegans* appear to be affected by a wide variety of genes, proteins, and neurobiologic systems that are known to mediate and/or support neuronal function (see Table 1). However, since the majority of these targets to date

Table 1 Some Addiction Related Targets/Systems/Genes in *C. elegans*

Drugs of Abuse	Category/System	Functional Class	Gene/Mutant	Measure/Effect	References
Alcohol	Aldehyde dehydrogenase	Aldehyde metabolism	*alh-6; alh-13*	EtOH effect on speed+	[39]
	BBS1	Fat storage and others	*bbs-1*	EtOH effect on: IR+; AFT+	[79]
	bHLH transcription factor	Lipid metabolism	*sbp-1*	EtOH effect on: IR−; AFT−	[79]
	C_2H_2 zinc-finger transcription factor	Gene repression; others	*pag-3*	EtOH effect on: AFT−	[79]
	Chloride intracellular channel (CLIC)	Cell tubulogenesis and others	*exc-4*	EtOH effect on: IR−	[80]
			exl-1	EtOH effect on: AFT+	
	Cholinergic signaling	Muscle contraction	*cha-1; unc-17*	EtOH-induced HC−	[52]
	CRF receptor-like GPCR	Peptide neurotransmission	*seb-3*	EtOH effect on: AFT−; W/D elicited tremors− ⋆GOF mutant: AFT+	[81]
	Delta-5 fatty acid desaturase	Delta-5 fatty acid synthesis	*fat-4*	EtOH effect on: AFT−	[53]
	Delta-6 fatty acid desaturase	Delta-6 fatty acid synthesis	*fat-3*	EtOH effect on: AFT−	[53]

(*Continued*)

Table 1 Some Addiction Related Targets/Systems/Genes in *C. elegans*—cont'd.

Drugs of Abuse	Category/System	Functional Class	Gene/Mutant	Measure/Effect	References
Alcohol (cont'd)	Diacylglycerol kinase	Synaptic transmission	*dgk-1*	EtOH effect on: IR−	[42,79]
	Dopamine receptor	Dopamine transmission	*dop-4 (D1-like)*	EtOH effect to + activity in liquid: +	[75]
	GABA vesicular transporter	GABA neurotransmission	*unc-47*	HC−	[52]
	Glutamic acid decarboxylase	GABA synthesis	*unc-25*	HC−	[52]
	Na^+/K^+ ATPase	Tolerance to muscle contraction	*eat-6*	Tolerance to EtOH-induced HC−	[52]
	NAD-dependent protein deacetylase	Transcription factor	*sir-2.1*	EtOH effect on: IR−; AFT+	[79]
	NALCN-related leak channel	Cation channel function	*nca-1; nca-2*	EtOH effect on: body bend rate+	[82]
	Neuropeptide receptor	Mediates NPY-like effects	*npr-1*	IR−; AFT+	[45,79]
	Nicotinic Acetylcholine receptor	Nicotine neurotransmission	*unc-63*	HC −	[52]
	Nuclear hormone receptor	Regulation of transcription	*nhr-49*	EtOH effect on: AFT−	[79]

Omega-3 fatty acid acyl desaturase	Polyunsaturated fatty acid synthesis	*fat-1*	EtOH effect on: IR+; AFT−	[53]
Potassium channel	EtOH mediated K+ current	*slo-1*	EtOH effect on: speed−; eggs−; IR−; AFT−	[32,42]
Proprotein convertase	Peptide secretion	*egl-3*	EtOH W/D-induced omega turns −	[83]
RAB3 GTP exchange factor	Synaptic transmission	*aex-3*	EtOH effect on speed: −	[84]
SM protein	Trafficking, synaptic function	*unc-18*	EtOH effect on swimming behavior−	[85]
Small molecular wt GTP-binding protein	Synaptic transmission	*rab-3*	EtOH effect on: speed−; movement to food−	[84]
SWI/SNF complexes	Chromatin remodeling	*swsn-9; swsn-4; swsn-1; swsn-2.1; swsn-2.2 swsn-3; swsn-6; pbrm-1; swsn-7; phf-10; let-526; dpff-1*	EtOH-induced HC and/or tolerance to HC; varied effects	[56]
Transcriptional corepressor	Transcription	*ctbp-1*	EtOH effect on: speed−; IR+; AFT−	[79]

(*Continued*)

Table 1 Some Addiction Related Targets/Systems/Genes in *C. elegans*—cont'd.

Drugs of Abuse	Category/System	Functional Class	Gene/Mutant	Measure/Effect	References
Alcohol (cont'd)	Triacylglycerol lipase	Hydrolase (predicted)	*lips-7*	EtOH effect on: IR−; AFT+	[79]
	TRPV channel	Cation transport	*osm-9*	Inhibits npr-1 enhanced AFT induced with EtOH	[45]
	Tryptophan hydroxylase	Serotonin synthesis	*tph-1*	EtOH preference−	[41]
	Tyrosine hydroxylase	Catecholamine synthesis	*cat-2*	EtOH-induced SDL: −	[76]
	Vesicular monoamine transporter	Monoamine packaging	*cat-1*	EtOH-induced SDL: −	[76]
	Voltage insensitive leak channel modulator	Modulates channel subunit protein levels; others	*unc-79*	EtOH-induced effects on: immobility− Swimming+	[37,82]
			unc-80	Swimming+	[82]
Nicotine	Aromatic amino acid decarboxylase	Monoamine synthesis	*bas-1*	NIC effect on gustatory plasticity−	[77]
	Cholinergic receptor subunit	Acetylcholine neurotransmission	*acr-5*	NIC preference− NIC cue conditioning−	[31]

	Cholinergic receptor subunit	Acetylcholine neurotransmission	*acr-15*	Nicotine preference− NIC cue conditioning− NIC locomotion stimulation−	[31] [62]
	Cholinergic receptor subunit	Acetylcholine neurotransmission	*acr-15*	NIC locomotion stimulation−	[62]
	Dopamine receptor	Dopamine neurotransmission	*dop-2*	Nicotine preference−	[31]
	Dopamine receptors	Dopamine neurotransmission	*dop-1/dop-2 (double mut)*	Nicotine preference−	[31]
	Phospholipase C beta	Second messenger system	*egl-8*	NIC locomotion stimulation−; dependent behavior−	[62]
	TRPC channels	Variable nonselective cation channel activity	*trp-1; trp-2*	NIC locomotion stimulation−	[62]
	Tyrosine hydroxylase	Dopamine neurotransmission	*cat-2*	NIC effect on gustatory plasticity−	[77]
Cocaine	Aromatic amino acid decarboxylase	Monoamine synthesis	*bas-1*	Cocaine-induced hypoactivity−	[68]

(Continued)

Table 1 Some Addiction Related Targets/Systems/Genes in *C. elegans*—cont'd.

Drugs of Abuse	Category/System	Functional Class	Gene/Mutant	Measure/Effect	References
Cocaine (cont'd)	Serotonin-gated chloride channel	Serotonin neurotransmission	*mod-1*	Cocaine-induced hypoactivity−	[68]
	Serotonin transporter	Serotonin neurotransmission	*mod-5*	Cocaine-induced hypoactivity−	[68]
	Tryptophan hydroxylase	Serotonin synthesis	*tph-1*	Cocaine-induced hypoactivity−	[68]
	Tyrosine hydroxylase	Dopamine neurotransmission	*cat-2*	Cue-conditioning to cocaine−	[20]
	Vesicular monoamine transporter	Monoamine neurotransmission	*cat-1*	Cue-conditioning to cocaine−	[20]
Methamphetamine	MAP kinase (MAP3K)	Neuronal signaling	*nsy-1*	MAP-induced toxicity−	[71]
	Tyrosine hydroxylase	Dopamine neurotransmission	*cat-2*	Cue-conditioning to MAP−	[20]
	Vesicular monoamine transporter	Monoamine neurotransmission	*cat-1*	Cue-conditioning to MAP−	[20]

Abbreviations: +, increased; −, reduced; AFT, acute functional tolerance; eggs, egg-laying behavior; IR, initial response to drug; HC, hypercontraction; MAP, methamphetamine; SDL, state-dependent learning; W/D, withdrawal. All observations were made from genetic manipulations using loss of function mutations or RNA interference, except *GOF gain of function mutation.

have been tested in EtOH models, but not with other drugs of abuse, additional studies are needed to determine if these molecular targets are selective for EtOH or represent common mechanisms for multiple drugs of abuse.

7. LEVERAGING *C. ELEGANS* FOR MODEL DEVELOPMENT AND DRUG DISCOVERY

The previous discussion indicates that *C. elegans* can be used to model aspects of drug addiction and identify systems and molecular mechanisms that mediate drug effects. Overall, the findings are surprisingly consistent with analogous findings in higher-level organisms—including humans—and suggest that the effects of addictive drugs are highly conserved. Additional study is needed to better characterize these models and provide a better understanding of how exposure to drugs of abuse can change *C. elegans* neurobiology—driving the animal to seek out further drug exposure. Fortunately, *C. elegans* has a tractable genome that enables both forward and reverse genetics approaches that can be applied to the study of addictions.[33] Additional opportunities for further *C. elegans* model development in the addiction field includes modeling the consequences of drug exposure during critical stages of development. Previous work has described the anatomical effects of embryonic EtOH exposure in *C. elegans* which has some parallels to fetal alcohol syndrome in humans.[86] However, few studies have examined the effects of drug exposure in development on future drug-associated behavior. Initiation of drug taking in humans typically begins in adolescence; thus, models that utilize the exclusive study of the effects of drugs in fully developed adults do not capture the impact of drugs on a developing nervous system. New models that focus on development, and/or take such factors into account, may better model the time course and progression of addiction in humans.

C. elegans are used for screening in a multitude of paradigms. Features such as low cost of maintenance, minimal space requirements, and the availability of image-analysis software for data collection affords advantages to *C. elegans* models in the development of high-throughput assays for medications screening purposes. However, although *C. elegans* are surprisingly highly conserved functionally, differences in some neurotransmitter systems and receptor pharmacology between *C. elegans* and humans could indicate limitations in this approach. For instance, although *C. elegans* make anandamide and 2-arachidonoylglycerol,[87] cannabinoid receptors have yet to be identified in *C. elegans*.[88] Such differences in molecular systems and pharmacology between *C. elegans* and vertebrates could be viewed as a limitation of the model and indeed could limit its application for certain drug classes. However, as previously discussed, *C. elegans* do respond to many of the common drugs of abuse (e.g., EtOH and the psychomotor stimulants) and the recent discovery of a class of opioid receptors in *C. elegans*[89] may provide a new avenue for drug-abuse modeling and treatment screening in *C. elegans*.

In order to evaluate the possible utility of *C. elegans* models in medications screening, drugs with known efficacy to treat addictions (such as naltrexone) need to be tested to establish predictive validity of the *C. elegans* models. It would be expected that some agents might prove more effective than others based on divergent pharmacology. However, diversity in pharmacology and molecular systems in *C. elegans* may also serve as an important advantage of this model system and provide additional information about how specific compounds may work to reduce drug intake. For instance, topiramate is a compound with a rich pharmacology and many possible mechanisms to reduce EtOH drinking. One possible mechanism is through actions at voltage-sensitive sodium channels,[90] which are absent in *C. elegans*.[91] Should topiramate prove ineffective in reducing the EtOH preference response in *C. elegans*, it would strengthen the argument that these channels may have a role in mediating EtOH drinking/seeking in vertebrates. Thus, the effects of agents to reduce drug intake in these models may be viewed in light of the molecular homology of the systems thought to mediate their respective effects in order to determine the possible influence of divergent pharmacology on the results. Such data might provide important information about the pharmacology and molecular substrates mediating the effects of established pharmacotherapies. Moreover, the advent of gene editing technology (such as CRISPR) offers the potential to replace *C. elegans* receptors with their human orthologs to possibly improve the translational impact and further increase the predictive validity of such models. This could significantly increase the utility of these models to identify new targets/candidate drugs for the treatment of addiction.

8. CONCLUSIONS

Addiction is a worldwide problem with severe consequences to health, relationships, crime, and economics at every level. Unfortunately, there are very few effective pharmacotherapies available to treat addicts. Animal models have been used to better understand the neurobiologic underpinnings of addictive behavior and *C. elegans* has emerged as a viable model system to study addiction. Various drug-induced behaviors in *C. elegans* have been utilized to identify genes and biologic systems that mediate specific behavioral responses (Table 1). Many of these genes/systems have orthologous representation in vertebrate animals and several have been implicated in human addiction. Thus far, EtOH is the best-studied drug of abuse in *C. elegans* behavioral models, and genes affecting alcohol-associated behavior in worms have been linked to polymorphisms in orthologous genes in humans with alcohol-related sensitivity and/or disorders.[40] In general, these genes fall into categories related to (1) alcohol metabolism (i.e., alh-6 and alh-13), (2) neurotransmitter/modulator function including acetylcholine (unc-63), monoamines (cat-1), dopamine (dop-4 and cat-2), serotonin (tph-1), GABA (unc-25), neuropeptide Y (npr-1), (3) cation channels (slo-1, nca-1, and nca-2), and (4) chromatin remodeling complexes (swsn-4 and swsn-9). These data suggest that a number of orthologous neurobiologic systems and molecular mediators of EtOH effects in humans are present and also involved in behavioral responses to EtOH in *C. elegans*. In addition, the HUMTH01-VNTR polymorphism in the human gene coding for tyrosine hydroxylase is associated with reduced odds of dependent smoking.[92] Although less information is currently available on the precise mechanisms and targets common to both *C. elegans* and humans associated with behavioral responses to other drugs of abuse, the consistent finding of dopamine system involvement (i.e., cat-2) in multiple classes of addictive drugs supports a conserved and central role for dopamine in the effects of drugs of abuse across animal species. The data generated thus far using *C. elegans* models are promising. However, further model development/refinement may enhance the validity of such models, and additional applications of the powerful molecular genetic methodologies used in *C. elegans* may enhance the utility of the species to aid in the identification of new targets and the development of addiction treatments.

ACKNOWLEDGMENT

This work was supported by the National Institute on Drug Abuse (NIDA) R21DA035468 (EAE).

REFERENCES

1. Justice UDo. The National Drug Threat Assessment; 2010
2. Edwards S, Koob GF. Experimental psychiatric illness and drug abuse models: from human to animal, an overview. *Methods Mol Biol.* 2012;829:31–48.

3. Koob GF, Volkow ND. Neurocircuitry of addiction. *Neuropsychopharmacology.* 2010;35(1): 217–238.
4. Phillips TJ, Reed C, Pastor R. Preclinical evidence implicating corticotropin-releasing factor signaling in ethanol consumption and neuroadaptation. *GenesBrainBehav.* 2015;14(1): 98–135.
5. Shorey-Kendrick LE, Ford MM, Allen DC, et al. Nicotinic receptors in non-human primates: analysis of genetic and functional conservation with humans. *Neuropharmacology.* 2015;96(Pt B):263–273.
6. Bell RL, Sable HJK, Colombo G, Hyytia P, Rodd ZA, Lumeng L. Animal models for medications development targeting alcohol abuse using selectively bred rat lines: neurobiological and pharmacological validity. *Pharmacol Biochem Behav.* 2012;103(1): 119–155.
7. Caille S, Clemens K, Stinus L, Cador M. Modeling nicotine addiction in rats. *Methods MolBiol.* 2012;829:243–256.
8. Niwa M, Yan YJ, Nabeshima T. Genes and molecules that can potentiate or attenuate psychostimulant dependence relevance of data from animal models to human addiction. *Ann NYAcad Sci.* 2008;1141:76–95.
9. Lu L, Shepard JD, Hall FS, Shaham Y. Effect of environmental stressors on opiate and psychostimulant reinforcement, reinstatement and discrimination in rats: a review. *Neurosci Biobehav Rev.* 2003;27(5):457–491.
10. Justinova Z, Goldberg SR, Heishman SJ, Tanda G. Self-administration of cannabinoids by experimental animals and human marijuana smokers. *Pharmacol Biochem Behav.* 2005;81(2):285–299.
11. Franklin KM, Hauser SR, Bell RL, Engleman EA. Caffeinated alcoholic beverages—an emerging trend in alcohol abuse. *J Addict ResTher.* 2013;suppl 4. pii: S4-012.
12. O'Dell LE, Khroyan TV. Rodent models of nicotine reward: what do they tell us about tobacco abuse in humans? *PharmacolBiochemBehav.* 2009;91(4):481–488.
13. O'Connor EC, Chapman K, Butler P, Mead AN. The predictive validity of the rat self-administration model for abuse liability. *Neurosci Biobehav Rev.* 2011;35(3):912–938.
14. Panlilio LV, Goldberg SR. Self-administration of drugs in animals and humans as a model and an investigative tool. *Addiction.* 2007;102(12):1863–1870.
15. Tzschentke TM. Measuring reward with the conditioned place preference (CPP) paradigm: update of the last decade. *AddictBiol.* 2007;12(3–4):227–462.
16. Huber R, Panksepp JB, Nathaniel T, Alcaro A, Panksepp J. Drug-sensitive reward in crayfish: an invertebrate model system for the study of SEEKING, reward, addiction, and withdrawal. *Neurosci Biobehav Rev.* 2011;35(9):1847–1853.
17. Kaun KR, Azanchi R, Maung Z, Hirsh J, Heberlein U. A *Drosophila* model for alcohol reward. *Nat Neurosci.* 2011;14(5):612–619.
18. Kandel ER. The molecular biology of memory storage: a dialogue between genes and synapses. *Science.* 2001;294(5544):1030–1038.
19. Hulme SE, Whitesides GM. Chemistry and the worm: *Caenorhabditis elegans* as a platform for integrating chemical and biological research. *Angew Chem Int Ed Engl.* 2011;50(21): 4774–4807.
20. Musselman HN, Neal-Beliveau B, Nass R, Engleman EA. Chemosensory cue conditioning with stimulants in a *Caenorhabditis elegans* animal model of addiction. *Behav Neurosci.* 2012;126(3):445–456.
21. *C. elegans* Sequencing Consortium.Genome sequence of the nematode *C. elegans*: a platform for investigating biology. *Science.* 1998;282(5396):2012–2018.
22. Antoshechkin I, Sternberg PW. The versatile worm: genetic and genomic resources for *Caenorhabditis elegans* research. *Nat Rev Genet.* 2007;8(7):518–532.
23. Chege PM, McColl G. *Caenorhabditis elegans*: a model to investigate oxidative stress and metal dyshomeostasis in Parkinson's disease. *Front Aging Neurosci.* 2014;6:89.

24. Bessa C, Maciel P, Rodrigues AJ. Using *C. elegans* to decipher the cellular and molecular mechanisms underlying neurodevelopmental disorders. *Mol Neurobiol.* 2013;48(3): 465–489.
25. Calahorro F, Ruiz-Rubio M. *Caenorhabditis elegans* as an experimental tool for the study of complex neurological diseases: Parkinson's disease, Alzheimer's disease and autism spectrum disorder. *Invert Neurosci.* 2011;11(2):73–83.
26. Koob GF, Roberts AJ, Kieffer BL, et al. Animal models of motivation for drinking in rodents with a focus on opioid receptor neuropharmacology. *Recent Dev Alcohol.* 2003;16:263–281.
27. Koob GF. Drugs of abuse—anatomy, pharmacology and function of reward pathways. *Trends Pharmacol Sci.* 1992;13(5):177–184.
28. McBride WJ, Rodd ZA, Bell RL, Lumeng L, Li TK. The alcohol-preferring (P) and high-alcohol-drinking (HAD) rats—animal models of alcoholism. *Alcohol.* 2014;48(3): 209–215.
29. Kalivas PW. Addiction as a pathology in prefrontal cortical regulation of corticostriatal habit circuitry. *Neurotox Res.* 2008;14(2–3):185–189.
30. Zhu G, Zhang F, Li W. Nematodes feel a craving—using *Caenorhabditis elegans* as a model to study alcohol addiction. *Neurosci Bull.* 2014;30(4):595–600.
31. Sellings L, Pereira S, Qian C, et al. Nicotine-motivated behavior in *Caenorhabditis elegans* requires the nicotinic acetylcholine receptor subunits acr-5 and acr-15. *Eur J Neurosci.* 2013;37(5):743–756.
32. Davis SJ, Scott LL, Hu K, Pierce-Shimomura JT. Conserved single residue in the BK potassium channel required for activation by alcohol and intoxication in *C. elegans*. *J Neurosci.* 2014;34(29):9562–9573.
33. Bettinger JC, Davies AG. The role of the BK channel in ethanol response behaviors: evidence from model organism and human studies. *Front Physiol.* 2014;5:346.
34. de Bono M, Maricq AV. Neuronal substrates of complex behaviors in *C. elegans*. *Annu Rev Neurosci.* 2005;28:451–501.
35. Bargmann CI. Chemosensation in *C. elegans*. *WormBook.* 2006;1–29.
36. Ardiel EL, Rankin CH. An elegant mind: learning and memory in *Caenorhabditis elegans*. *Learn Mem.* 2010;17(4):191–201.
37. Morgan PG, Sedensky MM. Mutations affecting sensitivity to ethanol in the nematode, *Caenorhabditis elegans*. *Alcohol Clin Exp Res.* 1995;19(6):1423–1429.
38. McIntire SL. Ethanol. *WormBook.* 2010;1–6.
39. Alaimo JT, Davis SJ, Song SS, et al. Ethanol metabolism and osmolarity modify behavioral responses to ethanol in *C. elegans*. *Alcohol Clin Exp Res.* 2012;36(11):1840–1850.
40. Grotewiel M, Bettinger JC. *Drosophila* and *Caenorhabditis elegans* as discovery platforms for genes involved in human alcohol use disorder. *Alcohol Clin Exp Res.* 2015;39:1292–1311.
41. Lee J, Jee C, McIntire SL. Ethanol preference in *C. elegans*. *Genes Brain Behav.* 2009;8(6): 578–585.
42. Davies AG, Pierce-Shimomura JT, Kim H, et al. A central role of the BK potassium channel in behavioral responses to ethanol in *C. elegans*. *Cell.* 2003;115(6):655–666.
43. Wang ZW, Saifee O, Nonet ML, Salkoff L. SLO-1 potassium channels control quantal content of neurotransmitter release at the *C. elegans* neuromuscular junction. *Neuron.* 2001;32(5):867–881.
44. Matson LM, Kasten CR, Boehm SL, Grahame NJ. Selectively bred crossed high-alcohol-preferring mice drink to intoxication and develop functional tolerance, but not locomotor sensitization during free-choice ethanol access. *Alcohol Clin Exp Res.* 2014;38(1):267–274.
45. Davies AG, Bettinger JC, Thiele TR, Judy ME, McIntire SL. Natural variation in the npr-1 gene modifies ethanol responses of wild strains of *C. elegans*. *Neuron.* 2004;42(5): 731–743.

46. de Bono M, Bargmann CI. Natural variation in a neuropeptide Y receptor homolog modifies social behavior and food response in *C. elegans*. *Cell.* 1998;94(5):679–689.
47. Thiele TE, Badia-Elder NE. A role for neuropeptide Y in alcohol intake control: evidence from human and animal research. *Physiol Behav.* 2003;79(1):95–101.
48. Leggio L, Addolorato G, Cippitelli A, Jerlhag E, Kampov-Polevoy AB, Swift RM. Role of feeding-related pathways in alcohol dependence: a focus on sweet preference, NPY, and ghrelin. *Alcohol Clin Exp Res.* 2011;35(2):194–202.
49. Weiss F, Porrino LJ. Behavioral neurobiology of alcohol addiction: recent advances and challenges. *J Neurosci.* 2002;22(9):3332–3337.
50. Edwards SM, Kenna GA, Swift RM, Leggio L. Current and promising pharmacotherapies, and novel research target areas in the treatment of alcohol dependence: a review. *Curr Pharm Des.* 2011;17(14):1323–1332.
51. Sparrow AM, Lowery-Gionta EG, Pleil KE, et al. Central neuropeptide Y modulates binge-like ethanol drinking in C57BL/6J mice via Y1 and Y2 receptors. *Neuropsychopharmacology.* 2012;37(6):1409–1421.
52. Hawkins EG, Martin I, Kondo LM, et al. A novel cholinergic action of alcohol and the development of tolerance to that effect in *Caenorhabditis elegans*. *Genetics.* 2015;199(1): 135–149.
53. Raabe RC, Mathies LD, Davies AG, Bettinger JC. The omega-3 fatty acid eicosapentaenoic acid is required for normal alcohol response behaviors in *C. elegans*. *Plos One.* 2014;9(8):e105999.
54. Vassoler FM, Byrnes EM, Pierce RC. The impact of exposure to addictive drugs on future generations: physiological and behavioral effects. *Neuropharmacology.* 2014;76 Pt B:269–275.
55. Walker DM, Cates HM, Heller EA, Nestler EJ. Regulation of chromatin states by drugs of abuse. *Curr Opin Neurobiol.* 2015;30:112–121.
56. Mathies LD, Blackwell GG, Austin MK, et al. SWI/SNF chromatin remodeling regulates alcohol response behaviors in *Caenorhabditis elegans* and is associated with alcohol dependence in humans. *Proc Natl Acad Sci USA.* 2015;112(10):3032–3037.
57. Jones AK, Sattelle DB. Functional genomics of the nicotinic acetylcholine receptor gene family of the nematode, *Caenorhabditis elegans*. *BioEssays.* 2004;26(1):39–49.
58. Rand JB. Acetylcholine. *WormBook.* 2007;1–21.
59. Waggoner LE, Dickinson KA, Poole DS, Tabuse Y, Miwa J, Schafer WR. Long-term nicotine adaptation in *Caenorhabditis elegans* involves PKC-dependent changes in nicotinic receptor abundance. *J Neurosci.* 2000;20(23):8802–8811.
60. Duerr JS, Gaskin J, Rand JB. Identified neurons in *C. elegans* coexpress vesicular transporters for acetylcholine and monoamines. *Am J Physiol Cell Physiol.* 2001;280(6): C1616–C1622.
61. Lewis JA, Wu CH, Levine JH, Berg H. Levamisole-resistant mutants of the nematode *Caenorhabditis elegans* appear to lack pharmacological acetylcholine receptors. *Neuroscience.* 1980;5(6):967–989.
62. Feng Z, Li W, Ward A, et al. A *C. elegans* model of nicotine-dependent behavior: regulation by TRP-family channels. *Cell.* 2006;127(3):621–633.
63. Talavera K, Gees M, Karashima Y, et al. Nicotine activates the chemosensory cation channel TRPA1. *Nat Neurosci.* 2009;12(10):1293–1299.
64. McDonald PW, Jessen T, Field JR, Blakely RD. Dopamine signaling architecture in *Caenorhabditis elegans*. *Cell Mol Neurobiol.* 2006;26(4–6):593–618.
65. LeBoeuf B, Correa P, Jee C, Garcia LR. *Caenorhabditis elegans* male sensory-motor neurons and dopaminergic support cells couple ejaculation and post-ejaculatory behaviors. *eLife.* 2014;3:e02938.
66. Jayanthi LD, Apparsundaram S, Malone MD, et al. The *Caenorhabditis elegans* gene T23G5.5 encodes an antidepressant- and cocaine-sensitive dopamine transporter. *Mol Pharmacol.* 1998;54(4):601–609.

67. Nass R, Hahn MK, Jessen T, McDonald PW, Carvelli L, Blakely RD. A genetic screen in *Caenorhabditis elegans* for dopamine neuron insensitivity to 6-hydroxydopamine identifies dopamine transporter mutants impacting transporter biosynthesis and trafficking. *J Neurochem.* 2005;94(3):774–785.
68. Ward A, Walker VJ, Feng Z, Xu XZ. Cocaine modulates locomotion behavior in *C. elegans. PLoS One.* 2009;4(6):e5946.
69. Chiu VM, Schenk JO. Mechanism of action of methamphetamine within the catecholamine and serotonin areas of the central nervous system. *Curr Drug Abuse Rev.* 2012;5(3): 227–242.
70. Yamamoto BK, Moszczynska A, Gudelsky GA. Amphetamine toxicities: classical and emerging mechanisms. *Ann NY Acad Sci.* 2010;1187:101–121.
71. Schreiber MA, McIntire SL. A *Caenorhabditis elegans* p38 MAP kinase pathway mutant protects from dopamine, methamphetamine, and MDMA toxicity. *Neurosci Lett.* 2011;498(1):99–103.
72. Sagasti A, Hisamoto N, Hyodo J, Tanaka-Hino M, Matsumoto K, Bargmann CI. The CaMKII UNC-43 activates the MAPKKK NSY-1 to execute a lateral signaling decision required for asymmetric olfactory neuron fates. *Cell.* 2001;105(2):221–232.
73. Tanaka-Hino M, Sagasti A, Hisamoto N, et al. SEK-1 MAPKK mediates Ca^{2+} signaling to determine neuronal asymmetric development in *Caenorhabditis elegans. EMBO Rep.* 2002;3(1):56–62.
74. Kim DH, Feinbaum R, Alloing G, et al. A conserved p38 MAP kinase pathway in *Caenorhabditis elegans* innate immunity. *Science.* 2002;297(5581):623–626.
75. Topper SM, Aguilar SC, Topper VY, Elbel E, Pierce-Shimomura JT. Alcohol disinhibition of behaviors in *C. elegans. PLoS One.* 2014;9(3):e92965.
76. Bettinger JC, McIntire SL. State-dependency in *C. elegans. Genes Brain Behav.* 2004;3(5): 266–272.
77. Matsuura T, Urushihata T. Chronic nicotine exposure augments gustatory plasticity in *Caenorhabditis elegans*: involvement of dopamine signaling. *Biosci Biotechnol Biochem.* 2015;79(3):462–469.
78. Zhou QY, Quaife CJ, Palmiter RD. Targeted disruption of the tyrosine hydroxylase gene reveals that catecholamines are required for mouse fetal development. *Nature.* 1995;374 (6523):640–643.
79. Bettinger JC, Leung K, Bolling MH, Goldsmith AD, Davies AG. Lipid environment modulates the development of acute tolerance to ethanol in *Caenorhabditis elegans. PLoS One.* 2012;7(5):e35192.
80. Bhandari P, Hill JS, Farris SP, et al. Chloride intracellular channels modulate acute ethanol behaviors in *Drosophila, Caenorhabditis elegans* and mice. *Genes Brain Behav.* 2012;11(4):387–397.
81. Jee C, Lee J, Lim JP, Parry D, Messing RO, McIntire SL. SEB-3, a CRF receptor-like GPCR, regulates locomotor activity states, stress responses and ethanol tolerance in *Caenorhabditis elegans. Genes Brain Behav.* 2013;12(2):250–262.
82. Speca DJ, Chihara D, Ashique AM, et al. Conserved role of unc-79 in ethanol responses in lightweight mutant mice. *PLoS Genet.* 2010;6(8):e1001057.
83. Mitchell P, Mould R, Dillon J, et al. A differential role for neuropeptides in acute and chronic adaptive responses to alcohol: behavioural and genetic analysis in *Caenorhabditis elegans. PLoS One.* 2010;5(5):e10422.
84. Kapfhamer D, Bettinger JC, Davies AG, et al. Loss of RAB-3/A in *Caenorhabditis elegans* and the mouse affects behavioral response to ethanol. *Genes Brain Behav.* 2008;7(6): 669–676.
85. Graham ME, Edwards MR, Holden-Dye L, Morgan A, Burgoyne RD, Barclay JW. UNC-18 modulates ethanol sensitivity in *Caenorhabditis elegans. Mol Biol Cell.* 2009;20(1): 43–55.

86. Davis JR, Li Y, Rankin CH. Effects of developmental exposure to ethanol on *Caenorhabditis elegans*. *Alcohol Clin Exp Res.* 2008;32(5):853–867.
87. Lehtonen M, Reisner K, Auriola S, Wong G, Callaway JC. Mass-spectrometric identification of anandamide and 2-arachidonoylglycerol in nematodes. *Chem Biodivers.* 2008;5(11):2431–2441.
88. Elphick MR, Egertova M. The phylogenetic distribution and evolutionary origins of endocannabinoid signalling. *Handb Exp Pharmacol.* 2005;168:283–297.
89. Cheong MC, Artyukhin AB, You YJ, Avery L. An opioid-like system regulating feeding behavior in *C. elegans*. *eLife.* 2015;4:e06683.
90. Johnson BA. Progress in the development of topiramate for treating alcohol dependence: from a hypothesis to a proof-of-concept study. *Alcohol Clin Exp Res.* 2004;28(8): 1137–1144.
91. Bargmann CI. Neurobiology of the *Caenorhabditis elegans* genome. *Science.* 1998;282 (5396):2028–2033.
92. Anney RJ, Olsson CA, Lotfi-Miri M, Patton GC, Williamson R. Nicotine dependence in a prospective population-based study of adolescents: the protective role of a functional tyrosine hydroxylase polymorphism. *Pharmacogenetics.* 2004;14(2):73–81.

INDEX

A

B

D

E

F

G

L

M

N

O

P

FSC
www.fsc.org
MIX
Papier | Fördert gute Waldnutzung
FSC® C083411

Kolibri 360 GmbH
Ferdinand-Jühlke-Straße 7
99095 Erfurt, Deutschland
produktsicherheit@kolibri360.de